Nonlinear and Adaptive Control

Tools and Algorithms for the User

Nonlinear and Adaptive Control

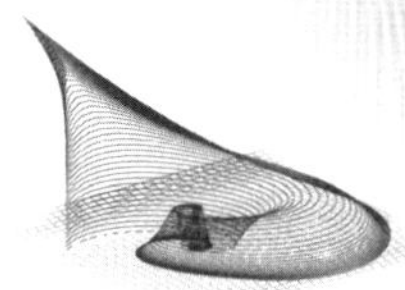

Tools and Algorithms for the User

Alessandro Astolfi

Imperial College London, UK

ICP

Imperial College Press

Published by

Imperial College Press
57 Shelton Street
Covent Garden
London WC2H 9HE

Distributed by

World Scientific Publishing Co. Pte. Ltd.
5 Toh Tuck Link, Singapore 596224
USA office: 27 Warren Street, Suite 401-402, Hackensack, NJ 07601
UK office: 57 Shelton Street, Covent Garden, London WC2H 9HE

British Library Cataloguing-in-Publication Data
A catalogue record for this book is available from the British Library.

NONLINEAR AND ADAPTIVE CONTROL
Tools and Algorithms for the User

ISBN 1-86094-617-8

Printed by FuIsland Offset Printing (S) Pte Ltd, Singapore

PREFACE

This book contains some of the main scientific contributions resulting from the research activities undertaken within the framework of the TMR Network "Nonlinear and adaptive control: tools and algorithms for the user" (NACO2, Contract Number HPRN-CT-1999-00046, www.ee.ic.ac.uk/naco2).

The support of the European Commission is gratefully acknowledged, and it is fair to say that without such support the research results reported in this book would have been only partly developed.

As already stated, the book contains a collection of diverse scientific results, ranging from purely theoretical results to engineering applications. This demonstrates the wide spectrum of activities undertaken within the NACO2 project and shows, once more, that nonlinear and adaptive control are located at the crossroads between mathematics and engineering and, as such, contribute to the development of such disciplines.

All chapters of the book either involve researchers from two or more nodes of the network (and other international researchers) or involve young researchers (G. Blankenstein, P. Castillo, H. DeBattista, A. de Rinaldis, P. Garcia, F. Grognard, G. Kaliora, D. Karagiannis, S. Velut) supported by the network. This highlights two main features of the NACO2 network: the importance of international collaborations and the high quality of training provided to the young researchers that have participated in the NACO2 programme.

The book is *ideally* divided into two parts. In the first (Chapters 1 to 4) recent results on the theory of nonlinear and adaptive control are presented. The second (Chapters 5 to 10) presents applications of nonlinear and adaptive control tools to engineering problems, ranging from mechanical engineering (Chapter 5, 6 and 7), to bioengineering (Chapters 8 and 9), to power engineering (Chapter 10). Notably, experimental results are reported in Chapters 6, 8 and 9.

I wish to thank all contributors to this book, and all researchers that have made the NACO2 programme an excellent and stimulating research adventure that, together with Prof. D.Q. Mayne, I had the honour to co-ordinate. I also wish to thank D. Karagiannis for his help in correcting the manuscript.

A. Astolfi
London

CONTENTS

CHAPTER 1

Observer-Based Solution to the Strictly Positive Real Problem

J. Collado[1], R. Lozano[2] and R. Johansson[3]

[1] *Department of Automatic Control, Cinvestav, P.O. Box 14-740, 07000 Mexico, D.F., Mexico, E-mail: jcollado@ctrl.cinvestav.mx*

[2] *UMR 6500, HEUDYASIC, BP 20529, 60205 Compiegne, cedex, France, E-mail: Rogelio.Lozano@hds.utc.fr*

[3] *Department of Automatic Control, Lund Institute of Technology, P.O. Box 118 SE-221 00 Lund, Sweden, E-mail: Rolf.Johansson@control.lth.se*

We study the extension of the class of linear time-invariant plants that may be transformed into SPR systems introducing an observer. It is shown that for open-loop stable systems a cascaded observer achieves the result. For open-loop unstable systems an observer-based feedback is required to succeed. In general any stabilizable and observable system may be transformed into an SPR system using cascade or cascade and feedback controllers. This overcomes the old conditions of minimum phase and relative degree one. The result is illustrated with some examples.

1. Introduction

The celebrated Kalman–Yakubovich–Popov (KYP) Lemma, also called Positive Real Lemma, gives algebraic equations for a square transfer matrix $Z(s)$ to be Strictly Positive Real (SPR). These algebraic equations are equivalent to an analytic condition in the frequency domain which is not easy to test. The solution of these algebraic equations provides a practical way to verify that a given transfer function is SPR.

The statement of the KYP Lemma is due to Popov who proposed the problem [18], although he did not propose a solution. Popov also introduced a Lyapunov function composed of the sum of a quadratic term and an integral term to solve the absolute stability problem for a class of nonlinear systems. Later, Yakubovich [23] established the equations and Kalman [9] further elaborated these results. Anderson [3] established the MIMO version of the KYP Lemma.

The standard assumptions on the state-space representation for the KYP Lemma are: a) minimality, b) relative degree one, c) minimum phase and d) the system must be square, *i.e.* the number of inputs and outputs should be the same.

It had been long recognized that the minimality condition may be weakened to simply stabilizability and observability. Indeed, Meyer [14] pointed at this relaxation of the minimality assumption explicitly, but they did not provide a proof for that statement. Implicitly, Rantzer [19], in his novel proof based on convexity properties and linear algebra, does not require minimality of the state space representation of $Z(s)$. It was not until recently [6] that the minimality relaxation was explicitly proved in an algebraic approach. It has also been pointed out that a state-space representation can have uncontrollable modes and still satisfy the conditions to be SPR provided that the uncontrollable modes are stable. Previous reviewers of this work pointed out that a proof was available in Yakubovich *et al.* [24], but in English it was published very recently. Other interesting properties of SPR systems and comparisons with other related results are presented in [2, 10, 13, 22].

This chapter presents a technique to render SPR any stabilizable and observable linear time-invariant system. The technique is based on a state observer and a feedback control law using the estimate of the state. Molander and Willems [15] solved the problem using state feedback under the assumption that the original system has relative degree one and is minimum-phase. In the context of nonlinear systems, Byrnes *et al.* [5] presented a solution to the problem using smooth state feedback provided that the system has relative degree one and is (weakly) minimum-phase. Furthermore the works of Kokotović *et al.* [11, 12], address the problem associated with the stabilization of a linear system in cascade with a globally asymptotically stable nonlinear system. The proposed solutions also require the system to be weakly minimum-phase and have relative degree one. Another interesting solution has been presented by Sun *et al.* [21] based on output feedback. They establish conditions to render the system Extended SPR (ESPR). This definition requires relative degree zero which means $D + D^T > 0$.

Some approaches have been proposed to overcome the condition on relative degree one. Barkana introduced a "parallel feedforward" in the context of adaptive control [4]. Another related idea is passification by means of shunting introduced by Fradkov and coworkers [1]. These two approaches represent derivation of a loop-transfer function with SPR properties for a control object without SPR properties by means of dynamic extensions or observers.

The present chapter addresses the problem of designing an observer and a controller so that the modified system becomes SPR. Since any LTI system with a state observer is not minimal, few studies have been made in the past [8] to define an output for nonminimal system in attempts to obtain SPR systems. The proposed method is described for both stable plants and unstable plants. In the case of stable plants the method reduces to an observer and defining a new output as a function of the estimated state. The new output has to satisfy an algebraic equation. In the unstable case we have in addition to introduce an estimated state feedback controller. The proposed approach does not require the original system to be minimum-phase nor to have relative degree one.

The chapter is organized as follows: Section 2 presents some definitions, Section 3 deals with the open-loop stable case, while the open-loop unstable case is addressed in Section 4. Some illustrative examples are given in Section 5 and concluding remarks are presented in Section 6.

2. Preliminaries

Consider a linear time-invariant m-inputs p-outputs system with transfer matrix $Z(s)$ and with a minimal realization given by

$$\begin{aligned} \dot{x} &= Ax + Bu \\ y &= Cx \end{aligned} \tag{1.1}$$

where $x \in \mathbb{R}^n$, $u \in \mathbb{R}^m$, $y \in \mathbb{R}^p$, $m \leq n$, $p \leq n$, and A, B, C are matrices of appropriate dimensions. Denote by $\mathbb{C}, \mathbb{C}_-$ and $\overset{\circ}{\mathbb{C}}_-$, the complex plane, the closed left complex plane and the open left complex plane, respectively. Denote by $\sigma(T)$ the set of eigenvalues of the square matrix T and let $\mathbb{R}_+$ represent the set of positive real numbers.

Definition 1: [2, 16] The transfer matrix $Z(s)$ is said to be PR if:

(i) All elements of $Z(s)$ are analytical in $\mathrm{Re}[s] > 0$, and
(ii) $Z(s) + Z^T(-s) \geq 0$ for all $\mathrm{Re}[s] > 0$.

Z(s) is said to be SPR if $Z(s - \varepsilon)$ is PR for some $\varepsilon > 0$.

For the scalar case, $m = p = 1$, the classical interpretation of $Z(s)$ being PR (SPR) is that its Nyquist plot lies entirely in the right complex plane (open right complex plane).

We will need in the sequel the following version of the KYP Lemma for strictly proper systems.

Lemma 1: *Let $Z(s) = C(sI - A)^{-1}B$ be a $m \times m$ transfer matrix such that $Z(s) + Z^T(-s)$ has normal rank m, where A is Hurwitz, (A, B) is stabilizable, and (C, A) is observable. Then $Z(s)$ is Strictly Positive Real (SPR) if and only if there exist symmetric positive definite matrices P and Q such that*

$$\begin{aligned} PA + A^T P &= -Q \\ PB &= C^T. \end{aligned}$$

3. Stable Case

Let us consider a linear time-invariant system described in standard state-space equations [20] as:

$$\Sigma_1 \begin{cases} \dot{x} = Ax + Bu \\ y = Cx. \end{cases} \tag{1.2}$$

Assumption 1: The A matrix is stable [10, 20], *i.e.* $\sigma(A) \subset \overset{\circ}{\mathbb{C}}_-$.

A full-order observer for the system Σ_1 is given by

$$\widehat{\Sigma} \begin{cases} \dot{\widehat{x}} = A\widehat{x} + Bu + LC\,(x - \widehat{x}) \\ z = M\widehat{x} \end{cases} \tag{1.3}$$

where the observer gain matrix L is such that

$$\sigma(A - LC) \subset \overset{\circ}{\mathbb{C}}_-. \tag{1.4}$$

The system (1.2) and the observer (1.3) may be written compactly as:

$$\Sigma_{1+\text{obs}} \left\{ \begin{bmatrix} \dot{x} \\ \dot{\widehat{x}} \end{bmatrix} = \begin{bmatrix} A & 0 \\ LC & A - LC \end{bmatrix} \begin{bmatrix} x \\ \widehat{x} \end{bmatrix} + \begin{bmatrix} B \\ B \end{bmatrix} u. \right. \tag{1.5}$$

Introducing the state estimation error as $\widetilde{x} \triangleq \widehat{x} - x$, the system $\Sigma_{1+\text{obs}}$ may be expressed as

$$\Sigma_0 \left\{ \begin{bmatrix} \dot{x} \\ \dot{\widetilde{x}} \end{bmatrix} = A_0 \begin{bmatrix} x \\ \widetilde{x} \end{bmatrix} + B_0 u \right. \tag{1.6}$$

where

$$A_0 \triangleq \begin{bmatrix} A & 0 \\ 0 & A - LC \end{bmatrix} \tag{1.7a}$$

and

$$B_0 \triangleq \begin{bmatrix} B \\ 0 \end{bmatrix}. \tag{1.7b}$$

Remark 1: The system Σ_0 is not minimal: all the modes associated to the block $(A - LC)$ are uncontrollable.

Since A and $A_L \triangleq A - LC$ are stable, then for all positive definite matrices Q_{11} and Q_{22}, there exist positive definite matrices P and P_L solving the Lyapunov equations

$$\begin{aligned} A^T P + PA &= -Q_{11} \\ A_L^T P_L + P_L A_L &= -Q_{22}. \end{aligned} \tag{1.8}$$

Now define

$$P_0 \triangleq \begin{bmatrix} P & P \\ P & \mu P_L \end{bmatrix} \tag{1.9}$$

where $\mu > 0$ will be determined later. Then

$$\begin{aligned} A_0^T P_0 + P_0 A_0 &= \begin{bmatrix} A & 0 \\ 0 & A - LC \end{bmatrix}^T \begin{bmatrix} P & P \\ P & \mu P_L \end{bmatrix} + \begin{bmatrix} P & P \\ P & \mu P_L \end{bmatrix} \begin{bmatrix} A & 0 \\ 0 & A - LC \end{bmatrix} \\ &= -Q_0. \end{aligned} \tag{1.10}$$

Note that block (1,1) corresponds to the first equation of (1.8), block (2,2) is a μ-scaled version of the second equation of (1.8) and the cross term is

$$A^T P + P(A - LC) = A^T P + PA - PLC = -Q_{11} - PLC.$$

Then

$$Q_0 = \begin{bmatrix} Q_{11} & Q_{11} + PLC \\ Q_{11} + C^T L^T P & \mu Q_{22} \end{bmatrix}. \tag{1.11}$$

The composite system Σ_0 (1.6) will satisfy the first equation of the KYP Lemma if $Q_0 > 0$ and $P_0 > 0$. Using the Schur complement formula [7], we obtain the following conditions for positive definiteness of Q_0 and P_0.
I) Conditions that guarantee that P_0 is positive definite are:

$$P_o > 0 \iff \begin{cases} \textbf{I.1} \;\; P > 0 \\ \textbf{I.2} \;\; \mu P_L - P > 0. \end{cases} \tag{1.12}$$

Condition **I.1** in (1.12) is satisfied in view of the Lyapunov Eqs. (1.8). Using the fact that for any given $R = R^T$ and $W = W^T$, not necessarily of the same dimensions, the following relation holds:

$$\begin{bmatrix} I & 0 \\ V^T R^{-1} & I \end{bmatrix} \begin{bmatrix} R & 0 \\ 0 & W - V^T R^{-1} V \end{bmatrix} \begin{bmatrix} I & R^{-1} V \\ 0 & I \end{bmatrix} = \begin{bmatrix} R & V \\ V^T & W \end{bmatrix}, \tag{1.13}$$

condition **I.2** in (1.12) is obtained by selecting in the above $R = V = P$ and $W = \mu P_L$. Condition **I.2** can also be expressed as $\mu P_L > P$. Further reduction is possible[a] if we use the following theorem.

Theorem 1: [17] Let H_1, H_2 be Hermitian matrices of the same dimensions, at least one of them being positive definite, say $H_1 > 0$. Then there exists a nonsingular matrix M such that

(i) $M^* H_1 M = I$, and
(ii) $M^* H_2 M = \mathrm{diag}(\rho_1, \rho_2, \ldots, \rho_n)$

where $\rho_i \in \mathbb{R}$ are eigenvalues of $H_1^{-1} H_2$.

Remark 2: The previous result says that given two Hermitian matrices of the same dimensions with one positive definite, they may be simultaneously diagonalized by means of a congruence transformation.

Applying Theorem 1 for $H_1 = \mu P_L$ and $H_2 = P$, the inequality $\mu P_L > P$ becomes $\mu I = M^* (\mu P_L) M > M^* P M = \mathrm{diag}(\lambda_1, \lambda_2, \ldots, \lambda_n)$, where $\lambda_i \in \mathbb{R}_+$ are eigenvalues of $P_L^{-1} P$. The condition **I.2** becomes

$$\mu I > \mathrm{diag}(\lambda_1, \lambda_2, \ldots, \lambda_n) \tag{1.14}$$

this condition being satisfied if

$$\mu > \mu_1 \triangleq \max_i \; \lambda_i. \tag{1.15}$$

II) Conditions that guarantee positive definiteness of Q_0 are:

$$Q_0 > 0 \iff \begin{cases} \textbf{II.1} \;\; Q_{11} > 0 \\ \textbf{II.2} \;\; \mu Q_{22} > \left(Q_{11} + C^T L^T P\right) Q_{11}^{-1} \left(Q_{11} + PLC\right). \end{cases} \tag{1.16}$$

Using Theorem 1, condition **II.2** may be reduced to a similar form as condition **I.2**. If we define

$$F \triangleq Q_{22}^{-1} \left(Q_{11} + C^T L^T P\right) Q_{11}^{-1} \left(Q_{11} + PLC\right)$$

having spectrum

$$\sigma(Q_{22}^{-1} \left(Q_{11} + C^T L^T P\right) Q_{11}^{-1} \left(Q_{11} + PLC\right)) = \{\nu_1, \nu_2, \ldots, \nu_n\} \in \mathbb{R}_+$$

then **II.2** is equivalent to

$$\mu > \mu_2 \triangleq \max_i \nu_i. \tag{1.17}$$

---

[a] Acknowledgment to Prof. V. Kharitonov who pointed out this simplification.

Combining conditions **I** and **II**, P_0 and Q_0 are positive definite if (see Eqs. (1.15) and (1.17))

$$\mu > \mu^* \triangleq \max\{\mu_1, \mu_2\}. \tag{1.18}$$

Remark 3: μ can always be chosen to satisfy the above inequality. Also note that the bound is tight, *i.e.* if $\mu = \mu^*$ then we can only guarantee $P_0 > 0$ or $Q_0 > 0$, but not both conditions.

We have proved the first part of the following theorem.

Theorem 2: Consider the stable transfer matrix $Z(s)$ with m-inputs and p-outputs and its state-space realization

$$\Sigma_1 \begin{cases} \dot{x} = Ax + Bu \\ y = Cx \end{cases}$$

where A is stable, the pair (A, B) is stabilizable and the pair (C, A) is observable. Then there exists an observer gain L given in (1.3) satisfying (1.4) such that the transfer matrix between u and the new output

$$z = M_0 \begin{bmatrix} x \\ \widetilde{x} \end{bmatrix} = M\widehat{x}, \qquad M = B^T P$$

is characterized by a state-space representation (A_0, B_0, M_0) which is SPR.

Proof: The proof of the first equation of the KYP Lemma is already done provided that $\mu > \mu^*$. If the new output z is defined as

$$\begin{aligned} z = M_0 \begin{bmatrix} x \\ \widetilde{x} \end{bmatrix} &= B_0^T P_0 \begin{bmatrix} x \\ \widetilde{x} \end{bmatrix} \\ &= \begin{bmatrix} B^T & 0 \end{bmatrix} \begin{bmatrix} P & P \\ P & \mu P_L \end{bmatrix} \begin{bmatrix} x \\ \widehat{x} - x \end{bmatrix} \\ &= B^T P x + B^T P (\widehat{x} - x) \\ &= B^T P \widehat{x} = M\widehat{x}, \end{aligned}$$

then the composed system (A_0, B_0, M_0), which is not minimal, satisfies the KYP Lemma equations, *i.e.*

$$A_0^T P_0 + P_0 A_0 = -Q_0 \quad \text{and} \quad M_0 = B_0^T P_0 = \begin{bmatrix} B^T P & B^T P \end{bmatrix}.$$

Not only the equation

$$z = M_0 \begin{bmatrix} x \\ \widetilde{x} \end{bmatrix} = \underbrace{B^T P}_{M} \widehat{x}$$

satisfies the second equation of the KYP Lemma, but also provides the output z as a function of the estimated state $\widehat{x}$, which is required to be implementable. □

The system (1.2) and the observer (1.3) can be combined to obtain the composite system (1.6). Note that A_0 in Eqs. (1.6) and (1.7a) satisfies the Lyapunov Eq. (1.10) where P_0 and Q_0 are positive definite. Therefore, if the new output z is defined as $z = M\widehat{x}$, the transfer function from u to z is SPR.

The diagram in Figure 1.1 shows the cascade compensator for the stable case.

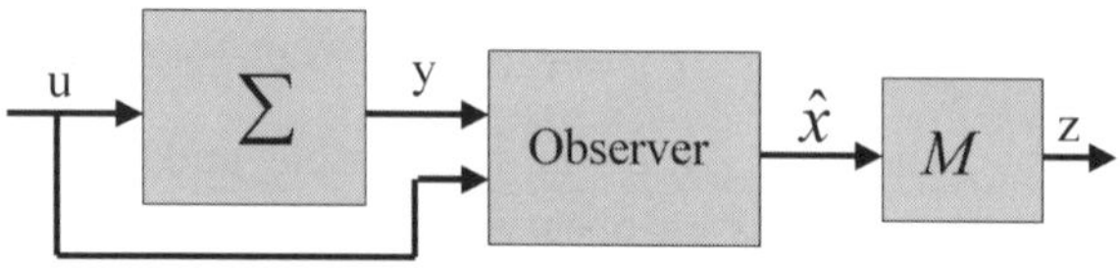

Figure 1.1. SPR transformation for open-loop stable systems.

4. Unstable Case

In this section we will remove Assumption 1 concerning the stability of the system. We will assume that system (1.2) is not stable, *i.e.* $\sigma(A) \nsubseteq \overset{\circ}{\mathbb{C}}_-$. Introduce a stabilizing control law based on the observer (1.3)

$$u = -K\widehat{x} + v \tag{1.19}$$

where v is a new input signal. The composed system becomes

$$\Sigma_K \left\{ \begin{bmatrix} \dot{x} \\ \dot{\widetilde{x}} \end{bmatrix} = \underbrace{\begin{bmatrix} A - BK & -BK \\ 0 & A - LC \end{bmatrix}}_{A_0} \begin{bmatrix} x \\ \widetilde{x} \end{bmatrix} + \underbrace{\begin{bmatrix} B \\ 0 \end{bmatrix}}_{B_0} v. \right.$$

Introduce the short hand notation $A_K = A - BK$ and $A_L = A - LC$. Again K and L are such that $\sigma(A - BK) \subset \overset{\circ}{\mathbb{C}}_-$ and $\sigma(A - LC) \subset \overset{\circ}{\mathbb{C}}_-$. Then for every positive definite Q_K and Q_L there exist $P_K > 0$ and $P_L > 0$ solving the Lyapunov equations

$$\begin{aligned} A_K^T P_K + P_K A_K &= -Q_K \\ A_L^T P_L + P_L A_L &= -Q_L. \end{aligned} \tag{1.20}$$

Define P_0 as in the stable case, *i.e.*

$$P_0 \triangleq \begin{bmatrix} P_K & P_K \\ P_K & \mu P_L \end{bmatrix}.$$

Then the block diagonal elements of the equation

$$A_0^T P_0 + P_0 A_0 = -Q_0$$

correspond to Eqs. (1.20). The off-diagonal block is

$$\begin{aligned} [-Q_0]_{1,2} &= -P_K BK + P_K A_L + A_K^T P_K \\ &= P_K (A - BK) + (A - BK)^T P_K - P_K LC \\ &= -Q_K - P_K LC. \end{aligned}$$

Obviously $[-Q_0]_{2,1} = [-Q_0]_{1,2}^T$. For stability of the feedback system it is required that $P_0 > 0$ and $Q_0 > 0$.

III) Conditions for positive definiteness of P_0:

$$P_0 > 0 \Longleftrightarrow \begin{cases} \textbf{III.1} & P_K > 0 \\ \textbf{III.2} & \mu P_L - P_K > 0. \end{cases}$$

Condition **III.1** is satisfied in view of the Lyapunov Eqs. (1.20). Let the spectrum of $P_L^{-1} P_K$ be $\{\eta_1, \eta_2, \ldots, \eta_n\} \in \mathbb{R}_+$. We then have that condition **III.2** is equivalent to

$$\mu > \mu_3 \triangleq \max_i \eta_i. \tag{1.21}$$

IV) For positive definiteness of Q_0 we require:

$$Q_0 > 0 \Longleftrightarrow \begin{cases} \textbf{IV.1} & Q_{11} > 0 \\ \textbf{IV.2} & \mu Q_L - \left(Q_K + C^T L^T P_K\right) Q_K^{-1} \left(Q_K + P_K LC\right) > 0. \end{cases}$$

Again let $\sigma(Q_L^{-1} \left(Q_K + C^T L^T P_K\right) Q_K^{-1} \left(Q_K + P_K LC\right)) = \{\rho_1, \ldots, \rho_n\} \in \mathbb{R}_+$, then condition **IV.2** is equivalent to

$$\mu > \mu_4 \triangleq \max_i \rho_i. \tag{1.22}$$

Combining (1.21) and (1.22), $P_0 > 0$ and $Q_0 > 0$ if and only if

$$\mu > \mu^* \triangleq \max \{\mu_3,\ \mu_4\}.$$

Now we may state the main result of this section.

Theorem 3: Given the strictly proper transfer matrix $Z(s)$ of dimensions $p \times m$ not identically zero, construct a stabilizable and observable realization (A, B, C). There exists a gain observer matrix L as in (1.3), an estimated state feedback K ($K = 0$ if A is Hurwitz), and a matrix M_0 which defines a new output such that the transfer matrix from v of Eq. (1.19) to the new output z is SPR.

Proof: The first equation of the KYP Lemma is just proved for sufficiently large μ, the second part is similar to the proof of Theorem 2 for stable systems. $\square$

Remark 4: Notice that in either, stable or unstable cases, the Lyapunov equation $A_0^T P_0 + P_0 A_0 = -Q_0$ does not have a positive definite solution P_0 of the form given in (1.9) for all $Q_0 > 0$ of the form given in (1.11). We are imposing the structure on the matrix P_0, therefore there exist positive definite P_0 and Q_0 only for μ sufficiently large.

Figure 1.2 shows the structure of the compensator proposed for the unstable case.

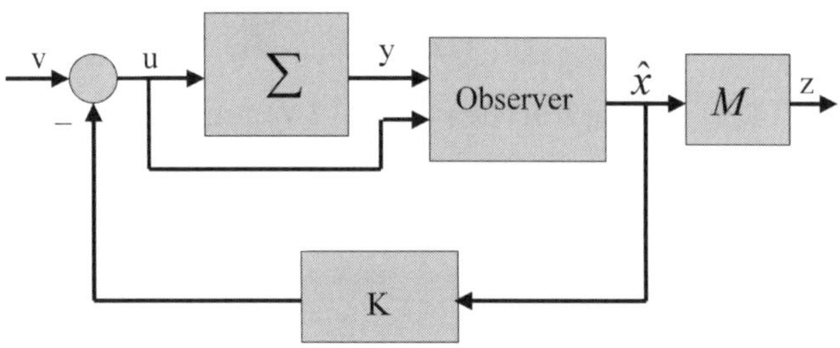

Figure 1.2. SPR transformation for open-loop unstable systems.

5. Illustrative Examples

In this section we will present three detailed examples. The first example is a relative degree two stable system, the second one presents a relative degree three unstable system and the third example is a nonminimum-phase unstable system with a nonminimal state-space representation. We construct a compensator which renders the new system SPR in the three cases.

Example 1: Consider the following transfer function

$$h(s) = \frac{1}{s^2 + 3s + 2} = \frac{1}{(s+1)(s+2)}$$

which has a minimal state-space representation

$$\Sigma_1 \begin{cases} \dot{x} = \begin{bmatrix} 0 & 1 \\ -2 & -3 \end{bmatrix} x + \begin{bmatrix} 0 \\ 1 \end{bmatrix} u \\ y = \begin{bmatrix} 1 & 0 \end{bmatrix} x. \end{cases}$$

A full-order observer for Σ_1 with eigenvalues at $\{-3, -4\}$ is

$$\widehat{\Sigma}_1 \left\{ \dot{\widehat{x}} = \begin{bmatrix} 0 & 1 \\ -2 & -3 \end{bmatrix} \widehat{x} + \begin{bmatrix} 0 \\ 1 \end{bmatrix} u + \underbrace{\begin{bmatrix} 4 \\ -2 \end{bmatrix}}_{L} \begin{bmatrix} 1 & 0 \end{bmatrix} (x - \widehat{x}) \,. \right.$$

If we choose $Q_{11} = 2I$, then the solution of the Lyapunov equation $A^T P + PA = -Q_{11}$ is

$$P = \begin{bmatrix} 2.5 & 0.5 \\ 0.5 & 0.5 \end{bmatrix} > 0.$$

For $Q_{22} = 2I$ the solution of the Lyapunov equation $A_L^T P_L + P_L A_L = -Q_{22}$ is

$$P_L = \begin{bmatrix} 0.25 & 0.0357 \\ 0.0357 & 0.3452 \end{bmatrix} > 0.$$

For the computation of μ^* we require the values:

$$\sigma(P_L^{-1} P) = \{10.0272, 1.1729\}$$

$$\sigma(Q_{22}^{-1} (Q_{11} + C^T L^T P) \, Q_{11}^{-1} (Q_{11} + PLC)) = \{44.5517, 0.9483\}$$

$$\begin{aligned} \mu^* &= \max \{\mu_1, \mu_2\} \\ &= \max \left\{ \max_i \lambda_i, \max_i \nu_i \right\} \\ &= \max \{10.0272, 44.5517\} \\ &= 44.5517. \end{aligned}$$

If $\mu = 45 > \mu^*$, then

$$P_0 = \left[\begin{array}{cc|cc} 2.5 & 0.5 & 2.5 & 0.5 \\ 0.5 & 0.5 & 0.5 & 0.5 \\ \hline 2.5 & 0.5 & 112.5 & 22.5 \\ 0.5 & 0.5 & 22.5 & 22.5 \end{array}\right]$$

$$\sigma(P_0) = \{0.3733, 2.5586, 17.1971, 117.871\}$$

$$Q_0 = \left[\begin{array}{cc|cc} 2 & 0 & 13 & 0 \\ 0 & 2 & 3 & 2 \\ \hline 13 & 3 & 90 & 0 \\ 0 & 2 & 0 & 90 \end{array}\right]$$

$$\sigma(Q_0) = \{0.0195, 1.9569, 90.0431, 91.9805\}.$$

The output system matrix becomes $M_0 = B_0^T P_0 = \begin{bmatrix} 0.5\ 0.5\ 0.5\ 0.5 \end{bmatrix}$ which gives us $M = \begin{bmatrix} 0.5\ 0.5 \end{bmatrix}$ and the transfer function between z and u is

$$H_{zu}(s) = \frac{0.5s + 0.5}{s^2 + 3s + 2} = \frac{0.5}{s+2}.$$

In this particular case, there was a cancellation in the final transfer function $H_{zu}(s)$, but this is not always the case. If we change the matrices Q_{11} and Q_{22}, we will get generically a second-order system in $H_{zu}(s)$.

Example 2: Consider the following transfer function

$$h(s) = \frac{1}{s^3 + 2s^2 - s - 2} = \frac{1}{(s+1)(s-1)(s+2)}$$

which has a minimal state-space representation

$$\Sigma_2 \begin{cases} \dot{x} = \begin{bmatrix} 0 & 1 & 0 \\ 0 & 0 & 1 \\ 2 & 1 & -2 \end{bmatrix} x + \begin{bmatrix} 0 \\ 0 \\ 1 \end{bmatrix} u \\ y = \begin{bmatrix} 1 & 0 & 0 \end{bmatrix} x. \end{cases}$$

A full-order observer for Σ_2 with eigenvalues at $\{-2, -3, -4\}$ is

$$\widehat{\Sigma}_2 \left\{ \dot{\widehat{x}} = \begin{bmatrix} 0 & 1 & 0 \\ 0 & 0 & 1 \\ 2 & 1 & -2 \end{bmatrix} \widehat{x} + \begin{bmatrix} 0 \\ 0 \\ 1 \end{bmatrix} u + \underbrace{\begin{bmatrix} 7 \\ 13 \\ 7 \end{bmatrix}}_{L} \begin{bmatrix} 1 & 0 & 0 \end{bmatrix} (x - \widehat{x}). \right.$$

If we assign the closed-loop eigenvalues at $\{-1, -1 \pm j\}$ we get $K = \begin{bmatrix} 4\ 5\ 1 \end{bmatrix}$. Choosing $Q_K = 2I$, then the solution of the Lyapunov equation $A_K^T P_K + P_K A_K = -Q_K$ is

$$P_K = \begin{bmatrix} 3.9 & 2.8 & 0.5 \\ 2.8 & 4.95 & 0.95 \\ 0.5 & 0.95 & 0.65 \end{bmatrix} > 0.$$

For $Q_L = 2I$ the solution of the Lyapunov equation $A_L^T P_L + P_L A_L = -Q_L$ is

$$P_L = \begin{bmatrix} 3.0958 & -1.2583 & -0.8625 \\ -1.2583 & 0.7500 & 0.2583 \\ -0.8625 & 0.2583 & 0.6292 \end{bmatrix} > 0.$$

For the computation of μ^* we require the values:

$$\sigma(P_L^{-1} P_K) = \{0.4098, 1.1196, 44.7432\}$$

$$\sigma(Q_L^{-1}\left(Q_K + C^T L^T P_K\right) Q_K^{-1}\left(Q_K + P_K LC\right)) = \{0.35694, 1, 3353.933\}$$

$$\mu^* = \max\{\mu_3, \mu_4\} = \max\{44.7432, 3353.933\} = 3353.933.$$

Now setting $\mu = 3400 > \mu^*$ yields

$$P_0 = \left[\begin{array}{ccc|ccc} 3.9 & 2.8 & 0.5 & 3.9 & 2.8 & 0.5 \\ 2.8 & 4.95 & 0.95 & 2.8 & 4.95 & 0.95 \\ 0.5 & 0.95 & 0.65 & 0.5 & 0.95 & 0.65 \\ \hline 3.9 & 2.8 & 0.5 & 10525.83 & -4278.33 & -2935.00 \\ 2.8 & 4.95 & 0.95 & -4278.33 & 2550 & 878.33 \\ 0.5 & 0.95 & 0.65 & -2935.00 & 878.33 & 2139.16 \end{array}\right]$$

and its spectrum is $\sigma(P_0) = \{0.4447, 1.6181, 7.3383, 529.6, 1464.4, 13221\}$,

$$Q_0 = \left[\begin{array}{ccc|ccc} 2 & 0 & 0 & 69.2 & 0 & 0 \\ 0 & 2 & 0 & 90.6 & 2 & 0 \\ 0 & 0 & 2 & 20.4 & 0 & 2 \\ \hline 69.2 & 90.6 & 20.4 & 6800 & 0 & 0 \\ 0 & 2 & 0 & 0 & 6800 & 0 \\ 0 & 0 & 2 & 0 & 0 & 6800 \end{array}\right]$$

and its spectrum is $\sigma(Q_0) = \{0.027, 1.9994, 1.9997, 6800, 6800, 6801.9\}$. The output matrix becomes $M_0 = \begin{bmatrix} 0.5 & 0.95 & 0.65 & 0.5 & 0.95 & 0.65 \end{bmatrix}$ and $M = \begin{bmatrix} 0.5 & 0.95 & 0.65 \end{bmatrix}$. The transfer function from the new input v to the new output z becomes

$$H_{zv}(s) = \frac{0.65s^2 + 0.95s + 0.5}{s^3 + 3s^2 + 4s + 2} = \frac{0.65\,(s + 0.731 \pm 0.4846j)}{(s+1)\,(s+1 \pm j)}.$$

The corresponding Nyquist diagrams are shown in Figure 1.3.

Example 3: Consider the following transfer function

$$h(s) = \frac{s-2}{s^2+s-2} = \frac{s-2}{(s-1)\,(s+2)}$$

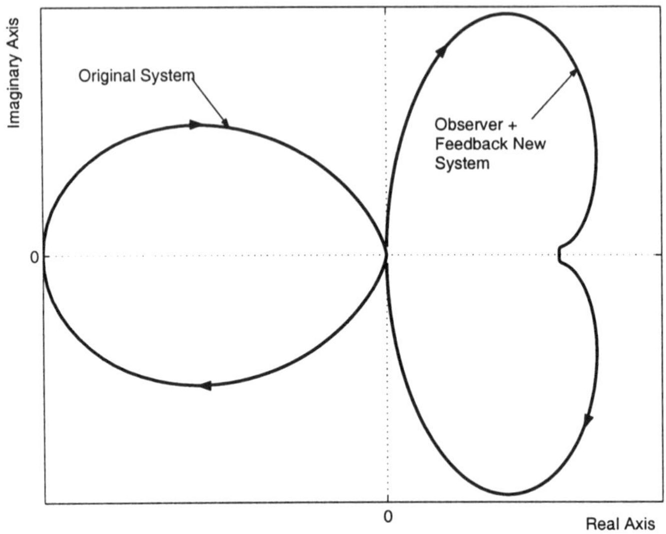

Figure 1.3. Nyquist diagrams for the second example.

which has a nonminimal, but stabilizable and observable state-space representation

$$\Sigma_3 \begin{cases} \dot{x} = \begin{bmatrix} 0 & 1 & 0 \\ 2 & -1 & -2 \\ 0 & 0 & -1 \end{bmatrix} x + \begin{bmatrix} 0 \\ 1 \\ 0 \end{bmatrix} u \\ y = \begin{bmatrix} -2 & 1 & 4 \end{bmatrix} x. \end{cases}$$

A full-order observer for Σ_3 with eigenvalues at $\{-2, -3, -4\}$ is

$$\widehat{\Sigma}_3 \begin{cases} \dot{\widehat{x}} = \begin{bmatrix} 0 & 1 & 0 \\ 2 & -1 & -2 \\ 0 & 0 & -1 \end{bmatrix} \widehat{x} + \begin{bmatrix} 0 \\ 1 \\ 0 \end{bmatrix} u + \underbrace{\begin{bmatrix} -13 \\ -7 \\ -3 \end{bmatrix}}_{L} \begin{bmatrix} -2 & 1 & 4 \end{bmatrix} (x - \widehat{x}) . \end{cases}$$

If we assign the closed-loop eigenvalues at $\{-1, -1 \pm j\}$ we get $K = \begin{bmatrix} 4 & 1 & -2 \end{bmatrix}$. Choosing $Q_K = 2I$, then the solution of the Lyapunov equation $A_K^T P_K + P_K A_K = -Q_K$ is

$$P_K = \begin{bmatrix} 2.5 & 0.5 & 0 \\ 0.5 & 0.75 & 0 \\ 0 & 0 & 1 \end{bmatrix} > 0.$$

For $Q_L = 2I$ the solution of the Lyapunov equation $A_L^T P_L + P_L A_L = -Q_L$ is

$$P_L = \begin{bmatrix} 3.7571 & -3.1286 & -9.8571 \\ -3.1286 & 3.2024 & 7.8619 \\ -9.8571 & 7.8619 & 27.9238 \end{bmatrix} > 0.$$

For the computation of μ^* we require the values:

$$\sigma(P_L^{-1} P_K) = \{0.0286, 0.6767, 20.1044\}$$

$$\sigma(Q_L^{-1}\left(Q_K + C^T L^T P_K\right) Q_K^{-1}\left(Q_K + P_K LC\right)) = \{0.0827, 1, 7626.24\}$$

$$\mu^* = \max\{\mu_3, \mu_4\} = \max\{20.1044, 7626.24\} = 7626.24.$$

Setting $\mu = 7700$ yields

$$P_0 = \left[\begin{array}{ccc|ccc} 2.5 & 0.5 & 0 & 2.5 & 0.5 & 0 \\ 0.5 & 0.75 & 0 & 0.5 & 0.75 & 0 \\ 0 & 0 & 1 & 0 & 0 & 1 \\ \hline 2.5 & 0.5 & 0 & 28930 & -24090 & -75900 \\ 0.5 & 0.75 & 0 & -24090 & 24658 & 60536 \\ 0 & 0 & 1 & -75900 & 60536 & 21501 \end{array}\right]$$

and $\sigma(P_0) = \{0.6171, 0.9999, 2.626, 968.03, 7577.86, 260055.76\}$,

$$Q_0 = \left[\begin{array}{ccc|ccc} 2 & 0 & 0 & 74 & 24 & 6 \\ 0 & 2 & 0 & -36 & -10 & 3 \\ 0 & 0 & 2 & -144 & -47 & -10 \\ \hline 74 & -36 & -144 & 15400 & 0 & 0 \\ 24 & -10 & -47 & 0 & 15400 & 0 \\ 6 & 3 & -10 & 0 & 0 & 15400 \end{array}\right]$$

and $\sigma(Q_0) = \{0.019, 1.9997, 1.9999, 15400, 15400, 15401.9874\}$. The output matrix becomes $M_0 = \begin{bmatrix} 0.5\ 0.75\ 0\ 0.5\ 0.75\ 0 \end{bmatrix}$ and $M = \begin{bmatrix} 0.5\ 0.75\ 0 \end{bmatrix}$. The transfer function from the new input v to the new output z becomes

$$H_{zv}(s) = \frac{0.75s + 0.5}{s^2 + 2s + 2} = \frac{0.75\,(s + 0.6667)}{(s + 1 \pm j)}.$$

The corresponding Nyquist diagrams are shown in Figure 1.4.

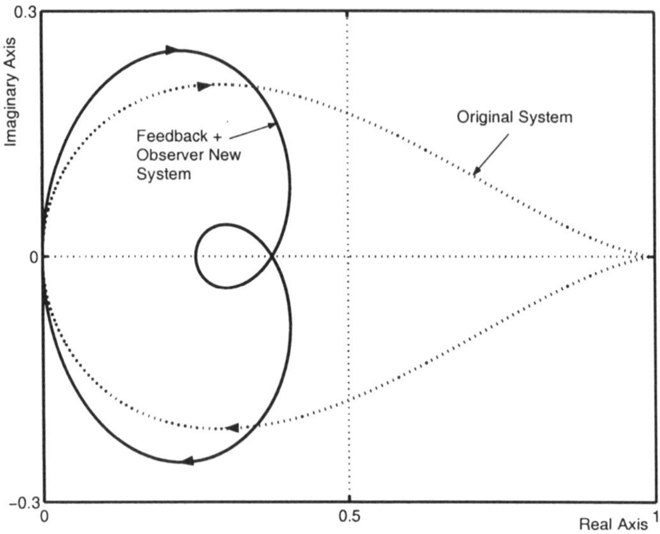

Figure 1.4. Nyquist diagrams for the third example.

6. Concluding Remarks

This chapter has presented a new approach to modify a linear time-invariant system so that the modified system is SPR. The proposed method applies to stable plants as well as unstable systems and does not require the system to be minimum-phase nor to have relative degree one. The original system may have a nonsquare transfer matrix, *i.e.* the number of inputs can be different from the number of outputs. We have proved that the Kalman–Yakubovich–Popov Lemma holds for a series connected system with an observer for stable open-loop systems. In the unstable case, an observer state feedback should be introduced in order to stabilize the system. Some examples failing the relative degree one or minimum phase conditions have been given to illustrate the procedure. Future work in this area includes studying the robustness of the proposed method with respect to parametric uncertainties.

Acknowledgments

This work was also partly supported by the Mexican-French collaboration project LAFMAA.

Bibliography

1. B.R. Andrievsky, A.L. Fradkov, and A.A. Stotsky. Shunt compensation for indirect sliding mode adaptive control. *Proc. 13th IFAC World Congress, San Francisco, California*, pages 193–198, 1996.
2. B.D.O. Anderson and S. Vongpanitlerd. *Network Analysis and Synthesis*. Prentice-Hall, Englewood Cliffs, 1973.
3. B.D.O. Anderson. A system theory criterion for positive real matrices. *SIAM J. Control*, 5(2):171–182, 1967.
4. I. Bar-Kana. Parallel feedforward and simplified adaptive control. *Int. J. Adaptive Control and Signal Processing*, 1:95–109, 1987.
5. C.I. Byrnes, A. Isidori, and J.C. Willems. Passivity, feedback equivalence, and the global stabilization of minimum phase nonlinear systems. *IEEE Trans. Automatic Control*, 36(11):1228–1240, 1991.
6. J. Collado, R. Lozano, and R. Johansson. On Kalman-Yakubovich-Popov Lemma for stabilizable systems. *IEEE Trans. Automatic Control*, 46(7):1089–1093, 2001.
7. R.A. Horn and C.R Johnson. *Matrix Analysis*. Cambridge University Press, reprint with corrections, Cambridge, 1996.
8. R. Johansson and A. Robertsson. Observer-based strict positive real (SPR) feedback control system design. *Automatica*, 38:1557–1564, 2002.
9. R.E. Kalman. Lyapunov functions for the problem of Lur'e in automatic control. *Proc. Nat. Acad. Sci.*, 49:201–205, 1963.
10. H.K. Khalil. *Nonlinear Systems*. Prentice-Hall, Upper Saddle River, 2nd ed., 1996.
11. P.V. Kokotović and H.J. Sussman. A positive real condition for global stabilization of nonlinear systems. *Systems & Control Letters*, 13(2):125–133, 1989.
12. M. Larsen and P.V. Kokotović. On passivation with dynamic output feedback. *IEEE Trans. Automatic Control*, 46(6):962–967, 2001.
13. R. Lozano, B. Brogliato, O. Egeland, and B. Maschke. *Dissipative Systems Analysis and Control*. Springer-Verlag, London, 2000.
14. K.R. Meyer. On the existence of Lyapunov functions for the problem of Lur'e. *SIAM J. Control*, 3(3), 1966.
15. P. Molander and J.C. Willems. Synthesis of state feedback control laws with a specified gain and phase margin. *IEEE Trans. Automatic Control*, 25:928–931, 1980.
16. K.S. Narendra and J.H. Taylor. *Frequency Domain Criteria for Absolute Stability*, Academic Press, New York, 1973.
17. V.V. Prasolov. *Problems and Theorems in Linear Algebra*. American Mathematical Society, Providence, 1994.
18. V.M. Popov. Hyperstability and optimality of automatic systems with several control functions. *Rev. Roumaine Sci. Tech Electrotech Engers.*, 9(4), 1961.
19. A. Rantzer. On the Kalman-Yakubovich-Popov Lemma. *Systems & Control Letters*, 28:7–10, 1996.
20. W. Rugh. *Linear Systems*. Prentice-Hall, Upper Saddle River, 2nd ed., 1996.

21. W. Sun, P.P. Khargonekar, and D. Shim. Solution to the positive real control problem for linear time-invariant systems. *IEEE Trans. Automatic Control*, 39(10):2034–2046, 1994.
22. M. Vidyasagar. *Nonlinear Systems Analysis*. Prentice-Hall, Englewood Cliffs, 2nd ed., 1993.
23. V.A. Yakubovich. Solution of certain matrix inequalities in the stability theory of nonlinear control systems (English translation). *Soviet. Math. Dokl.*, 3:620–623, 1962.
24. V.A Yakubovich, G.A. Leonov, and A.K. Gelig. Stability of stationary sets in control systems with discontinuous nonlinearities. *World Scientific*, 2004. Previously available in Russian under the title "Stability of nonlinear systems with non-unique equilibria", Nauka, 1978.

CHAPTER 2

Nonlinear Control of Feedforward Systems

G. Kaliora and A. Astolfi

Department of Electrical and Electronic Engineering, Imperial College London, SW7 2AZ, United Kingdom, E-mail: {g.kaliora,a.astolfi}@imperial.ac.uk

The stabilization problem for a class of nonlinear feedforward systems is solved using bounded control. It is shown that when the lower subsystem of the cascade is *input-to-state stable* and the upper subsystem not exponentially unstable, global asymptotic stability can be achieved via a simple static feedback having bounded amplitude that requires knowledge of the "upper" part of the state only. This is made possible by invoking the bounded real lemma and a generalization of the small gain theorem. Thus, stabilization is achieved with typical saturation functions, saturations of constant sign, or quantized control. Moreover, the problem of asymptotic stabilization of a stable linear system with bounded outputs is solved by means of dynamic feedback. Finally, a new class of stabilizing control laws for a chain of integrators with input saturation is proposed. Some robustness issues are also addressed and the theory is illustrated with examples on the stabilization of physical systems.

1. Introduction

In this chapter we study the problem of asymptotic stabilization with bounded control of stable cascades described by equations of the form

$$\begin{aligned} \dot{z} &= Jz + h(\xi) + r(\xi)u \\ \dot{\xi} &= f(\xi) + g(\xi)u \end{aligned} \tag{2.1}$$

and some related problems. See Section 2 for the precise formulation and the standing assumptions.

Nonlinear control with saturated signals is a problem that although well studied, see *e.g.* [10,32,33,35] and the references therein, still gathers a lot of interest [9,18,19,22]. Limitations on available energy impose bounded actuator input while it is also very common that due to sensors limitations the outputs of the system are bounded. System (2.1) belongs to the family of systems in feedforward form. This class of systems can be stabilized using the *forwarding approach* or one of its modifications, see *e.g.* [2,10,21,25,29]. On the other hand *nonlinear small gain theorem* based approaches have also been used for the stabilization of these systems [1, 18, 33, 36]. Finally, system (2.1) can (under some special assumptions) also be studied from an *absolute stability* [38] point of view.

Forwarding is a systematic tool for the stabilization of general cascades, a special case of which is the form described by Eqs. (2.1). This methodology requires, in general, the (approximate) solution of a partial differential equation and tends to generate complex control laws. Moreover, although forwarding tackles successfully saturated inputs, it is not a low amplitude design, so it does not impose restrictions on the control amplitude. The Control Lyapunov Function approach can also provide control laws for stabilization in the presence of input constraints, with the use of *universal formulae* [15]. Relevant results are general, however, a large number of the studies on low amplitude designs are typically based on small gain considerations. They also require full state feedback, and in some cases only semiglobal results are provided [7, 17, 27, 32, 34].

From a structural point of view, for systems described by a generalized linear [17, 33, 35] or nonlinear [9, 18] chain of integrators, the control laws consist of a generalization of the nested saturations scheme of [35] or linear combinations of saturations [33]. These designs also make use of passivity, in the sense that, at each step of the procedure, the feedback consists of functions of the state for which the system is always of relative degree one. See also [1] where Teel's nested saturation scheme is robustified against unmodeled dynamics. On the other hand, when linear versions of system (2.1) are considered, an analysis based on absolute stability can be easily implemented and can lead to simple control laws. This way of thinking also provides flexibility and robustness against some classes of perturbations. This is made possible because no phase restriction is imposed, unlike in passivity-based designs.

The results of this chapter are motivated by the observation that under the assumption that the lower subsystem of (2.1) is *input-to-state stable*[a] and locally exponentially stable, an absolute stability point of view can be used in the design of stabilizing saturated controllers for linear as well as nonlinear systems. In a more precise formulation, the results stem from a

---

[a]This condition can be relaxed to ISS with restrictions or to global asymptotic stability.

general result on the L_2 *stability of feedback interconnections* found in [26] and from the *linear bounded real lemma* [8]. The proposed design requires partial state feedback only, and it bears no connection to passivity arguments. As a matter of fact, it will be shown that, in a very clear and natural framework, a number of stabilization issues for system (2.1), such as global asymptotic and input-to-state stabilization (possibly with restrictions), robust stabilization, stabilization with bounded outputs, can be addressed. More specifically, the main contribution of this chapter is the presentation of a new class of bounded control laws for system (2.1).

Following this general result, the linear bounded real lemma and the generalized small gain theorem of [26] are used to solve the following problems in a unified way.

– Robust stabilization of a particular class of systems (2.1) with partial state feedback in the presence of time delays. Under the present framework, these perturbations can be accommodated in a natural way, unlike the case where passivity-based controllers are used. See for example [20] where some robustness issues of the nested saturations scheme have been studied.

– Asymptotic stabilization with control of constant sign.

– *Practical stabilization* with quantized control, *i.e.* it will be shown that a control input taking values in a discrete set can drive the state of the closed-loop system in an arbitrarily small neighborhood of the origin.

In addition, the problem of global asymptotic stabilization for stable linear systems with bounded output is solved via dynamic linear feedback. Stabilization with feedback of perturbed and bounded outputs was achieved via time varying control in [6,14,23] and via dynamic control, that includes state observation, in [16,22]. The dynamic law presented here is not based on state estimation and it is applicable to minimum and nonminimum-phase systems, providing a partial answer to the question raised in [16] about the output stabilizability of systems with unstable zeros in the presence of saturated outputs.

Another byproduct of the main result is a new globally asymptotically stabilizing control law for a chain of integrators in the presence of input saturation which is obtained with recursive application of the main result. This is conceptually and structurally different from the ones of [9,18,33,35]. Moreover, the stabilization of mechanical systems is addressed as an application of the main results. In particular, the *Translational Oscillator with a Rotational Actuator*, commonly referred to as TORA [3], is globally asymptotically stabilized by output feedback. Various constructive nonlinear control methodologies have been tested on this system, see for example [11, 22, 29], while in [24] the problem was set in an Euler–Lagrange framework and a passivity-based output feedback controller was proposed. With the exception of this last reference, all proposed controllers either require the whole state, or utilize some kind of state observation, to achieve global asymptotic stability. In this chapter it will be shown that a simple dynamic output feedback controller of dimension one can globally asymptotically stabilize the TORA, which may be compared with the elaborate stabilization scheme of [11]. Also a preliminary result on the stabilization of underactuated ships moving on a linear course is presented.

As mentioned above, the majority of the results that are presented in this chapter are established from an interconnections point of view, *i.e.* they are proven with the application of a generalization of the small gain theorem. However, all of them can be phrased in Lyapunov stability and invariance principle arguments.

In what follows the construction of the bounded control signals – or the mathematical description of a bounded output – will be achieved with the use of saturation functions. More specifically we shall use three different types of such nonlinearities, all belonging to the sector $[0, 1]$[b], the simplest of which, denoted with $\sigma_s(\cdot)$, $\sigma_+(\cdot)$ and $\sigma_q(\cdot)$, are

$$\begin{aligned} \sigma_s(y) = \operatorname{sign}(y)\min\{|y|, 1\}, \quad \sigma_+(y) = \max\{0, \sigma_s(y)\} \\ \sigma_q(y) = \begin{cases} 0 & \text{for} \quad |y| < 1 \\ \operatorname{sign}(y) & \text{elsewhere.} \end{cases} \end{aligned} \tag{2.2}$$

For the rest of the chapter we use the general symbol $\sigma(\cdot)$ to denote any of the functions (2.2).

The chapter is organized as follows. In Section 2 we present a preliminary result on the stabilization of a cascade consisting of an ISS-LES system driving an integrator. In the same section we formulate the two main problems that are addressed and solved. In Section 3 we present and prove two useful lemmas about the solvability of a matrix inequality. In Section 4 we elaborate on our main result on the stabilization of a nonlinear feedforward system with a bounded, partial state feedback control law. Motivated by the results in Sections 3 and 4, a dynamic control law that solves the problem of asymptotic stabilization of a linear stable SISO system with bounded output is presented in Section 5. In Section 6 we give some applications of our main result. Finally, in Section 7 we provide some conclusions.

---

[b] A nonlinear function $\sigma(y)$ is said to belong to the sector $[k_1, k_2]$ if for all y, $k_1 \leq \frac{\sigma(y)}{y} \leq k_2$.

Comment: With the exception of the saturation functions defined in (2.2), it is assumed that all mappings and functions are at least $\mathcal{C}^1$, throughout the chapter. Note that the saturation functions (2.2) are *piecewise* $\mathcal{C}^1$. Moreover, whenever linear approximations are used these are always considered at the origin and for functions and mappings that are $\mathcal{C}^1$ at the origin. It will become clear that all statements that involve $\sigma_s(\cdot)$ can be applied iteratively.

Notation: The symbol $\|s\|$ is used to denote the Euclidean norm of a vector s.

2. A Motivating Result and Problem Formulation

In this section we show how stabilization of a simple cascade can be obtained using bounded partial state feedback, and we formally state the problems studied in the chapter. Consider a system described by equations of the form

$$\begin{aligned} \dot{z} &= h(\xi) \\ \dot{\xi} &= f(\xi) + g(\xi)u, \end{aligned} \tag{2.3}$$

with $z \in \mathbb{R}$, $\xi \in \mathbb{R}^n$ and $u \in \mathbb{R}$ and assume that the lower subsystem is locally exponentially stable (LES) and input-to-state stable (ISS) with respect to u. We now show that this cascaded system can be stabilized with a simple bounded feedback law that requires knowledge of z only. The rationale behind this result is straightforward. To begin with note that if ϵ is sufficiently small, by LES and ISS of the ξ-subsystem, any trajectory of the closed-loop system will converge to the slice $\|\xi\| < \delta_\epsilon$, where δ_ϵ can be made arbitrarily small reducing ϵ. Note also that $z(t)$ is bounded for all bounded t. On the slice $\|\xi\| < \delta_\epsilon$ the system can be approximated by a linear time invariant system given by the equations

$$\begin{bmatrix} \dot{z} \\ \dot{\xi} \end{bmatrix} = \begin{bmatrix} 0 & H \\ 0 & F \end{bmatrix} \begin{bmatrix} z \\ \xi \end{bmatrix} + \begin{bmatrix} 0 \\ G \end{bmatrix} u, \tag{2.4}$$

where

$$H \triangleq \left. \frac{\partial h(\xi)}{\partial \xi} \right|_{\xi=0}, \quad F \triangleq \left. \frac{\partial f(\xi)}{\partial \xi} \right|_{\xi=0}, \quad G \triangleq g(0).$$

Then the following result can be established.

Proposition 1: *Consider system (2.3) with $f(0) = 0$ and $h(0) = 0$ and the nonlinearity $\sigma_s(\cdot)$. Suppose that*

(H1) the system $\dot{\xi} = f(\xi) + g(\xi)u$ *is ISS with respect to the input* u, *and* $\dot{\xi} = f(\xi)$ *is LES;*

(H2) $HF^{-1}G < 0$.

Then there exist $\kappa^\star > 0$ *and* $\epsilon^\star > 0$, *such that for any* $\kappa \in (0, \kappa^\star)$ *and* $\epsilon \in (0, \epsilon^\star)$ *the closed-loop system*

$$\begin{aligned} \dot{z} &= h(\xi) \\ \dot{\xi} &= f(\xi) - g(\xi)\epsilon\sigma_s(\tfrac{\kappa}{\epsilon}z) \end{aligned} \tag{2.5}$$

is GAS-LES.

Proof: Consider, first, the linear approximation of system (2.3) which is given by system (2.4). Next, rewrite the control law as,

$$u = -\epsilon\sigma_s(\frac{\kappa z}{\epsilon}) = -\kappa z - \left(\epsilon\sigma_s(\frac{\kappa z}{\epsilon}) - \kappa z\right) = -\kappa z - \psi(\kappa z),$$

where $\psi(\kappa z)$ denotes a nonlinearity acting on κz that belongs to the sector $[-1, 0]$. Consider now the system

$$\begin{bmatrix} \dot{z} \\ \dot{\xi} \end{bmatrix} = \begin{bmatrix} 0 & H \\ -\kappa G & F \end{bmatrix} \begin{bmatrix} z \\ \xi \end{bmatrix} - \begin{bmatrix} 0 \\ G \end{bmatrix} \psi(\kappa z), \tag{2.6}$$

and note that, for sufficiently small κ the poles of

$$\begin{aligned} \mathcal{G}_n(s) &= [\kappa \;\; 0] \left(sI - \begin{bmatrix} 0 & H \\ -\kappa G & F \end{bmatrix} \right)^{-1} \begin{bmatrix} 0 \\ G \end{bmatrix} \\ &= \left(\tfrac{\kappa G(sI-F)^{-1}H)}{s} \right) \left(1 + \tfrac{\kappa G(sI-F)^{-1}H)}{s} \right)^{-1} \end{aligned}$$

lie in the left-half of the complex plane, and that

$$\left| \frac{\kappa G(j\omega I - F)^{-1}H}{j\omega} \right| \leq \left| 1 + \frac{\kappa G(j\omega I - F)^{-1}H}{j\omega} \right|. \tag{2.7}$$

Hence, the system

$$\begin{aligned} \begin{bmatrix} \dot{z} \\ \dot{\xi} \end{bmatrix} &= \begin{bmatrix} 0 & H \\ -\kappa G & F \end{bmatrix} \begin{bmatrix} z \\ \xi \end{bmatrix} - \begin{bmatrix} 0 \\ G \end{bmatrix} w, \\ \eta &= \kappa z \end{aligned} \tag{2.8}$$

with input w and output η is asymptotically (exponentially) stable and has an L_2-gain not larger than one. Therefore, by the circle criterion (or the small gain theorem) we conclude GAS-LES of system (2.6).

Consider now system (2.5) and note that, by LES and the ISS property of the ξ-subsystem, if ϵ is sufficiently small[c], there exists a finite time $t_o > 0$,

---

[c] Recall that $|\sigma_s(\cdot)| \leq 1$.

such that for all $t \geq t_o$, $\|\xi(t)\| \leq c_1\epsilon$, for some positive number c_1. Now rewrite system (2.5) as

$$\begin{bmatrix} \dot{z} \\ \dot{\xi} \end{bmatrix} = \begin{bmatrix} 0 & H + \delta_H(\xi) \\ -\kappa G & F + \delta_F(\xi, u) \end{bmatrix} \begin{bmatrix} z \\ \xi \end{bmatrix} - \begin{bmatrix} 0 \\ G \end{bmatrix} \psi(\kappa z), \tag{2.9}$$

and note that there exist positive constants c_H and c_F, such that for all $t \geq t_o$

$$\|\delta_H(\xi(t))\| \leq c_H\epsilon$$
$$\|\delta_F(\xi(t), u(t))\| \leq c_F\epsilon.$$

System (2.9) can be regarded as a perturbed linear system with perturbations $\delta_H(\cdot)$ and $\delta_F(\cdot)$ that can be rendered asymptotically arbitrarily small reducing ϵ. Note also that the perturbations are such that, if ϵ is sufficiently small, all but one of the eigenvalues of the family of systems (2.9) with $\kappa = 0$ are in the left part of the complex plane, with the remaining eigenvalue at the origin. We conclude that there exists $\epsilon^\star > 0$ such that for all $\epsilon \in (0, \epsilon^\star)$, and for all κ sufficiently small, every element in the family of transfer functions

$$\mathcal{G}(s) = [\kappa \;\; 0] \left(sI - \begin{bmatrix} 0 & H + \Delta_H \\ -\kappa G & F + \Delta_F \end{bmatrix} \right)^{-1} \begin{bmatrix} 0 \\ G \end{bmatrix}$$

with $\|\Delta_H\| \leq c_H\epsilon$ and $\|\Delta_F\| \leq c_F\epsilon$ has L_2-gain not larger than one. As a result, by the small gain theorem (or the circle criterion), system (2.5) is GAS-LES. □

Remark 1: Note that, if the pair $\{F, G\}$ is controllable, system (2.3) is controllable if and only if $HF^{-1}G \neq 0$. Moreover, Hypothesis (H2) is not restrictive. In fact, if $HF^{-1}G > 0$ the result of Proposition 1 holds with $\kappa^\star < 0$.

We remark that ϵ is the level of saturation, whereas κ is the feedback gain, or in other words, κz is the appropriate output that needs to be fed back. An interesting extension of Proposition 1 would be the iterative application of the methodology proposed. Indeed this is possible, as it will be discussed in the following sections, where, it will be proven that the closed-loop system (2.5) is also *ISS with restrictions* with respect to a new external input.

The result of Proposition 1 can be interpreted as a consequence of the circle criterion, hence this facilitates the handling of a series of system uncertainties, such as time delays. While it is known that passivity-based designs may be inadequate in the presence of delays, the result in Proposition 1 is robust against (constant) time delays in the input or output path, as summarized in the following corollary.

Corollary 1: *Consider system (2.3) and a positive constant τ. Under the assumptions of Proposition 1, there exists a positive $\kappa_1^\star = \kappa_1^\star(\tau)$ and an $\epsilon^\star > 0$ such that for all $\kappa_1 \in (0, \kappa_1^\star)$ and $\epsilon \in (0, \epsilon^\star)$ the control law*

$$u = -\epsilon\sigma_s\left(\frac{\kappa_1}{\epsilon}z(t-\tau)\right) \tag{2.10}$$

globally asymptotically (locally exponentially) stabilizes system (2.3).

Proof: Note that, as before, $|\sigma_s(\frac{\kappa_1}{\epsilon}z(t-\tau))| \leq 1$ for any positive constant τ, thus if ϵ is small enough ξ will eventually be such that $\|\xi\| < \delta_\epsilon$ for some small enough constant $\delta_\epsilon > 0$. In this slice of the state space we consider the system

$$\begin{aligned} \dot{z}(t) &= H\xi(t) \\ \dot{\xi}(t) &= F\xi(t) - G\epsilon\sigma_s(\tfrac{\kappa_1}{\epsilon}z(t-\tau)). \end{aligned} \tag{2.11}$$

If $G_\xi(s)$ is the transfer function of the open-loop ξ-subsystem with output $H\xi$, then the transfer function of the system

$$\begin{aligned} \dot{z}(t) &= H\xi(t) \\ \dot{\xi}(t) &= F\xi(t) + Gu \\ \eta &= z(t-\tau) \end{aligned}$$

is $\frac{e^{-s\tau}G_\xi(s)}{s}$. Note now that the Nyquist diagram of $\frac{G_\xi(s)}{s}$ is bounded from the left by a vertical line, say through the point $(-\frac{1}{\kappa}, 0)$. Then there exists a positive number $\kappa_1 < \kappa$ such that the Nyquist diagram of $\frac{e^{-s\tau}G_\xi(s)}{s}$ is also bounded from the left by a vertical line through the point $(-\frac{1}{\kappa_1}, 0)$. To see this, note that the term $e^{-j\omega\tau}$ does not modify the amplitude of $\frac{G_\xi(j\omega)}{j\omega}$ and does not introduce any phase shift for $\omega \to 0$. The conclusion, therefore, follows as an application of the circle criterion. □

Remark 2: Corollary 1 provides a "delay dependent" stability result, *i.e.* the closed-loop system is not asymptotically stable for any τ, but only for $0 \leq \tau \leq \tau^\star$. However, unlike other delay dependent criteria, the result in Corollary 1 is constructive, *i.e.* for any delay τ an appropriate stabilizing feedback (2.10) can be found.

We are now ready to state formally the stabilization problems dealt with in this chapter.

Partial state feedback stabilization problem. Consider a system described by equations of the form

$$\begin{aligned} \dot{z} &= Jz + h(\xi) + r(\xi)u \\ \dot{\xi} &= f(\xi) + g(\xi)u, \end{aligned} \tag{2.12}$$

where $z \in \mathbb{R}^p$, $\xi \in \mathbb{R}^n$ and $u \in \mathbb{R}$ and suppose the following.

(A1) The system $\dot{\xi} = f(\xi) + g(\xi)u$ is ISS with respect to u, and $\dot{\xi} = f(\xi)$ is LES;

(A2) $J + J' \leq 0$[d].

Find (if possible) a positive constant ϵ and an output

$$\eta = Kz \tag{2.13}$$

such that (2.12) in closed loop with the control law

$$u = -\epsilon\sigma(\frac{1}{\epsilon}\eta) + v \tag{2.14}$$

is LES and ISS with restrictions with respect to v.
Regarding this problem we define the following matrices

$$\begin{gathered} H \triangleq \left.\frac{\partial h(\xi)}{\partial \xi}\right|_{\xi=0} \in \mathbb{R}^{p\times n}, \quad F \triangleq \left.\frac{\partial f(\xi)}{\partial \xi}\right|_{\xi=0} \in \mathbb{R}^{n\times n}; \\ R \triangleq r(0) \in \mathbb{R}^{p\times 1}, \; G \triangleq g(0) \in \mathbb{R}^{n\times 1}, \end{gathered} \tag{2.15}$$

$$A \triangleq \begin{bmatrix} J & H \\ \mathbf{0} & F \end{bmatrix}, \quad B \triangleq \begin{bmatrix} R \\ G \end{bmatrix}, \quad C \triangleq [I \; \mathbf{0}], \tag{2.16}$$

and the approximation of (2.12) for small ξ, given by

$$\begin{bmatrix} \dot{z} \\ \dot{\xi} \end{bmatrix} = A \begin{bmatrix} z \\ \xi \end{bmatrix} + Bu. \tag{2.17}$$

Remark 3: Note that, as proved in [30], if the subsystem $\dot{\xi} = f(\xi)$ is GAS-LES, assumption (A1) is without loss of generality because the control can always be rescaled appropriately, provided that the whole state is measurable. However, if this rescaling is undesired[e], the minimal assumption under which the partial state feedback stabilization problem above is

[d]Assumption (A2) can be replaced by $J'S + SJ \leq 0$ for some $S = S' > 0$.

[e]This is the case when not the whole state is measurable, or when the requirements on the control signal amplitude cannot be fulfilled when a feedback transformation is applied.

solvable, in the context of this work, is that the system is *ISS with some restriction* [18]. For example the system $\dot{x} = -x^3 + (1 + x^3)u$ is not ISS, but it is ISS with the restriction $|u| < 1$.

The second problem that will be solved in the chapter is the problem of asymptotic stabilization of a linear stable SISO system $\dot{z} = Jz + \bar{H}w$ when the available output is subject to saturation. This is formally stated as follows.

Bounded output stabilization problem. Consider a nonlinearity $\sigma(\cdot)$ and a system described by equations of the form

$$\begin{aligned} \dot{z} &= Jz + \bar{H}w \\ y &= \sigma(Kz), \end{aligned} \tag{2.18}$$

with $z \in \mathbb{R}^p$, $w \in \mathbb{R}$ and $y \in \mathbb{R}$. Suppose (A2) holds. Find (if possible) a dynamic control law

$$\begin{aligned} \dot{\xi} &= F\xi - Gy \\ w &= \Gamma\xi \end{aligned} \tag{2.19}$$

such that the closed-loop system (2.18)–(2.19) is GAS-LES.

3. Two Useful Lemmas

In this section we present two lemmas that are instrumental in proving the main results of the chapter. They are both related to the existence of solutions for a special *matrix inequality*. Note that the proofs of both these lemmas are constructive, *i.e.* we provide a family of solutions of the considered matrix inequality.

Lemma 1: *Let A, B, and C be defined as in (2.16) and suppose $\{A, B\}$ is controllable, $F \in \mathbb{R}^{n\times n}$ is a Hurwitz matrix and $J \in \mathbb{R}^{p\times p}$ is such that (A2) holds. Then, there exist $P \in \mathbb{R}^{(n+p)\times(n+p)}$ and $K \in \mathbb{R}^{1\times p}$ such that*

$$\begin{aligned} (KC - B'P)'(KC - B'P) + PA + A'P &\leq 0 \\ P = P' &> 0, \end{aligned} \tag{2.20}$$

and $A_{cl} = A - BKC$ is Hurwitz.

Proof: Let P be defined as

$$P \triangleq \begin{bmatrix} \chi I & Y \\ Y' & P_\xi \end{bmatrix} \tag{2.21}$$

with χ a positive constant and $P_\xi = P_\xi' > 0$ to be selected. As a result, the inequality (2.20) rewrites

$$\begin{bmatrix} \chi(J+J') & \chi H+YF+J'Y \\ \chi H'+F'Y'+Y'J & Y'H+H'Y+P_\xi F+F'P_\xi \end{bmatrix} + \begin{bmatrix} K'-\chi R-YG \\ -Y'R-P_\xi G \end{bmatrix} \begin{bmatrix} K'-\chi R-YG \\ -Y'R-P_\xi G \end{bmatrix}' \leq 0. \tag{2.22}$$

Setting

$$K = \chi R' + G'Y' \tag{2.23}$$

the problem is translated into finding matrices Y and P_ξ such that (2.22) holds. To this end, note that Y can always be selected such that

$$\chi H + YF + J'Y = 0, \tag{2.24}$$

for all H, F, J and all positive constants χ. With K and Y defined by (2.23) and (2.24), (2.22) reduces to

$$\begin{bmatrix} \chi(J+J') & 0 \\ 0 & T+P_\xi F+F'P_\xi+P_\xi GG'P_\xi \end{bmatrix} \leq 0,$$

where

$$\begin{aligned} T &= Y'H + H'Y + Y'RR'Y + Y'RG'P_\xi + P_\xi GR'Y \\ &= Y'(H + RG'P_\xi) + (H' + P_\xi GR')Y + Y'RR'Y. \end{aligned}$$

Since $\chi(J + J') \leq 0$, the problem is reduced to finding a P_ξ such that

$$T + P_\xi F + F'P_\xi + P_\xi GG'P_\xi \leq 0. \tag{2.25}$$

To solve this problem, let $\bar{P}_\xi$ be such that $\bar{P}_\xi F' + F\bar{P}_\xi = -(GG' + I)$. Then, setting $P_\xi = \bar{P}_\xi^{-1}$ yields

$$P_\xi F + F'P_\xi + P_\xi GG'P_\xi = -P_\xi P_\xi < 0. \tag{2.26}$$

Hence, it is sufficient to show that T can be made arbitrarily small. To this end, notice that the solution of (2.24) is

$$Y = \chi\bar{Y} \tag{2.27}$$

where $\bar{Y}$ is the solution of

$$H + \bar{Y}F + J'\bar{Y} = 0. \tag{2.28}$$

Therefore

$$T = \chi(\bar{Y}'(H + RG'P_\xi) + (H' + P_\xi GR')\bar{Y}) + \chi^2(\bar{Y}'RR'\bar{Y})$$

and this can be made arbitrarily small by a proper selection of $\chi > 0$.

Besides, P constructed as above is positive definite. To prove this, note that, following standard decomposition arguments, P is positive definite if and only if $P_\xi - \chi \bar{Y}'\bar{Y}$ is positive definite, which is true for a positive definite P_ξ and small enough χ. Therefore there exists a positive χ such that condition (2.20) holds.

To complete the proof we need to show that $A_{cl} = A - BKC$ is Hurwitz. To this end, observe that inequality (2.20) is equivalent to

$$A'_{cl}P + PA_{cl} + PBB'P + C'K'KC \leq 0, \tag{2.29}$$

which yields

$$P^{-1}A'_{cl} + A_{cl}P^{-1} \leq -BB' \leq 0. \tag{2.30}$$

On the other hand, it is trivial to check that if $\{A, B\}$ is controllable $\{B', A_{cl}\}$ is observable. From that and from inequality (2.30), according to [39], it is concluded that A_{cl} is Hurwitz. □

Remark 4: Inequality (2.29) arises in the non-standard H_∞ control problem [8] described by the equations

$$\begin{aligned} \dot{x} &= Ax + Bu + Bw \\ z &= u \\ y &= Cx, \end{aligned} \tag{2.31}$$

where A, B and C are as in (2.16), u is the control input, w is the exogenous input, z is the penalty variable and y is the measurement. Lemma 1 expresses the fact that there exists a static output feedback control law $u = -Ky$ rendering system (2.31) asymptotically stable and with a L_2-gain from w to z less than or equal to one. Note that if J has eigenvalues on the $j\omega$ axis then $\gamma = 1$ is the smallest achievable L_2-gain for system (2.31), *i.e.* any static or dynamic output feedback stabilizing controller yields a closed-loop system with L_2-gain larger or equal to one.

Lemma 2: *Let $J \in \mathbb{R}^{p\times p}$, $\bar{H} \in \mathbb{R}^{p\times 1}$ and $K \in \mathbb{R}^{1\times p}$ be known matrices such that (A2) holds, $\{J', \bar{H}\}$ is controllable and $\{K, J\}$ is observable. Then, there exist $P \in \mathbb{R}^{(2p)\times(2p)}$, $G \in \mathbb{R}^{p\times 1}$, $\Gamma \in \mathbb{R}^{1\times p}$ and a Hurwitz matrix $F \in \mathbb{R}^{p\times p}$ such that (2.20) holds, with $R = 0$, $H = \bar{H}\Gamma$ and A, B, and C as in (2.16), and the matrix $A_{cl} = A - BKC$ is Hurwitz.*

Proof: Partition P as in (2.21) and repeat the first steps of the proof of Lemma 1. However, note that we are looking now for F, G and Γ. Let F be a Hurwitz matrix with distinct eigenvalues, and L be such that $\mathbf{spec}(J' + \bar{H}L) = \mathbf{spec}(-F)$. Note that such an L exists because of controllability of the pair $\{J', \bar{H}\}$. Then there exists a nonsingular matrix X such that

$$J'X + \bar{H}LX + XF = 0.$$

Therefore, setting $Y = \chi X$, for some positive χ, solves the Sylvester equation (2.24) with $H = \bar{H}LX$. Next, set $\Gamma = LX$, $G = Y^{-1}K'$ and let $\bar{P}_\xi$ be the positive definite matrix that solves the Lyapunov equation

$$\bar{P}_\xi F' + F\bar{P}_\xi = -\left(X^{-1}K'K(X^{-1})' + I\right).$$

Choosing $P_\xi = \chi^2 \bar{P}_\xi^{-1}$, it is easy to verify that the first of inequalities (2.20) holds for a large enough $\chi > 0$. On the other hand, with the above selections for P_ξ and Y, the matrix P, is positive definite for a large enough $\chi > 0$. Observe, now, that inequality (2.20), or the equivalent inequality (2.29), yields

$$A'_{cl}P + PA_{cl} \leq -(PBB'P + C'K'KC) \leq -C'K'KC \leq 0, \tag{2.32}$$

and that observability of the pair $\{K, J\}$ implies detectability of the pair $\{KC, A_{cl}\}$. As a result, by [39], A_{cl} is Hurwitz. □

Remark 5: In conjunction with what was stated in Remark 4, consider the non-standard H_∞ control problem described by the equations

$$\begin{aligned} \dot{x} &= Jx + \bar{H}u \\ z &= Kx \\ y &= Kx + w, \end{aligned} \tag{2.33}$$

where J is such that (A2) holds, u is the control input, w is the exogenous input, z is the penalty variable and y is the measurement. Lemma 2 expresses the fact that there exists a dynamic output feedback control law, of the same dimension as system (2.33), described by equations of the form

$$\begin{aligned} u &= \Gamma\xi \\ \dot{\xi} &= F\xi - Gy, \end{aligned} \tag{2.34}$$

such that the closed-loop system (2.33)–(2.34) is asymptotically stable and with an L_2-gain from w to z less than or equal to one. Note that if J has eigenvalues on the $j\omega$ axis then an L_2-gain equal to one is the smallest achievable gain, with any output feedback.

Remark 6: The results in Lemmas 1 and 2 can be trivially given a multivariable control extension. Namely, under the assumptions of Lemma 1, for $G \in \mathbb{R}^{n\times m}$ and $R \in \mathbb{R}^{p\times m}$ there exist a matrix $K \in \mathbb{R}^{m\times p}$ and a positive definite matrix $P \in \mathbb{R}^{(n+p)\times(n+p)}$ such that (2.20) holds and $A_{cl} = A - BKC$ is Hurwitz. Similarly, under the assumptions of Lemma 2, for $\bar{H} \in \mathbb{R}^{p\times m}$ and $K \in \mathbb{R}^{q\times p}$ (system with m inputs and q outputs) there exist matrices $G \in \mathbb{R}^{p\times q}$, $\Gamma \in \mathbb{R}^{m\times p}$ and a positive definite $P \in \mathbb{R}^{2p\times 2p}$ such that (2.20) holds.

4. Stabilization with Bounded Partial State Feedback

In this section we provide our main result on the stabilization, with partial state feedback bounded control, of systems described by Eqs. (2.12).

Proposition 2: *Consider a nonlinearity $\sigma(\cdot)$ belonging to the sector $[0,1]$ and the system described by the equations*

$$\begin{aligned} \dot{z} &= Jz + h(\xi) + r(\xi)u \\ \dot{\xi} &= f(\xi) + g(\xi)u, \end{aligned} \tag{2.35}$$

with $z \in \mathbb{R}^p$, $\xi \in \mathbb{R}^n$, $u \in \mathbb{R}$ and $f(0) = 0$, $h(0) = 0$. Suppose (A1) and (A2) hold and moreover, assume the following.

(C1) The linear approximation of (2.35) is controllable.

Then there exists $\epsilon^\star > 0$ and a matrix $K \in \mathbb{R}^{1\times p}$ such that if $\epsilon \in (0, \epsilon^\star)$, the static partial state feedback control law

$$u = -\epsilon\sigma(\frac{1}{\epsilon}Kz) \tag{2.36}$$

globally stabilizes system (2.35). Moreover, if

(C2) all trajectories $z(t)$ of $\dot{z}(t) = Jz(t)$ such that $\sigma(Kz(t)) = 0$, for all $t \geq 0$, are such that $\lim_{t\to\infty} z(t) = 0$,

then (2.36) globally asymptotically (locally exponentially) stabilizes system (2.35). Furthermore, the system

$$\begin{aligned} \dot{z} &= Jz + h(\xi) - r(\xi)\epsilon\sigma_s(\tfrac{1}{\epsilon}Kz) + r(\xi)w \\ \dot{\xi} &= f(\xi) - g(\xi)\epsilon\sigma_s(\tfrac{1}{\epsilon}Kz) + g(\xi)w, \end{aligned} \tag{2.37}$$

is ISS with respect to the new input w, with the restriction $|w| \leq \mu$, with $\mu < \epsilon$.

Remark 7: The second claim of Proposition 2 holds with the choice of the "nonlinearity" $\sigma(s) = 0$, for all $s \in \mathbb{R}$. This is due to (C2), which, in this case, implies that system (2.35) is the interconnection of two asymptotically stable systems, possessing bounded trajectories and operating in open loop.

Remark 8: A similar result has been proven in [10] on the basis of the results in [36]. Note, however that the result of [10] requires, in general, full state feedback, and that the result in Proposition 2 is based on a different construction. As a result Proposition 2 can also be used in the design of output feedback control laws (see Section 5), in the design of quantized or constant sign controllers (see Corollary 2 and Corollary 3) and when dealing with some robustness problems (see Corollary 1). In fact, the assumptions on the system in [10] are different to the assumptions in Proposition 2. Therein, the construction uses the fact that the pair $\{J, R\}$ is stabilizable, while a cross-term corresponding to $h(\xi)$ of system (2.35) is assumed to be of order at least two. Under these assumptions the feedback used in [10] is of the form (2.36), but this time K is such that $J - RK$ is Hurwitz. From what will become clear from the proof of Proposition 2 and the examples presented in the rest of the chapter, it is obvious that the two results are not addressing the same problem. For example, Proposition 2 also covers the case where J is skew symmetric and $R = 0$, *i.e.* the pair $\{J, R\}$ is not stabilizable, and the upper subsystem is driven entirely by ξ.

Proof: As discussed earlier, because of (A1), there exists $\epsilon^\star > 0$ such that if $\epsilon \in (0, \epsilon^\star)$, the state of the closed-loop system (2.35)–(2.36) will in finite time enter a small enough "slice" where $\|\xi\| < \delta_\epsilon$, for an arbitrarily small $\delta_\epsilon > 0$. There, we can consider the approximation of system (2.35) for small $\|\xi\|$, as explained in the proof of Proposition 1, in other words, it suffices to study the stabilization with bounded control problem for system (2.17) to obtain stabilization results for the nonlinear system (2.12).

Denoting $x = [z' \quad \xi']'$, the state-space equations of the cascade (2.12) and the output described in the partial state feedback stabilization problem (Eq. (2.13)) are written as

$$\begin{aligned} \dot{x} &= Ax + Bu \\ \eta &= KCx. \end{aligned} \tag{2.38}$$

Let K be a matrix such that the linear feedback $u_l = -KCx$ exponentially stabilizes system (2.17). The proposed control law (2.36) can be written as

$$u = -\epsilon\sigma(\frac{1}{\epsilon}\eta) = -\left[KCx + \left(\epsilon\sigma(\frac{1}{\epsilon}\eta) - KCx\right)\right] = -KCx - \psi(\eta),$$

where $\psi(\eta)$ is a new nonlinearity restricted to the sector $[-1, 0]$. Note that up to now, K is some matrix that sets A_{cl} to be Hurwitz. However, to prove stability in the presence of the nonlinearity $\psi(\eta)$ a special "stabilizing"

K has to be selected. To this end, note that system (2.38)–(2.36) can be regarded as the feedback interconnection of the system

$$\begin{aligned} \dot{x} &= A_{cl}x + Bv \\ \eta &= KCx, \end{aligned} \tag{2.39}$$

where $A_{cl} \triangleq A - BKC$, with $v = -\psi(\eta)$. Moreover, the L_2-gain of $\psi(\eta)$ is not larger than one, hence, selecting K satisfying inequality (2.20) for some $P > 0$, yields[f], by Assumption (C2) and the *generalized small gain theorem* in [26], an asymptotically stable closed-loop system. Moreover, A_{cl} is a Hurwitz matrix, from Lemma 1.

To complete the proof of Proposition 2, we need to prove the ISS property of system (2.37)[g]. First notice, that if $\epsilon \in (0, \epsilon^\star)$, for any w such that $|w| < \epsilon$, in finite time, all trajectories of the nonlinear system (2.37) will eventually be such that $\|\xi(t)\| < \delta'_\epsilon$ for all $t \geq \bar{t}$. Therein we consider the approximation of system (2.37) for small $\|\xi\|$

$$\dot{x} = Ax - B\epsilon\sigma_s(\frac{\eta}{\epsilon}) + Bw, \tag{2.40}$$

and we prove that it is ISS with some restriction on w. To this end, consider the positive definite function $V = x'Px$, with P as defined in (2.21). Along the trajectories of (2.40) one has

$$\dot{V} = x'(A'P + PA)x - 2x'PB\left(\epsilon\sigma_s(\frac{\eta}{\epsilon}) - w\right), \tag{2.41}$$

where $\eta = Kz$. With simple calculations, using (2.24) and (2.27), it is easy to see that

$$x'(A'P + PA)x = \chi z'(J + J')z - \xi'\left(Q_\xi - \chi(H'\bar{Y} + \bar{Y}'H)\right)\xi, \tag{2.42}$$

where Q_ξ is a positive definite matrix and P_ξ is the positive definite solution of the Lyapunov equation $F'P_\xi + P_\xi F = -Q_\xi$. Note that P_ξ and Q_ξ are as in the proof of Lemma 1. Note also that, by Assumption (A2), $J + J'$ is negative semi-definite and that

$$PB = \begin{bmatrix} \chi R + YG \\ Y'R + P_\xi G \end{bmatrix} = \begin{bmatrix} K' \\ Y'R + P_\xi G \end{bmatrix}. \tag{2.43}$$

---

[f]Recall that, by Remark 4 and Lemma 1, system (2.39) has a L_2-gain (H_∞-norm) less than or equal to one.
[g]Note that the symmetric nonlinearity $\sigma_s(s)$ is used.

As a result, by simple manipulations, Eq. (2.41) becomes

$$\begin{aligned}\dot{V} &= \chi z'(J+J')z - \xi'\left(Q_\xi - \chi(H'\bar{Y}+\bar{Y}'H)\right)\xi \\ &\quad -2\xi'(Y'R+P_\xi G)\left(\epsilon\sigma_s(\tfrac{\eta}{\epsilon})-w\right) \\ &\quad -2\eta\left(\epsilon\sigma_s(\tfrac{\eta}{\epsilon})-w\right)+\left[\epsilon\sigma_s(\tfrac{\eta}{\epsilon})-w\right]^2-\left[\epsilon\sigma_s(\tfrac{\eta}{\epsilon})-w\right]^2,\end{aligned}$$

hence

$$\begin{aligned}&\dot{V} \leq -\left[\epsilon\sigma_s(\tfrac{1}{\epsilon}\eta)-w\right]\left(2\eta-\left[\epsilon\sigma_s(\tfrac{1}{\epsilon}\eta)-w\right]\right) - \\ &-\left[\epsilon\sigma_s(\tfrac{1}{\epsilon}\eta)-w,\ \xi'\right]\begin{bmatrix} 1 & \chi R'\bar{Y}+G'P_\xi \\ \chi\bar{Y}'R+P_\xi G & Q_\xi-\chi(H'\bar{Y}+\bar{Y}'H)\end{bmatrix}\begin{bmatrix}\epsilon\sigma_s(\tfrac{1}{\epsilon}\eta)-w \\ \xi\end{bmatrix}.\end{aligned} \tag{2.44}$$

The matrix

$$M = \begin{bmatrix} 1 & \chi R'\bar{Y}+G'P_\xi \\ \chi\bar{Y}'R+P_\xi G & Q_\xi-\chi(H'\bar{Y}+\bar{Y}'H)\end{bmatrix} \tag{2.45}$$

is positive definite by construction, as shown in the proof of Lemma 1. Under the restriction $|w| < \epsilon$, we see that the following implications hold

$$\begin{aligned}|\eta| > |w| &\Rightarrow -\left[\epsilon\sigma_s(\tfrac{1}{\epsilon}\eta)-w\right]\left(2\eta-\left[\epsilon\sigma_s(\tfrac{1}{\epsilon}\eta)-w\right]\right) < 0 \\ &\Rightarrow \dot{V} \leq 0\,.\end{aligned}$$

This means that the system (2.40) with output η is *input-to-output stable* with some (nonzero) restriction on w. Using the result in [31], to prove ISS it is sufficient to show that the pair $\{KC, A\}$ is detectable. To this end, note that the matrix $A'_{cl} = A' - C'K'B'$ is Hurwitz, therefore the pair $\{A', C'K'\}$ is stabilizable. Thus, by [39], the pair $\{KC, A\}$ is detectable. This completes the proof of Proposition 2. □

Remark 9: In the light of Remark 6, if we consider a system of the form (2.35) where $u \in \mathbb{R}^m$, then, under the assumptions of Proposition 2, there exists a matrix $K \in \mathbb{R}^{m\times p}$ with $K = [k_1,\ k_2,\ \cdots,\ k_m]'$ such that the control law

$$u = -\begin{bmatrix}\epsilon_1\sigma_s(\frac{1}{\epsilon_1}k_1 z) \\ \epsilon_2\sigma_s(\frac{1}{\epsilon_2}k_2 z) \\ \vdots \\ \epsilon_m\sigma_s(\frac{1}{\epsilon_m}k_m z)\end{bmatrix} \tag{2.46}$$

globally asymptotically stabilizes the underlying system.

Remark 10: System (2.35) with output η is not, in general, minimum-phase, nor with relative degree one. This fact distinguishes the present stabilization method from a family of other nonlinear control results that rely on some passivity property of the system, see for example [10], or even the results in [35] and [33].

It is easy to see that cascades with a simple integrator for the upper system (see also Proposition 1) belong to the class of systems described by Eqs. (2.35) with $J = 0$ and $r(\xi) = 0$. In this case we can name the "desired output" mentioned in the stabilization problem as $\eta = \kappa z$, where κ is as described in Proposition 1. In general, when integrators are present, special attention has to be given to the choice of the nonlinearity $\sigma(\cdot)$. Note for example, that using the nonnegative nonlinearity $\sigma_+(\cdot)$ for the system (2.3) we cannot achieve GAS, since there are no isolated equilibria (the trajectories of the system can converge to any point $[z_-,\ 0]$, where $z_- \leq 0$). However, when J is a full rank matrix, the equilibrium is always uniquely defined, hence GAS can be achieved. On the other hand, if the aim is not to globally asymptotically stabilize system (2.35) but to *practically stabilize* it, *i.e.* to achieve convergence to a small enough neighborhood of the origin, then the saturation function could be like $\sigma_q(\cdot)$ of (2.2). This discussion can be formally summarized as follows.

Corollary 2: *Consider system (2.35). Suppose conditions (A1) and (A2) hold. Suppose moreover that J is a full rank matrix. Then there exists $\epsilon^\star > 0$ and a matrix $K \in \mathbb{R}^{1\times p}$ such that if $\epsilon \in (0, \epsilon^\star)$, the static partial state feedback control law*

$$u = -\epsilon\sigma_+(\frac{1}{\epsilon}Kz) \quad \left(or \quad u = \epsilon\sigma_+(-\frac{1}{\epsilon}Kz) \right) \tag{2.47}$$

globally asymptotically stabilizes system (2.35). Moreover, $u(t) \leq 0$ (or $u(t) \geq 0$[h]) for all $t \geq 0$.

Proof: Note that if J is a full rank matrix, then the linear approximation of system (2.35) is controllable as long as the ξ-subsystem is controllable, and the matrices H and R are not both zero. Also, the half space defined

[h]Note that $\sigma_+(-\eta) = -\sigma_-(\eta)$, where $\sigma_-(\cdot)$ is defined in similar way to $\sigma_+(\cdot)$, but is equal to zero for all $\eta > 0$. Like the nonlinearities (2.2), $\sigma_-(\cdot)$ also belongs to the sector $[0, 1]$.

by

$$\mathbb{S} \triangleq \{z \in \mathbb{R}^p : \sigma_+(\frac{1}{\epsilon}Kz) = 0\} = \{z \in \mathbb{R}^p : Kz \leq 0\}$$

$$\left(\text{or} \quad \mathbb{S} \triangleq \{z \in \mathbb{R}^p : \sigma_+(-\frac{1}{\epsilon}Kz) = 0\} = \{z \in \mathbb{R}^p : Kz \geq 0\}\right)$$

contains the point $z = 0$ but does not contain any neighborhood of $z = 0$. Therefore the only trajectory of $\dot{z} = Jz$ contained in $\mathbb{S}$ is such that $\lim_{t\to\infty} z(t) = 0$[i]. As a result, conditions (C1) and (C2) are satisfied, and the result follows from Proposition 2. □

Corollary 3: *Consider system (2.3). Suppose that assumptions (H1) and (H2) of Proposition 1 hold. Then there exist* $\kappa^\star > 0$, $\epsilon^\star > 0$ *and* $t_o \in \mathbb{R}_+$ *such that for any* $\kappa \in (0, \kappa^\star)$ *and* $\epsilon \in (0, \epsilon^\star)$ *all trajectories* $(z(t), \xi(t))$ *of the closed-loop system*

$$\begin{aligned} \dot{z} &= h(\xi) \\ \dot{\xi} &= f(\xi) - g(\xi)\epsilon\sigma_q(\tfrac{\kappa}{\epsilon}z) \end{aligned} \tag{2.48}$$

are such that

$$\lim_{t\to\infty} \xi(t) = 0, \quad \textit{and} \ |z(t)| \leq \epsilon/\kappa, \, \forall t \geq t_o.$$

Proof: As in the proof of Propositions 1 and 2, we focus on the approximated system for small $\|\xi\|$. For such a system consider the Lyapunov function

$$V = [z,\ \xi'] \begin{bmatrix} \chi & Y \\ Y' & P_\xi \end{bmatrix} \begin{bmatrix} z \\ \xi \end{bmatrix} \tag{2.49}$$

with $P_\xi = P'_\xi > 0$ such that $F'P_\xi + P_\xi F = -P_\xi GG'P_\xi - P_\xi P_\xi = -Q_\xi$[j], $Q_\xi = Q'_\xi > 0$ and $Y = -\chi HF^{-1}$. Along the trajectories of (2.48) one has:

$$\dot{V} = -\xi'(Q_\xi + \chi(H'HF^{-1} + F^{-T}H'H))\xi - 2\kappa z\epsilon\sigma_q(\frac{\kappa}{\epsilon}z) - 2\xi'P_\xi G\epsilon\sigma_q(\frac{\kappa}{\epsilon}z), \tag{2.50}$$

with $\kappa = -\chi HF^{-1}G > 0$. Consider now the following two exclusive cases.

---

[i]This is due to the fact that because $\det J \neq 0$, the system $\dot{z} = Jz$ has no trajectory with a component of the form $z_i(t) = c$, with $c \neq 0$.

[j]See also Eq. (2.26) in the proof of Lemma 1.

- $|z| \geq \epsilon/\kappa$. In this case $\sigma_q(\frac{\kappa}{\epsilon}z) \neq 0$, hence, using the fact that $-y\sigma(y) \leq -[\sigma(y)]^2$, $\dot{V} \leq -W_1(x)$, with

$$W_1(x) = [\epsilon\sigma_q(\frac{\kappa}{\epsilon}z),\ \xi'] \begin{bmatrix} 1 & G'P_\xi \\ P_\xi G & Q_\xi + \chi(H'HF^{-1} + F^{-T}H'H) \end{bmatrix} \begin{bmatrix} \epsilon\sigma_q(\frac{\kappa}{\epsilon}z) \\ \xi \end{bmatrix}.$$

  It is easy to see that, for a small enough χ, $W_1(x) > 0$.
- $|z| < \epsilon/\kappa$. In this case $\sigma_q(\frac{\kappa}{\epsilon}z) = 0$ and

$$\dot{V} = -\xi'(Q_\xi + \chi(H'HF^{-1} + F^{-T}H'H))\xi = -W_2(\xi).$$

  Note that the matrix $Q_\xi+\chi(H'HF^{-1}+F^{-T}H'H)$ can be made positive definite with an appropriate choice of Q_ξ and a small enough $\chi > 0$.

From the above, we can see that $\dot{V}(x)$ is bounded from above by a negative semidefinite function, namely

$$\dot{V}(x) \leq -\min\{W_1(x), W_2(\xi)\} \leq 0.$$

As a result, by LaSalle's invariance principle, the trajectories of system (2.48) are bounded and asymptotically converging to the set

$$\left\{z \in \mathbb{R} : \sigma_q(\frac{\kappa}{\epsilon}z) = 0\right\} \times \{\xi = 0\} = \left\{z \in \mathbb{R} : |z| < \frac{\epsilon}{\kappa}\right\} \times \{\xi = 0\}. \qquad \square$$

The extension of Corollary 3 for system (2.35) is straightforward and is omitted here for the sake of brevity, see [12].

In Figures 2.1 and 2.2 we illustrate the conclusion of Corollaries 2 and 3 with some simulation results for a fourth-order system with states $z_1, \ldots, z_4$ and control u. The open-loop eigenvalues are at $\pm 2j$, $\pm j$. The "chattering" of the control signal observed in the top graphs of Figure 2.2 can be reduced if, instead of the simple quantized nonlinearity $\sigma_q(\cdot)$ of (2.2), we use a nonlinearity with *hysteresis*. In the bottom graphs of Figure 2.2 we show the improved simulation results, where hysteresis has been implemented.

5. Stabilization with Sensors Saturations

In this section we consider the asymptotic stabilization of linear stable systems for which the measured output is subject to a constraint, for example the case where the measurement device has some range limitations. Consider a SISO linear system with saturated output, namely

$$\begin{aligned} \dot{z} &= Jz + \bar{H}w \\ \eta &= \sigma(Kz), \end{aligned} \tag{2.51}$$

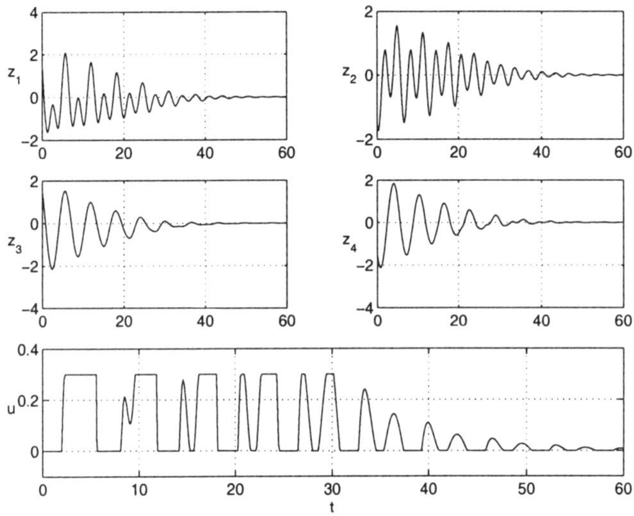

Figure 2.1. A fourth-order linear system with two pairs of open-loop imaginary eigenvalues, in closed loop with a positive control law of the form (2.47).

with J such that (A2) is satisfied (*i.e.* $J + J' \leq 0$). The goal of this section is to show that system (2.51) is globally asymptotically stabilizable by dynamic output feedback, as illustrated in the following proposition.

Proposition 3: *Consider system (2.51) with $z \in \mathbb{R}^p$, $w \in \mathbb{R}$, $\eta \in \mathbb{R}$, and J and $\sigma(\cdot)$ such that Assumptions (A2) and (C2) hold. Assume that the pair $\{J', \bar{H}\}$ is controllable and the pair $\{K, J\}$ is observable. Then there exist matrices $\Gamma \in \mathbb{R}^{1\times p}$, $G \in \mathbb{R}^{p\times 1}$ and a Hurwitz matrix $F \in \mathbb{R}^{p\times p}$ such that (2.51) in closed loop with the dynamic controller*

$$\begin{aligned} \dot{\xi} &= F\xi - G\eta \\ w &= \Gamma\xi \end{aligned} \tag{2.52}$$

is globally asymptotically (locally exponentially) stable.

Proof: It is trivial to verify that the closed loop system (2.51)–(2.52) is described by equations of the form

$$\begin{aligned} \dot{z} &= Jz + \bar{H}\Gamma\xi \\ \dot{\xi} &= F\xi + Gu \\ u &= -\sigma(Kz), \end{aligned} \tag{2.53}$$

i.e. it is the feedback interconnection of a system of the form (2.17) with $H = \bar{H}\Gamma$ and $R = 0$, and the nonlinear feedback $u = -\sigma(Kz)$. Hence,

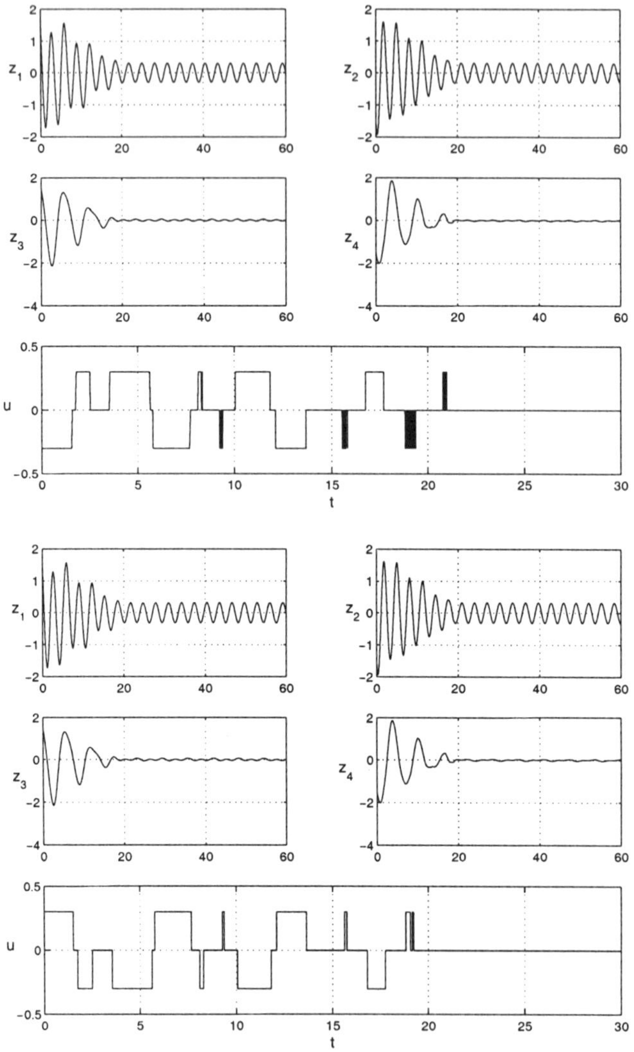

Figure 2.2. A fourth-order linear system with two pairs of open-loop imaginary eigenvalues, in closed loop with a control law of the form (2.36) with a quantized saturation function $\sigma_q(s)$ (top graphs) and with a quantized control law with hysteresis (bottom graphs).

selecting Γ, G and a Hurwitz matrix F as in the proof of Lemma 2 and using arguments similar to those in the proof of Proposition 2, it follows that the interconnection is globally asymptotically stable. □

Remark 11: Proposition 3 can be easily extended, using the same arguments as in the proof of Corollary 2 and Corollary 3, to the case $\sigma(\cdot) = \sigma_+(\cdot)$, provided that $\det(J) \neq 0$, or to the case $\sigma(\cdot) = \sigma_q(\cdot)$, if one is interested in practical, rather than asymptotic, stability.

It should be noted that the result of Proposition 3 is not restricted by the sign of the system zeros, *i.e.* it is applicable to both minimum and nonminimum-phase systems. In light of Remarks 6 and 9 it is also applicable to MIMO systems. Other extensions and discussions on the bounded output stabilization problem are discussed in detail in [13].

6. Applications

In this section we consider some applications of the main results of Section 4, namely the global stabilization of a chain of integrators with bounded input, the global asymptotic stabilization of linear null controllable systems by positive (negative) control, the global asymptotic stabilization of the benchmark TORA system and the global asymptotic stabilization of underactuated ships moving on a linear course.

6.1. *Stabilization of a chain of integrators with bounded control revisited*

The problem of global asymptotic stabilization of a chain of integrators with bounded control has been extensively studied by several researchers. In this section we revisit it, and in light of the results of Propositions 1 and 2, we present a novel stabilizing bounded control law, complete with some remarks on its robustness.

Proposition 4: *Consider the system*

$$\begin{aligned} \dot{x}_1 &= x_2 \\ &\vdots \\ \dot{x}_{n-1} &= x_n \\ \dot{x}_n &= u. \end{aligned} \tag{2.54}$$

There exist positive numbers $\lambda_1, \lambda_2, \ldots, \lambda_{n-1}, \lambda_n$ such that, for any $\epsilon > 0$, system (2.54) in closed loop with

$$u = -\frac{\epsilon}{2}\sigma_s(\frac{2\lambda_n}{\epsilon}x_n) - \frac{\epsilon}{4}\sigma_s(\frac{4\lambda_{n-1}}{\epsilon}x_{n-1}) - \cdots - \frac{\epsilon}{2^n}\sigma_s(\frac{2^n\lambda_1}{\epsilon}x_1) + w \tag{2.55}$$

is LES and ISS with the restriction $|w| < \frac{\epsilon}{2^{n+1}}$. Moreover, if $w = 0$, $|u| < \epsilon$.

Proof: The proof can be carried out iteratively. To this end, set $u = -\frac{\epsilon}{2}\sigma_s(\frac{2\lambda_n}{\epsilon}x_n) + v_{n-1}$ and note that the system

$$\begin{aligned} \dot{x}_{n-1} &= x_n \\ \dot{x}_n &= -\tfrac{\epsilon}{2}\sigma_s(\tfrac{2\lambda_n}{\epsilon}x_n) + v_{n-1} \end{aligned} \tag{2.56}$$

satisfies the assumptions of Proposition 1[k] for every $\lambda_n > 0$. It is also obvious that the last equation of (2.56) represents an ISS system with the restriction $|v_{n-1}| < \frac{\epsilon}{2}$. As a result, there exists a positive λ_{n-1} such that

$$v_{n-1} = -\frac{\epsilon}{4}\sigma_s(\frac{4\lambda_{n-1}}{\epsilon}x_{n-1}) + v_{n-2}$$

achieves input-to-state stability of system (2.56), with the restriction $|v_{n-2}| < \frac{\epsilon}{4}$, and local exponential stability for $v_{n-2} = 0$, according to Propositions 1 and 2.

The proof is then completed by recursive application of Proposition 1. We remark, that at each step the positive constant $\lambda_i \in (0, \lambda_i^\star)$ that will achieve absolute stability (see the proof of Proposition 2 or the proof of Proposition 1) will automatically belong to the set of positive λ_i that would achieve exponential stability, if linear feedback was used. Also, we can see that, at each step i, the transfer function of the system

$$\begin{aligned} \dot{x}_i &= x_{i+1} \\ &\vdots \\ \dot{x}_n &= -\lambda_n x_n - \cdots - \lambda_{i+1}x_{i+1} + v_i \end{aligned}$$

from the input v_i to the output $y_i = x_i$ will have one eigenvalue at the origin, $n - i$ eigenvalues on the left half complex plane and no zeros. Using the root locus we can see that for a small enough positive λ_i, the feedback $v_i = -\lambda_i x_i$ will achieve exponential stability. Finally, by a trivial property of the geometric series, if $w = 0$,

$$|u| \leq \frac{\epsilon}{2} + \frac{\epsilon}{4} + \frac{\epsilon}{8} + \cdots + \frac{\epsilon}{2^n} + \frac{\epsilon}{2^{n+1}} < \epsilon, \tag{2.57}$$

and ϵ can be arbitrarily selected. □

Remark 12: The design option that the saturation levels should follow the geometric series $\epsilon/2, \epsilon/2^2, \ldots, \epsilon/2^n$ is academic, namely it is considered for

[k]In fact, the lower subsystem of system (2.56) is ISS *with restrictions*. See also Remark 3.

the case of an infinite chain of integrators because of the property (2.57). In practical situations, one can use the feedback

$$u = -\epsilon_n \sigma_s(\frac{\lambda_n}{\epsilon_n} x_n) - \cdots - \epsilon_2 \sigma_s(\frac{\lambda_2}{\epsilon_2} x_2) - \epsilon_1 \sigma_s(\frac{\lambda_1}{\epsilon_1} x_1),$$

where, if u_{max} is the maximum available control energy, the constants $\epsilon_1, \epsilon_2, \ldots, \epsilon_n$ must be such that

$$\begin{aligned} \epsilon_n &> \epsilon_{n-1} + \cdots + \epsilon_2 + \epsilon_1 \\ &\vdots \\ \epsilon_2 &> \epsilon_1 \\ \epsilon_1 &> 0 \\ u_{max} &\geq \epsilon_n + \cdots + \epsilon_2 + \epsilon_1. \end{aligned} \tag{2.58}$$

The feasibility of the above system of inequalities is trivial, since we know at least one solution, for example, $\epsilon_i = \frac{1}{2^{n-i+1}} u_{max}$. Replacing the last inequality in (2.58) with the equality constraint $\epsilon_n + \cdots + \epsilon_2 + \epsilon_1 = u_{max}$ we can treat the problem of finding the appropriate set of ϵ_i as an optimization problem. This approach allows us to increase the saturation level in the feedback of the upper component x_1 enhancing the overall performance of the closed-loop system.

Remark 13: System (2.54) is a special case of the class of systems studied in [19]. Therein, a similar construction has been performed. However, in the proposed design the *saturating gains*, namely $\frac{\epsilon}{2}$, $\frac{\epsilon}{4}$, ..., are constants, whereas in [19] the gains are functions of the state and have to satisfy some nontrivial conditions. Finally, for large values of $\|x\|$ the *saturating gains* in [19] tend to zero, and this is not the case for the control law (2.55).

The result in Proposition 4 can be easily extended to a larger class of systems, namely nonlinear chains of integrators described by equations of the form

$$\begin{aligned} \dot{x}_1 &= \phi_1(x_2) \\ &\vdots \\ \dot{x}_{n-1} &= \phi_{n-1}(x_n) \\ \dot{x}_n &= \phi_n(u) \end{aligned} \tag{2.59}$$

with $d\phi_i(0) > 0$, for all $i = 1, \ldots, n$. For illustration purposes consider the system described by the equations

$$\dot{x}_1 = \sin(x_2),\ \dot{x}_2 = \sin(x_3),\ \dot{x}_3 = \sin(x_4),\ \dot{x}_4 = \sin(u). \tag{2.60}$$

In Figure 2.3 the response of system (2.60) in closed loop with

$$u = -\frac{\epsilon}{2}\sigma_s(\frac{2\lambda_4}{\epsilon}x_4) - \frac{\epsilon}{4}\sigma_s(\frac{4\lambda_3}{\epsilon}x_3) - \frac{\epsilon}{8}\sigma_s(\frac{8\lambda_2}{\epsilon}x_2) - \frac{\epsilon}{16}\sigma_s(\frac{16\lambda_1}{\epsilon}x_1) \quad (2.61)$$

is presented. For this particular case, global asymptotic stability can be achieved if $\epsilon \in (0, \epsilon^\star]$ with $\epsilon^\star < \frac{\pi}{2}$. In the particular simulations we used $\epsilon = \frac{\pi}{4}$ and $[\lambda_1, \lambda_2, \lambda_3, \lambda_4] = [0.008, 0.108, 0.540, 1.20]$.

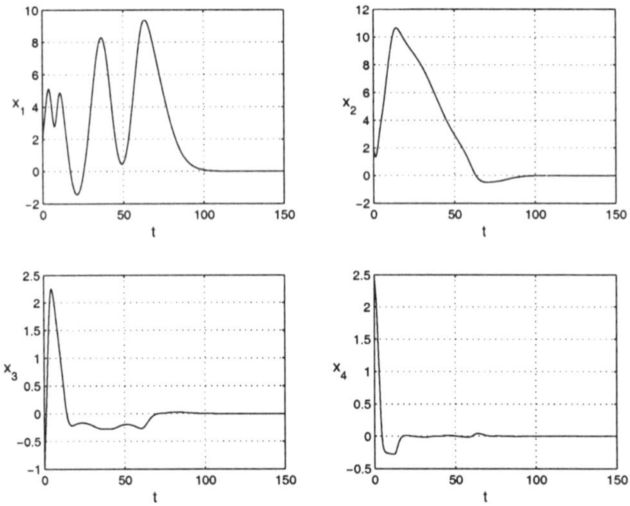

Figure 2.3. State histories of the closed-loop system (2.60)–(2.61).

Remark 14: Output feedback stabilization of system (2.54) with output $\eta = x_1$ can be addressed by a straightforward application of Proposition 4 and [33]. Finally, chains of integrators of the form

$$\dot{x}_1 = \alpha_1 x_2, \ \cdots, \ \dot{x}_{n-1} = \alpha_{n-1}x_n, \ \dot{x}_n = \alpha_n u, \quad (2.62)$$

where $0 < \underline{\alpha}_j \leq \alpha_j \leq \overline{\alpha}_j$, $j = 1, \ldots, n$ and the limits $\underline{\alpha}_j$, $\overline{\alpha}_j$ are known, can be treated following the steps of the proof of Proposition 4. Robust stabilization of system (2.62) in the presence of uncertain system parameters has also been studied in [18]. The nested saturation scheme employed there also required some nontrivial algebraic conditions to be satisfied.

6.2. *Asymptotic stabilizability by control of constant sign*

In this section we present a general result on the asymptotic stabilizability of linear stable systems with bounded control of constant sign, that is a consequence of Proposition 2 or Corollary 2.

Proposition 5: *Any stable and controllable linear system*

$$\dot{x} = Ax + Bu, \tag{2.63}$$

with A such that $\det(A) \neq 0$, *is* asymptotically stabilizable *by positive (or negative) control.*

Proof: Note first that because $\det(A) \neq 0$, the matrix A has no zero eigenvalue. It can be verified that, under the assumptions of Proposition 5, the system (2.63) can be written, in a set of suitable coordinates, in the form

$$\begin{aligned} \dot{z} &= Jz + H\xi + Ru \\ \dot{\xi} &= F\xi + Gu, \end{aligned} \tag{2.64}$$

where $z \in \mathbb{R}^p$, $\xi \in \mathbb{R}^m$, $p + m = n$, and $J + J' \leq 0$, $\det(J) \neq 0$. The last equation of the cascade (2.64) represents the asymptotically stable part of (2.63), if there is any, *i.e.* F is Hurwitz. In the case that such an asymptotically stable part does not exist, it is easy to verify that the system $\dot{z} = Jz + Ru$ is *globally asymptotically stabilized* by the control law

$$u = -\frac{\epsilon}{2}\sigma_+(\frac{2}{\epsilon}Kz) \quad \left(\text{or} \quad u = \frac{\epsilon}{2}\sigma_+(-\frac{2}{\epsilon}Kz)\right) \tag{2.65}$$

for all $\epsilon > 0$ and for some appropriately chosen[l] $K \in \mathbb{R}^{1\times p}$. According to Corollary 2, a similar control law[m] would also stabilize the cascade (2.64) if the asymptotically stable part exists. □

Remark 15: The results in Proposition 5 and Corollary 2 should be examined in the light of what established in [28], where it was proven that a linear system $\dot{x} = Ax + Bu$ is locally controllable at the origin with $u(t) \in [0, 1]$, for all t, if and only if the pair $\{A, B\}$ is controllable in the ordinary sense and all eigenvalues of A have nonzero imaginary parts.

---

[l]In fact, in this case a "good" saturated linear feedback can be obtained by invoking standard passivity arguments.
[m]Not necessarily with the same K.

Using [28] and [4] it is easy to show that a linear system is asymptotically controllable with positive (or negative) bounded control if and only if $\mathbf{spec}(A) \subset \mathbf{C}^- \cup \{\mathbf{C^0} \setminus \{0\}\}$.

6.3. *Asymptotic stabilization of the TORA*

In this section we apply the results of Section 4 to solve the asymptotic stabilization problem for the TORA [3]. After appropriate normalized transformations [37], the dynamics of the system are described by the equations

$$\begin{aligned} \ddot{x}_d + x_d &= \gamma(\dot{\phi}^2 \sin\phi - \ddot{\phi}\cos\phi) \\ \ddot{\phi} &= u - \gamma \ddot{x}_d \cos\phi, \end{aligned} \tag{2.66}$$

where x_d is the translational position, $v_d = \dot{x}_d$ the translational velocity, ϕ the angular position, $\omega = \dot{\phi}$ the angular velocity and $0 < \gamma < 1$ a constant depending on the physical parameters of the device. The presence of the term in ω^2 in the model makes the stabilization of the system an intricate problem, especially considering that an ideal control law would utilize measurements of the translational and angular positions only. It is shown in [11] that, via a coordinates transformation, system (2.66) can be written in the form

$$\begin{aligned} \dot{x}_1 &= x_2 \\ \dot{x}_2 &= -x_1 + \gamma \sin x_3 \\ \dot{x}_3 &= \frac{1}{\psi(x_3)} x_4 \\ \dot{x}_4 &= \frac{\psi(x_3)}{1 - \gamma^2 \cos^2 x_3} u + \frac{\gamma(x_1 - \gamma \sin x_3)\cos x_3}{1 - \gamma^2 \cos^2 x_3} \psi(x_3), \end{aligned} \tag{2.67}$$

where

$$\psi(x_3) = -\sqrt{\frac{1 - \gamma^2 \cos^2 x_3}{1 - \gamma^2}}.$$

The measured variables are x_1 and x_3, which are functions of the translational and angular positions only. In [11] global output feedback stabilization and tracking was achieved with a combination of a nonlinear observer and backstepping. Here, we propose a simpler output feedback scheme. First, consider the preliminary feedback transformation

$$u = u(v, x_1, x_3) = \frac{1 - \gamma^2 \cos^2 x_3}{\psi(x_3)} v - \gamma(x_1 - \gamma \sin x_3)\cos x_3, \tag{2.68}$$

and the subsystem

$$\begin{aligned} \dot{x}_3 &= \alpha(x_3) x_4 \\ \dot{x}_4 &= v, \end{aligned} \tag{2.69}$$

with $\alpha(x_3) = \frac{1}{\psi(x_3)}$ and output $y = x_3$. Stabilization of (2.69) can be achieved using the dynamic output feedback

$$\begin{aligned} \dot{\hat{\theta}} &= -(\lambda(x_3)+1)\hat{\theta} - (\lambda(x_3)+1)\beta(x_3) - x_3\alpha(x_3) + (b+1)w \\ v &= -x_3\alpha(x_3) - \hat{\theta} - \beta(x_3) + w, \end{aligned} \tag{2.70}$$

where w is a new control variable to be used in the next step, $\beta(x_3)$ and $\lambda(x_3)$ are functions satisfying

$$\frac{\partial \beta(x_3)}{\partial x_3}\alpha(x_3) = \lambda(x_3), \quad \lambda(x_3) > 0 \tag{2.71}$$

and b is a constant to be selected later. Asymptotic stabilization can be proven considering the new coordinate $z = \hat{\theta} + \beta(x_3) - x_4$, and noting that system (2.69)–(2.70) can be written as

$$\begin{aligned} \dot{x}_3 &= \alpha(x_3)x_4 \\ \dot{x}_4 &= -x_3\alpha(x_3) - x_4 - z + w \\ \dot{z} &= -\lambda(x_3)z + bw, \end{aligned} \tag{2.72}$$

which is LES-ISS. Condition (2.71) can be seen either as a differential equation for the definition of $\beta(x_3)$, if $\lambda(x_3)$ is selected by the designer, or as the definition of $\lambda(x_3)$, if $\beta(x_3)$ is selected. For example, $\beta(x_3) = -\beta x_3$ with $\beta > 0$ is a simple choice. Consider now the cascade

$$\begin{aligned} \dot{x}_1 &= x_2 \\ \dot{x}_2 &= -x_1 + \gamma \sin x_3 \\ \dot{x}_3 &= \alpha(x_3)x_4 \\ \dot{x}_4 &= -x_3\alpha(x_3) - x_4 - z + w \\ \dot{z} &= -\lambda(x_3)z + bw, \end{aligned} \tag{2.73}$$

that results from the first two of Eqs. (2.67) and the system (2.72) and note that it satisfies assumptions (A1) and (A2), hence can be asymptotically stabilized with bounded control of the form

$$w = -\epsilon\sigma_s\left(\frac{1}{\epsilon}K\begin{bmatrix} x_1 \\ x_2 \end{bmatrix}\right), \tag{2.74}$$

with K selected as in Lemma 1[n]. However, to obtain an output feedback controller, K has to be of the form $K = [k_1 \ \ 0]$, for some k_1. We now show that such a K exists. To this end, consider the approximation of (2.73) for

[n]Note that using the saturation functions $\sigma_+(\cdot)$ or $\sigma_-(\cdot)$ would also yield GAS.

small $\|[x_3\ x_4\ z]'\|$, and define the matrices

$$J = \begin{bmatrix} 0 & 1 \\ -1 & 0 \end{bmatrix}, \quad H = \begin{bmatrix} 0 & 0 & 0 \\ \gamma & 0 & 0 \end{bmatrix}, \quad F = \begin{bmatrix} 0 & \alpha & 0 \\ -\alpha & -1 & -1 \\ 0 & 0 & -\lambda \end{bmatrix}, \quad G = \begin{bmatrix} 0 \\ 1 \\ b \end{bmatrix}, \tag{2.75}$$

where $\alpha = \alpha(0)$ and $\lambda = \lambda(0)$. From Lemma 1

$$K = \chi G'\bar{Y} = \chi[\bar{Y}_{1,2} + b\bar{Y}_{1,3} \quad \bar{Y}_{2,2} + b\bar{Y}_{2,3}]$$

where χ is a positive constant, $\bar{Y}$ is the matrix that solves the Sylvester equation $H + \bar{Y}F + J'\bar{Y} = 0$ and $\bar{Y}_{i,j}$ is the (i,j) entry of $\bar{Y}$. Selecting $b = -\bar{Y}_{2,2}/\bar{Y}_{2,3}$ yields $K = [k_1\ \ 0]$, hence $w = w(x_1)$. This is possible, as it can be shown that with $\gamma \neq 0$ and $\alpha = -1$, $\bar{Y}_{2,3} \neq 0$ for all λ. The result is summarized in the following statement.

Proposition 6: *Consider system (2.67) and a nonlinearity $\sigma(\cdot) = \sigma_s(\cdot)$ or $\sigma(\cdot) = \sigma_+(\cdot)$, belonging to the sector $[0,1]$. There exist constants b, k_1, $\epsilon \in \mathbb{R}$, with $\epsilon > 0$ and a positive function $\lambda(x_3)$ such that system (2.67) in closed loop with the dynamic output feedback controller*

$$\begin{aligned} \dot{\hat{\theta}} &= -(\lambda+1)\hat{\theta} - (\lambda+1)\beta(x_3) - x_3\alpha(x_3) - (b-1)\epsilon\sigma\left(\frac{1}{\epsilon}k_1x_1\right) \\ v &= -x_3\alpha(x_3) - \hat{\theta} - \beta(x_3) - \epsilon\sigma\left(\frac{1}{\epsilon}k_1x_1\right) \\ u &= \frac{1-\gamma^2\cos^2 x_3}{1-\psi(x_3)}v - \gamma(x_1 - \gamma\sin x_3)\cos x_3 \end{aligned} \tag{2.76}$$

is GAS (LES).

The control law (2.76) is much simpler in structure and implementation than the output feedback designs proposed in [11] or [22], while in [29] only state feedback is considered. In Figure 2.4 some simulation results of the closed loop with the proposed controller are depicted. For the simulations we have used, as in [29], $\gamma = 0.1$, so that the results are directly comparable with the ones given in this reference. It can be concluded that full state feedback does not outperform the output feedback presented here.

6.4. *Stabilization of underactuated ships on a linear course*

In this section we apply the result of Proposition 2 for the global asymptotic stabilization of a normalized model of an underactuated ship moving on a linear course. The model examined is taken from [5], were the authors

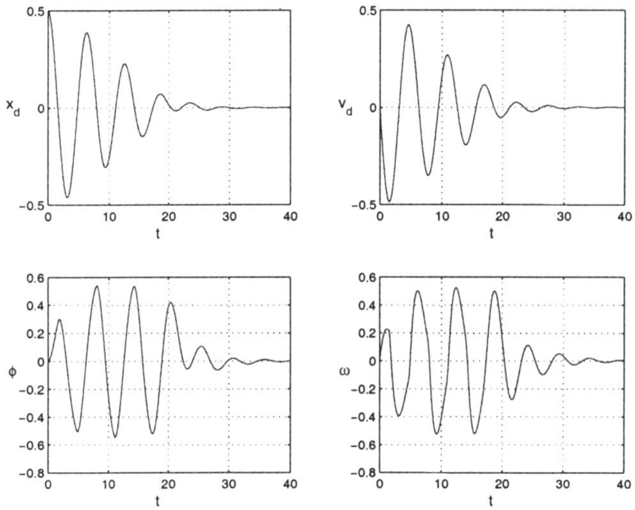

Figure 2.4. State histories of the closed-loop system (2.67)–(2.76).

designed state and output feedback controllers based on the backstepping technique and nonlinear observers. Their controllers achieve global tracking of a straight line in the presence of non-vanishing environmental disturbances, that occur due to wave, wind and ocean current. Such a model is given by

$$\begin{aligned} \dot{y} &= u\sin(\psi) + \cos(\psi)\nu \\ \dot{\psi} &= r \\ \dot{\nu} &= -\frac{m_1 u}{m_2} r - \frac{d_2}{m_2}\nu - \frac{d_{\nu 2}}{m_2}|\nu|\nu + \frac{1}{m_2}\tau_{w\nu}(t) \\ \dot{r} &= \frac{(m_1 - m_2)u}{m_3}\nu - \frac{d_3}{m_3} r - \frac{d_{r2}}{m_2}|r|r + \frac{1}{m_3}\tau_r + \frac{1}{m_3}\tau_{wr}(t) \end{aligned} \tag{2.77}$$

where, y, ν are the sway displacement (deviation from the course on the axis vertical to the ship axis) and velocity and ψ, r are the yaw angle and velocity. The forward speed, that is controlled independently by the main thruster control system, is given by u, and is considered constant, or slowly varying. The control action is represented by τ_r, the torque applied to the ship rudder. The positive constants m_i, $i = 1, 2, 3$ denote the ship inertia with respect to the three axis, including added mass, and the positive constants d_2, d_3, $d_{\nu 2}$, d_{r2} denote the hydrodynamic damping in sway and yaw. The terms $\tau_{w\nu}(t)$, $\tau_{wr}(t)$ represent the environmental disturbance moments and are considered to be bounded.

System (2.77) is in block feedforward form, *i.e.* we can distinguish the interconnection of the subsystem of $[\nu \ r]'$ with the integrator $\dot{\psi} = r$ and, at the next step, the interconnection of the subsystem of $[\psi \ \nu \ r]'$ with the subsystem $\dot{y} = u\sin(\psi) + \cos(\psi)\nu$. The inertia, m_2, around the second axis of the ship is always larger than the inertia, m_1, around the first axis which implies that the linear part of the subsystem of $[\nu \ r]'$ is exponentially stable ($m_1 - m_2 < 0$). In addition, the nonlinear damping terms $-\frac{d_{\nu 2}}{m_2}|\nu|\nu$ and $-\frac{d_{r2}}{m_2}|r|r$ do not "disturb" this stability property, so it is easy to verify that all assumptions of Proposition 2 are satisfied for the cascade of the subsystem of $[\nu \ r]'$ with the integrator $\dot{\psi} = r$. A bounded control law – feedback of ψ only – can be designed. Next, repeating the procedure once more, we obtain a stabilizing controller for the cascade (2.77), namely

$$\tau_r = -\frac{\epsilon}{2}\sigma_s(\frac{2\lambda_1}{\epsilon}\psi) - \frac{\epsilon}{2.1}\sigma_s(\frac{2.1\lambda_2}{\epsilon}y), \tag{2.78}$$

where λ_1 and λ_2 are suitably chosen. Note that we have used the arguments in Remark 12 to enhance the performance of the controller. To illustrate the properties of the closed-loop system (2.77)–(2.78) via simulations we consider a simplified situation where $m_1 = 1$, $m_2 = 2$, $m_3 = 1$, $d_2 = 2$, $d_3 = 2$, $d_{\nu 2} = 0.1$, $d_{r2} = 0.1$, $\tau_{w\nu}(t) = 0$ and $\tau_{wr}(t) = 0$, and the nominal forward speed is $u = 1$. For this set of parameters, appropriate gains for the controller (2.78) are $(\lambda_1, \lambda_2) = (1.4, 0.86)$ and $\epsilon = 5$. In Figure 2.5 we depict the state histories and the control action of the closed-loop system (2.77)–(2.78).

7. Conclusions

The problem of stabilization of a class of cascaded systems with bounded control has been addressed and solved using the linear bounded real lemma and a generalized version of the small gain theorem. Globally asymptotically stabilizing control laws that require only partial state feedback have been designed. These control laws make use of typical saturating functions, constant sign saturations or quantizations and they exhibit a simple structure, however, in some cases, they require the amplitude of the control signal to be kept small enough.

The main results are applied to the global stabilization problem for a chain of integrators subject to input saturation, yielding a control law that is significantly different from existing results and also to the global stabilization of the nonlinear benchmark system of TORA and to the stabilization of underactuated ships moving on a linear course. At the same

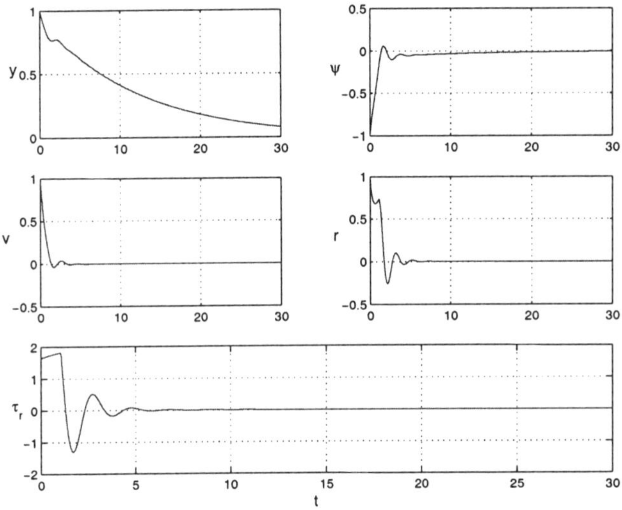

Figure 2.5. State histories and control action of the closed-loop system (2.77)–(2.78).

time, the new stabilization scheme provides motivation for a dynamic output feedback stabilization methodology, which can accommodate saturated outputs. This dynamic solution is clearly different from observation-based schemes available in the literature.

Bibliography

1. M. Arcak, A.R. Teel, and P.V. Kokotović. Robust nonlinear control of feedforward systems with unmodeled dynamics. *Automatica*, 37:265–272, 2001.
2. A. Astolfi, R. Ortega, and R. Sepulchre. Passivity-based control of non-linear systems. In *Control of Complex Systems, K. Astrom, P. Albertos, M. Blanke, A. Isidori, W. Schaufelberger and R. Sanz (Eds.)*, Springer-Verlag, 2001.
3. R.T. Bupp and D.S. Bernstein. A benchmark problem for nonlinear control design: problem statement, experimental testbed and passive nonlinear compensation. *Proc. American Control Conference, Seattle, Washington*, pages 4363–4367, 1995.
4. P. De Leenheer and D. Aeyels. Stabilization of positive linear systems. *Systems & Control Letters*, 44(4):259–271, 2001.
5. K.D. Do, Z.P. Jiang, and J. Pan. Robust global stabilization of underactuated ships on a linear course: state and output feedback. *Int. J. Control*, 76(1):1–17, 2003.
6. R. Freeman. Time-varying feedback for the global stabilization of nonlinear systems with measurement disturbances. *Proc. European Control Conference, Brussels, Belgium*, 1997.

7. G. Garcia and S. Tarbouriech. Stabilization with eigenvalues placement of a norm bounded uncertain system by bounded inputs. *Int. J. Robust and Nonlinear Control*, 9:599–615, 1999.
8. M. Green and D.J.N. Limebeer. *Linear Robust Control.* Prentice-Hall, 1995.
9. F. Grognard, R. Sepulchre, and G. Bastin. Global stabilization of feedforward systems with exponentially unstable Jacobian linearization. *Systems & Control Letters*, 37(2):107–115, 1999.
10. A. Isidori. *Nonlinear Control Systems II.* Springer-Verlag, 1999.
11. Z.P. Jiang and I. Kanellakopoulos. Global output-feedback tracking for a benchmark nonlinear system. *IEEE Trans. Automatic Control*, 45(5):1023–1027, 2000.
12. G. Kaliora. *Control of nonlinear systems with bounded signals.* Ph.D. Thesis, Imperial College, University of London, 2002.
13. G. Kaliora and A. Astolfi. Output feedback asymptotic stabilization with bounded measurements. *Proc. 4th Asian Control Conference*, 2002.
14. G. Kreisselmeier. Stabilization of linear systems in the presence of output measurement saturation. *Systems & Control Letters*, 29(1):27–30, 1996.
15. Y. Lin and E.D. Sontag. A universal formula for stabilization with bounded controls. *Systems & Control Letters*, 16:393–397, 1991.
16. Z. Lin and T. Hu. Semi-global stabilization of linear systems subject to output saturation. *Systems & Control Letters*, 43(3):211–217, 2001.
17. Z. Lin and A. Saberi. Semi-global exponential stabilization of linear systems subject to "input saturation" via linear feedbacks. *Systems & Control Letters*, 21:225–239, 1993.
18. L. Marconi and A. Isidori. Robust global stabilization of a class of uncertain feedforward nonlinear systems. *Systems & Control Letters*, 41(4):281–290, 2000.
19. F. Mazenc. Stabilization of feedforward systems approximated by a nonlinear chain of integrators. *Systems & Control Letters*, 32(4):223–229, 1997.
20. F. Mazenc, S. Mondie, and S.I. Niculescu. Global asymptotic stabilization for chains of integrators with a delay in the input. *Proc. 40th IEEE Conf. Decision and Control, Orlando, Florida*, pages 1843–1848, 2001.
21. F. Mazenc and L. Praly. Adding integrations, saturated controls and stabilization for feedforward systems. *IEEE Trans. Automatic Control*, 41(11):1559–1578, 1996.
22. F. Mazenc and J.C. Vivalda. Global asymptotic output feedback stabilization of feedforward systems. *Proc. European Control Conference, Porto, Portugal*, pages 2564–2569, 2001.
23. D. Nesic and E. Sontag. Output stabilization of nonlinear systems: linear systems with positive outputs as a case study. *Proc. 37th IEEE Conf. Decision and Control, Tampa, Florida*, pages 885–890, 1998.
24. R. Ortega, A. Loria, P.J. Nicklasson, and H. Sira-Ramirez. *Passivity-based Control of Euler–Lagrange Systems.* Springer-Verlag, 1998.
25. L. Praly, R. Ortega, and G. Kaliora. Stabilization of nonlinear systems via forwarding mod{L_gV}. *IEEE Trans. Automatic Control*, 46(9):1461–1466, 2001.

26. A. Rapaport and A. Astolfi. A remark on the stability of interconnected nonlinear systems. *IEEE Trans. Automatic Control*, 49(1):120–124, 2004.
27. A. Saberi, J. Han, and A. Stoorvogel. Constrained stablization problems for linear plants. *Automatica*, 38:639–654, 2002.
28. S. Saperstone and J. Yorke. Controllability of linear oscillatory systems using positive controls. *SIAM J. Control*, 9(2):253–262, 1971.
29. R. Sepulchre, M. Janković, and P.V. Kokotović. *Constructive Nonlinear Control.* Springer-Verlag, 1996.
30. E. Sontag. Further facts about input to state stabilization. *IEEE Trans. Automatic Control*, 35(4):473–476, 1990.
31. E. Sontag. The ISS philosophy as a unifying framework for stability-like behavior. In *Nonlinear Control in the year 2000, A.Isidori, F. Lamnabhi, W. Respondek (Eds.)*, 2001.
32. A. Stoorvogel and A. Saberi. Output regulation for linear plants with actuators subject to amplitude and rate constraints. *Int. J. Robust and Nonlinear Control*, 9:631–657, 1999.
33. H. Sussmann, E. Sontag, and Y. Yang. A general result on the stabilization of linear systems using bounded controls. *IEEE Trans. Automatic Control*, 39(12):2411–2425, 1994.
34. M. Sznaier, R. Suarez, S. Miani, and J. Alvarez-Ramirez. Optimal l^{∞} disturbance attenuation and global stabilization of linear systems with bounded control. *Int. J. Robust and Nonlinear Control*, 9:659–675, 1999.
35. A.R. Teel. Global stabilization and restricted tracking for multiple integrators with bounded controls. *Systems & Control Letters*, 18:165–171, 1992.
36. A.R. Teel. A nonlinear small gain theorem for the analysis of control systems with saturation. *IEEE Trans. Automatic Control*, 11(9):1256–1270, 1996.
37. C.J. Wan, D.S. Bernstein, and V.T. Coppola. Global stabilization of the oscillating eccentric rotor. *Nonlinear Dynamics*, 10:49–62, 1996.
38. J.L. Willems. *Stability Theory of Dynamical Systems.* Nelson, 1970.
39. W.M. Wonham. *Linear Multivariable Control.* Springer-Verlag, 1995.

CHAPTER 3

Output Feedback Stabilization of a Class of Uncertain Systems

D. Karagiannis[1], A. Astolfi[1] and R. Ortega[2]

[1] *Department of Electrical and Electronic Engineering, Imperial College London, SW7 2AZ, United Kingdom, E-mail: {d.karagiannis,a.astolfi}@imperial.ac.uk*

[2] *Laboratoire des Signaux et Systèmes, Supélec, 91192 Gif-sur-Yvette, France, E-mail: Romeo.Ortega@lss.supelec.fr*

The problem of global output feedback stabilization for a class of nonlinear systems whose zero dynamics are not necessarily stable is addressed in this chapter. It is shown that, using a novel observer design tool together with standard backstepping and small-gain techniques, it is possible to design a stabilizing output feedback controller which ensures robustness with respect to dynamic uncertainties. The proposed stabilization method generalizes existing tools in several directions. The method is illustrated by means of an academic example and is applied to the stabilization of the benchmark translational oscillator/rotational actuator (TORA) system with measurement of the rotational position.

1. Introduction

The problem of output feedback stabilization of nonlinear systems has been an active area of research in recent years. Several control methodologies have been proposed, which achieve global or semiglobal results by exploiting certain feedback structures. In particular, the class of systems in "lower triangular" form has received special attention, see *e.g.* [13, 14, 16] and references therein.

In [13] a recursive method, known as *tuning functions design*, has been introduced and has been used extensively on systems with parametric uncertainties. In [8] this method has been combined with the *nonlinear small-gain theorem* [10] and the notion of *input-to-state stability* [6,20,21] to tackle systems with unstructured dynamic uncertainties, described by equations of the form

$$\begin{aligned}
\dot{\eta} &= f(\eta, y) \\
\dot{x}_1 &= x_2 + \Delta_1(\eta, y) \\
&\vdots \\
\dot{x}_i &= x_{i+1} + \Delta_i(\eta, y) \\
&\vdots \\
\dot{x}_n &= u + \Delta_n(\eta, y) \\
y &= x_1,
\end{aligned} \tag{3.1}$$

where $(\eta, x_1, \ldots, x_n) \in \mathbb{R}^m \times \mathbb{R} \times \cdots \times \mathbb{R}$ is the state of the system, y is the measurable output, u is the control input and $\Delta_i(\cdot)$ are uncertain functions.

Note that a particular case of the form (3.1) is the well-known *output feedback form* [13, 14, 16], where the functions $\Delta_i(\cdot)$ are replaced by the *structured* uncertainty $\phi_i(y)^T \eta$, where η is a vector of unknown parameters (*i.e.* $\dot{\eta} = 0$). Special instances of the system (3.1) have also been studied in [15], [17] and [12]. In [15] the matrix $\partial f / \partial \eta$ is constant and Hurwitz, the perturbation functions $\Delta_i(\cdot), i = 1, \ldots, n-1$ depend only on the output y and $\Delta_n(\cdot)$ is linear in η. In [12] the $\Delta_i(\cdot)$ are allowed to depend on the unmeasured states $x_2, \ldots, x_i$, but they must satisfy a *global Lipschitz* condition, which is slightly relaxed into a *linear growth* condition in [19]. In [17] a Lipschitz-like condition with an output-dependent upper bound is used instead.

A common hypothesis in the aforementioned methods is that the zero dynamics of the considered systems possess some strong stability property, *i.e.* they are globally asymptotically stable (GAS) or input-to-state stable (ISS). A method by means of which it is possible to relax this assumption has been recently proposed in [7] and has been shown to achieve *semiglobal practical* stability for systems that are possibly nonminimum-phase.[a] A global result for *weakly* minimum-phase systems in output feedback form has been reported in [4].

The purpose of this chapter is to partially extend the results of [8]. In particular, we relax the hypothesis that the η-subsystem is ISS with respect to y and replace it with an input-to-state stabilizability condition (see Assumption 2).

The chapter is organized as follows. In Section 2 we define the considered

[a]Note that in [7] the functions $\Delta_i(\cdot)$ may also depend on the unmeasured states $x_2, \ldots, x_i$.

class of systems and the assumptions under which the proposed method will be applicable. In Section 3 we propose a reduced-order observer for the unmeasured states and design the output feedback controller by combining a backstepping construction with a small-gain condition. Some special cases are considered in Section 4. In Section 5 we apply the proposed method to two examples including the benchmark translational oscillator/rotational actuator (TORA) system and in Section 6 we provide some conclusions.

2. Problem Description

Consider a class of uncertain nonlinear systems described by equations of the form

$$\begin{aligned} \dot{\eta} &= F(y)\eta + G(y) + \Delta_0(\eta, y) \\ \dot{x}_1 &= x_2 + \phi_1(y)^T\eta + \Delta_1(\eta, y) \\ &\vdots \\ \dot{x}_i &= x_{i+1} + \phi_i(y)^T\eta + \Delta_i(\eta, y) \\ &\vdots \\ \dot{x}_n &= u + \phi_n(y)^T\eta + \Delta_n(\eta, y) \\ y &= x_1 \end{aligned} \tag{3.2}$$

with state $(\eta, x_1, \ldots, x_n) \in \mathbb{R}^m \times \mathbb{R} \times \cdots \times \mathbb{R}$, input u and output y, where $\Delta_i(\cdot)$ are unknown perturbation functions. We assume that the origin is an equilibrium for the system (3.2) with $u = 0$, *i.e.* $\Delta_i(0,0) = 0$ and $G(0) = 0$, and all functions are sufficiently smooth and we pose the following stabilization problem.

Output feedback stabilization problem. For the system (3.2), find (if possible) a dynamic output feedback control law described by equations of the form

$$\begin{aligned} \dot{\xi} &= \pi(\xi, y) \\ u &= \alpha(\xi, y) \end{aligned} \tag{3.3}$$

with $\xi \in \mathbb{R}^p$ such that the closed-loop system (3.2)–(3.3) is globally asymptotically stable. Note that the system (3.2) has relative degree n and its zero dynamics are given by

$$\dot{\eta} = F(0)\eta + \Delta_0(\eta, 0),$$

hence they are not necessarily stable. Moreover, the functions $\Delta_i(\cdot)$ need not be bounded. However, the following conditions must hold.

Assumption 1: There exist positive definite, locally quadratic and smooth functions $\rho_{i1}(\cdot)$ and $\rho_{i2}(\cdot)$, $i = 0, \ldots, n$, such that

$$|\Delta_i(\eta, y)|^2 \leq \rho_{i1}(|\eta|) + \rho_{i2}(|y|). \tag{3.4}$$

Assumption 2: There exists a smooth function $y^*(\eta)$ such that the system

$$\dot{\eta} = F(y^*(\eta + d_1) + d_2)\eta + G(y^*(\eta + d_1) + d_2) + \Delta_0(\eta, y^*(\eta + d_1) + d_2)$$

is ISS with respect to d_1 and d_2, *i.e.* there exists a positive definite and proper function $V_1(\eta)$ such that

$$\dot{V}_1 \leq -\kappa_{11}(|\eta|) + \gamma_{11}(|d_1|) + \gamma_{12}(|d_2|),$$

where $\kappa_{11}(\cdot)$, $\gamma_{11}(\cdot)$ and $\gamma_{12}(\cdot)$ are smooth class-$\mathcal{K}_\infty$ functions.

Remark 1: In [8] it is assumed that the η-subsystem is ISS with respect to y, *i.e.* Assumption 2 holds for $y^* = 0$. It must be noted that in [8] the functions $\rho_{i1}(\cdot)$, $\rho_{i2}(\cdot)$ of Assumption 1 are multiplied by *unknown* coefficients, which are estimated online using standard Lyapunov techniques.

Remark 2: Assumption 2 is a *robust stabilizability* condition on the internal dynamics. In the linear case, *i.e.* when the matrix F is constant and the vectors G and Δ_0 are linear functions, it is always satisfied, if the pair $(F + \partial\Delta_0/\partial\eta, G + \partial\Delta_0/\partial y)$ is stabilizable, or if the pair (F, G) is stabilizable and $\partial\Delta_0/\partial\eta$ and $\partial\Delta_0/\partial y$ are sufficiently small.

3. Output Feedback Stabilization

In this section a solution to the output feedback stabilization problem is proposed based on a reduced-order observer and a combination of backstepping and small-gain ideas. In particular, it is shown that the closed-loop system can be described as an interconnection of ISS subsystems, whose gains can be tuned to satisfy the small-gain theorem.

3.1. *Reduced-order observer design*

To begin with, we will construct an observer for the unmeasured states η and $x_2, \ldots, x_n$. To this end, define the estimation errors

$$\begin{aligned} z_1 &= \hat{\eta} - \eta + \beta_1(y) \\ z_2 &= \hat{x}_2 - x_2 + \beta_2(y) \\ &\vdots \\ z_n &= \hat{x}_n - x_n + \beta_n(y) \end{aligned}$$

and the update laws

$$\dot{\hat{\eta}} = F(y)\left(\hat{\eta} + \beta_1(y)\right) + G(y) - \frac{\partial \beta_1}{\partial y}\left[\hat{x}_2 + \beta_2(y) + \phi_1(y)^T\left(\hat{\eta} + \beta_1(y)\right)\right]$$

$$\dot{\hat{x}}_2 = \hat{x}_3 + \beta_3(y) + \phi_2(y)^T\left(\hat{\eta} + \beta_1(y)\right) - \frac{\partial \beta_2}{\partial y}\left[\hat{x}_2 + \beta_2(y) + \phi_1(y)^T\left(\hat{\eta} + \beta_1(y)\right)\right]$$

$$\vdots$$

$$\dot{\hat{x}}_n = u + \phi_n(y)^T\left(\hat{\eta} + \beta_1(y)\right) - \frac{\partial \beta_n}{\partial y}\left[\hat{x}_2 + \beta_2(y) + \phi_1(y)^T\left(\hat{\eta} + \beta_1(y)\right)\right],$$

where $\beta_i(y)$ are continuous functions yet to be defined. The "error dynamics" are described by the system

$$\dot{z} = A(y)z - \Delta_r(\eta, y) + \Delta_1(\eta, y)\frac{\partial \beta}{\partial y}, \tag{3.5}$$

where

$$z = \begin{bmatrix} z_1 \\ z_2 \\ \vdots \\ z_n \end{bmatrix}, \qquad \Delta_r = \begin{bmatrix} \Delta_0 \\ \Delta_2 \\ \vdots \\ \Delta_n \end{bmatrix}, \qquad \beta = \begin{bmatrix} \beta_1 \\ \beta_2 \\ \vdots \\ \beta_n \end{bmatrix}$$

and

$$A(y) = \left[\begin{array}{c|ccc} \left(F(y) - \frac{\partial \beta_1}{\partial y}\phi_1(y)^T\right) & -\frac{\partial \beta_1}{\partial y} & 0 \cdots & 0 \\ \hline \left(\phi_2(y)^T - \frac{\partial \beta_2}{\partial y}\phi_1(y)^T\right) & -\frac{\partial \beta_2}{\partial y} & 1 \cdots & 0 \\ \vdots & \vdots & \ddots & \\ \left(\phi_{n-1}(y)^T - \frac{\partial \beta_{n-1}}{\partial y}\phi_1(y)^T\right) & -\frac{\partial \beta_{n-1}}{\partial y} & 0 \cdots & 1 \\ \left(\phi_n(y)^T - \frac{\partial \beta_n}{\partial y}\phi_1(y)^T\right) & -\frac{\partial \beta_n}{\partial y} & 0 \cdots & 0 \end{array}\right]. \tag{3.6}$$

In addition to the estimation error z, we define the *output error*

$$\tilde{y} = y - y^*,$$

where

$$y^* = y^*(\hat{\eta} + \beta_1(y)) = y^*(\eta + z_1)$$

verifies Assumption 2.

Consider now the function $V_2(z) = z^T P z$, where P is a constant, positive definite matrix, and its time-derivative along the trajectories of (3.5), namely

$$\dot{V}_2 = z^T \left(A(y)^T P + PA(y)\right) z - 2\Delta_r(\eta, y)^T Pz + 2\Delta_1(\eta, y) \frac{\partial \beta}{\partial y}^T Pz.$$

Define the matrix

$$B(y) = I + \frac{\partial \beta}{\partial y} \frac{\partial \beta}{\partial y}^T$$

and note that

$$\begin{aligned} \dot{V}_2 \leq\ & z^T \left(A(y)^T P + PA(y) + \frac{PB(y)P}{\gamma(y)} \right) z \\ & + \gamma(y) \left(|\Delta_r(\eta, y)|^2 + |\Delta_1(\eta, y)|^2 \right), \end{aligned}$$

for any function $\gamma(y) > 0$. From Assumption 1 and the definition of $\tilde{y}$ it is possible to select functions $\gamma_{21}(\cdot)$, $\gamma_{22}(\cdot)$ and $\gamma_{23}(\cdot)$ such that

$$\begin{aligned} \dot{V}_2 \leq\ & z^T \left(A(y)^T P + PA(y) + \frac{PB(y)P}{\gamma(y)} \right) z \\ & + \gamma_{21}(|\eta|) + \gamma_{22}(|\tilde{y}|) + \gamma(y)\gamma_{23}(|z_1|). \end{aligned} \tag{3.7}$$

Now consider the following condition.

Assumption 3: There exist functions $\beta(y)$, $\gamma(y)$, a positive definite matrix P and a class-$\mathcal{K}_\infty$ function $\kappa_{21}(\cdot)$ such that, for any y,

$$z^T \left(A(y)^T P + PA(y) + \frac{PB(y)P}{\gamma(y)} \right) z + \gamma(y)\gamma_{23}(|z_1|) \leq -\kappa_{21}(|z|). \tag{3.8}$$

Remark 3: Assumption 3 is a *robust detectability* condition on the system (3.2) and can be considered as dual to Assumption 2. In fact, in the linear case, it is a necessary and sufficient condition for detectability when $\Delta_i = 0$ (see Section 4.4).

Remark 4: The main restriction in the condition (3.8) is the presence of the term $\gamma_{23}(|z_1|)$ which stems from the dependence of the perturbations on y^*. If the system is minimum-phase, then Assumption 2 is satisfied with $y^* = 0$, hence $\gamma_{23}(|z_1|) = 0$ in (3.7). Then the inequality (3.8) can be satisfied by making the matrix $A(y)$ negative definite and taking $\gamma(y)$ sufficiently large.

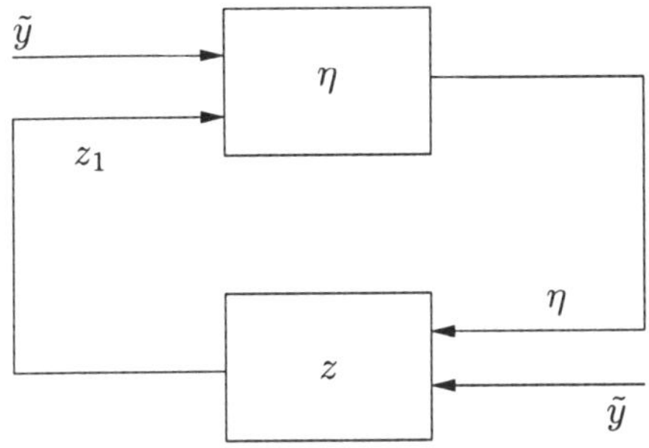

Figure 3.1. Block diagram of the interconnected systems (3.9)–(3.11).

3.2. *Small-gain condition*

Consider again the η-subsystem, which is described by the equation

$$\dot{\eta} = F(y^*(\eta + z_1) + \tilde{y})\eta + G(y^*(\eta + z_1) + \tilde{y}) + \Delta_0(\eta, y^*(\eta + z_1) + \tilde{y}), \quad (3.9)$$

and note that from Assumption 2 we have

$$\dot{V}_1 \leq -\kappa_{11}(|\eta|) + \gamma_{11}(|z_1|) + \gamma_{12}(|\tilde{y}|). \quad (3.10)$$

Moreover, from Assumption 3 and condition (3.7) we conclude that the system

$$\dot{z} = A(y^*(\eta+z_1)+\tilde{y})z - \Delta_r(\eta, y^*(\eta+z_1)+\tilde{y}) + \Delta_1(\eta, y^*(\eta+z_1)+\tilde{y})\frac{\partial \beta}{\partial y} \quad (3.11)$$

is ISS with respect to η and $\tilde{y}$, *i.e.*

$$\dot{V}_2 \leq -\kappa_{21}(|z|) + \gamma_{21}(|\eta|) + \gamma_{22}(|\tilde{y}|). \quad (3.12)$$

Thus we have assumed that each of the systems (3.9) and (3.11) can be rendered ISS by selecting the functions $\beta(y)$ and $y^*(\eta+z_1)$ appropriately. In the following we will consider the stability of their interconnection (depicted in Figure 3.1) by means of the Lyapunov formulation of the nonlinear small-gain theorem [10].

To this end, define class-$\mathcal{K}_\infty$ functions κ_1, κ_2, γ_1, γ_2 such that

$$\begin{aligned} \gamma_2^{-1} \circ \gamma_{21}(|\eta|) \leq V_1(\eta) \leq \kappa_1^{-1} \circ \kappa_{11}(|\eta|) \\ \gamma_1^{-1} \circ \gamma_{11}(|z_1|) \leq V_2(z) \leq \kappa_2^{-1} \circ \kappa_{21}(|z|) \end{aligned} \quad (3.13)$$

and note that the conditions (3.10) and (3.12) can be written as

$$\begin{aligned} \dot{V}_1 &\leq -\kappa_1(V_1) + \gamma_1(V_2) + \gamma_{12}(|\tilde{y}|) \\ \dot{V}_2 &\leq -\kappa_2(V_2) + \gamma_2(V_1) + \gamma_{22}(|\tilde{y}|). \end{aligned}$$

The following theorem states the main result of this chapter.

Theorem 1: Consider a system described by equations of the form (3.2) and such that Assumptions 1, 2 and 3 hold. Let κ_1, κ_2, γ_1 and γ_2 be class-$\mathcal{K}_\infty$ functions satisfying (3.13) with V_1 as in Assumption 2 and $V_2 = z^T P z$ as in Assumption 3 and suppose that there exist constants $0 < \varepsilon_1 < 1$ and $0 < \varepsilon_2 < 1$ such that

$$\frac{1}{1-\varepsilon_1}\kappa_1^{-1} \circ \gamma_1 \circ \left(\frac{1}{1-\varepsilon_2}\kappa_2^{-1} \circ \gamma_2(r)\right) < r, \tag{3.14}$$

for all $r > 0$. Then the system (3.9)–(3.11) with input $\tilde{y}$ is ISS. If, in addition, the ISS gain of this system is locally linear, then there exists a dynamic output feedback control law, described by equations of the form (3.3), such that the closed-loop system (3.2)–(3.3) is globally asymptotically stable.

Remark 5: Theorem 1 states that it is possible to globally asymptotically stabilize the system (3.2), where $\Delta_i(\cdot)$ satisfy the growth condition (3.4), provided three subproblems are solvable. The first problem is the robust stabilization of the η-subsystem with input y (Assumption 2). The second problem is the input-to-state stabilization of the observer dynamics with respect to η (Assumption 3). The third problem is the stabilization of the interconnection of the two subsystems, which can be achieved by satisfying the small-gain condition (3.14). This "reduction" idea is also the basis of the methodology proposed in [7], although therein an entirely different route is followed.

Proof: From condition (3.14) and the nonlinear small-gain theorem [10, Theorem 3.1] the system (3.9)–(3.11) with input $\tilde{y}$ is ISS. Since the gain of this system is locally linear, it suffices to prove that there exists a continuous control law $u(y, \hat{x}_2, \ldots, \hat{x}_n, \hat{\eta})$ such that the gain of the system with state $(\tilde{y}, \hat{x}_2, \ldots, \hat{x}_n, \hat{\eta})$, output $\tilde{y}$ and input (η, z) can be arbitrarily assigned. This can be achieved using a standard backstepping construction, which can be described by the following recursive procedure.

Step 1: Consider the dynamics of $\tilde{y}$, which are described by the equation

$$\begin{aligned}\dot{\tilde{y}} &= \hat{x}_2 + \beta_2(y) - z_2 + \phi_1(y)^T\left(\hat{\eta} + \beta_1(y) - z_1\right) + \Delta_1(\eta, y)\\ &- \frac{\partial y^*}{\partial(\eta + z_1)}\left[F(y)\left(\hat{\eta} + \beta_1(y)\right) + G(y) + \frac{\partial \beta_1}{\partial y}\left(-z_2 - \phi_1(y)^T z_1 + \Delta_1(\eta, y)\right)\right].\end{aligned}$$

Note that the term $\partial y^*/\partial(\eta + z_1)$ is known. Consider $\hat{x}_2$ as a virtual control input and define the error $\tilde{x}_2 = \hat{x}_2 - x_2^*$, where

$$x_2^* = \lambda_1(y,\hat{\eta}) - \beta_2(y) - \phi_1(y)^T(\hat{\eta} + \beta_1(y)) + \frac{\partial y^*}{\partial(\eta + z_1)}\left[F(y)(\hat{\eta} + \beta_1(y)) + G(y)\right],$$

for some function $\lambda_1(\cdot)$ yet to be defined.

Step 2: The dynamics of $\tilde{x}_2$ are given by

$$\begin{aligned}\dot{\tilde{x}}_2 = {} & \hat{x}_3 + \beta_3(y) + \phi_2(y)^T(\hat{\eta} + \beta_1(y)) \\ & -\frac{\partial \beta_2}{\partial y}\left[\hat{x}_2 + \beta_2(y) + \phi_1(y)^T(\hat{\eta} + \beta_1(y))\right] - \frac{\partial x_2^*}{\partial \hat{\eta}}\dot{\hat{\eta}} \\ & -\frac{\partial x_2^*}{\partial y}\left[\hat{x}_2 + \beta_2(y) - z_2 + \phi_1(y)^T(\hat{\eta} + \beta_1(y) - z_1) + \Delta_1(\eta, y)\right].\end{aligned}$$

Consider $\hat{x}_3$ as a virtual control input and define the error $\tilde{x}_3 = \hat{x}_3 - x_3^*$, where

$$\begin{aligned}x_3^* = {} & \lambda_2(y,\hat{x}_2,\hat{\eta}) - \beta_3(y) - \phi_2(y)^T(\hat{\eta} + \beta_1(y)) \\ & + \left(\frac{\partial \beta_2}{\partial y} + \frac{\partial x_2^*}{\partial y}\right)\left[\hat{x}_2 + \beta_2(y) + \phi_1(y)^T(\hat{\eta} + \beta_1(y))\right] + \frac{\partial x_2^*}{\partial \hat{\eta}}\dot{\hat{\eta}}.\end{aligned}$$

Continuing with this (by now classical) step-by-step design philosophy through the dynamics of $\tilde{x}_3, \ldots, \tilde{x}_n$, the control u appears.

Step n: The dynamics of $\tilde{x}_n$ are given by

$$\begin{aligned}\dot{\tilde{x}}_n = {} & u + \phi_n(y)^T(\hat{\eta} + \beta_1(y)) - \frac{\partial \beta_n}{\partial y}\left[\hat{x}_2 + \beta_2(y) + \phi_1(y)^T(\hat{\eta} + \beta_1(y))\right] \\ & -\frac{\partial x_n^*}{\partial y}\left[\hat{x}_2 + \beta_2(y) - z_2 + \phi_1(y)^T(\hat{\eta} + \beta_1(y) - z_1) + \Delta_1(\eta, y)\right] \\ & -\sum_{i=2}^{n-1}\frac{\partial x_n^*}{\partial \hat{x}_i}\dot{\hat{x}}_i - \frac{\partial x_n^*}{\partial \hat{\eta}}\dot{\hat{\eta}}.\end{aligned}$$

Finally, we select the control law u as

$$\begin{aligned}u = {} & \lambda_n(y,\hat{x}_2,\ldots,\hat{x}_n,\hat{\eta}) - \phi_n(y)^T(\hat{\eta} + \beta_1(y)) \\ & + \left(\frac{\partial \beta_n}{\partial y} + \frac{\partial x_n^*}{\partial y}\right)\left[\hat{x}_2 + \beta_2(y) + \phi_1(y)^T(\hat{\eta} + \beta_1(y))\right] \\ & + \sum_{i=2}^{n-1}\frac{\partial x_n^*}{\partial \hat{x}_i}\dot{\hat{x}}_i + \frac{\partial x_n^*}{\partial \hat{\eta}}\dot{\hat{\eta}}.\end{aligned} \tag{3.15}$$

Note that the $\tilde{x}$-subsystem is described by the equations

$$\begin{aligned}
\dot{\tilde{x}}_1 &= \lambda_1(y,\hat{\eta}) + \tilde{x}_2 + \left(\frac{\partial y^*}{\partial(\eta+z_1)}\frac{\partial \beta_1}{\partial y} - 1\right)\left(z_2 + \phi_1(y)^T z_1 - \Delta_1(\eta,y)\right)\\
\dot{\tilde{x}}_2 &= \lambda_2(y,\hat{x}_2,\hat{\eta}) + \tilde{x}_3 + \frac{\partial x_2^*}{\partial y}\left(z_2 + \phi_1(y)^T z_1 - \Delta_1(\eta,y)\right)\\
&\vdots\\
\dot{\tilde{x}}_n &= \lambda_n(y,\hat{x}_2,\ldots,\hat{x}_n,\hat{\eta}) + \frac{\partial x_n^*}{\partial y}\left(z_2 + \phi_1(y)^T z_1 - \Delta_1(\eta,y)\right).
\end{aligned} \tag{3.16}$$

Consider now the function $W(\tilde{x}) = |\tilde{x}|^2$, whose time-derivative along the trajectories of (3.16) is given by

$$\begin{aligned}
\dot{W} = {} & 2\tilde{y}\lambda_1(y,\hat{\eta}) + 2\tilde{y}\tilde{x}_2 + 2\tilde{y}\left(\frac{\partial y^*}{\partial(\eta+z_1)}\frac{\partial \beta_1}{\partial y} - 1\right)\\
& \times \left(z_2 + \phi_1(y)^T z_1 - \Delta_1(\eta,y)\right)\\
& + 2\tilde{x}_2\lambda_2(y,\hat{x}_2,\hat{\eta}) + 2\tilde{x}_2\tilde{x}_3\\
& + 2\tilde{x}_2\frac{\partial x_2^*}{\partial y}\left(z_2 + \phi_1(y)^T z_1 - \Delta_1(\eta,y)\right)\\
& + \cdots + 2\tilde{x}_n\lambda_n(y,\hat{x}_2,\ldots,\hat{x}_n,\hat{\eta})\\
& + 2\tilde{x}_n\frac{\partial x_n^*}{\partial y}\left(z_2 + \phi_1(y)^T z_1 - \Delta_1(\eta,y)\right).
\end{aligned}$$

Then, we can select the functions $\lambda_i(\cdot)$ in such a way that, for some positive constant ε and some function $\alpha_2(\cdot)$ of class-$\mathcal{K}_\infty$,

$$\dot{W} \leq -\varepsilon W - \alpha_1(W) + \alpha_2(|\eta|,|z|), \tag{3.17}$$

where $\alpha_1(\cdot)$ is *any* smooth function of class-$\mathcal{K}_\infty$. Using Assumption 1, both $\alpha_1(\cdot)$ and $\alpha_2(\cdot)$ can be made locally linear.

Finally, as said previously, by hypotheses and by application of the gain assignment technique as in [8], an appropriate choice of $\alpha_1(\cdot)$ completes the proof of Theorem 1. □

4. Special Cases

In this section, we discuss the applicability of Theorem 1 for special cases of systems described by equations of the form (3.2). It is worth noting that Theorem 1 is more general than some of the results in [4,8,13–15], although unknown parameters are also present therein.

Parametric uncertainty can be treated, in the present framework, either by incorporating the unknown parameters into the perturbation terms Δ_i, or (in the linear parameterization case) by including them in the vector η. While the former (similar to [15]) requires only that the parameter vector belongs to a known bounded set, the latter implies that the origin may not be an equilibrium for the system (3.2) and so a somewhat different formulation is needed (see, for instance, the approach in [11]).

4.1. *Systems without zero dynamics*

Consider the system (3.2), where η is an empty vector, *i.e.* there are no zero dynamics, and suppose that Assumption 1 holds. Note that, in this case, Assumption 2 is trivially satisfied with $y^* = 0$ (*i.e.* $\tilde{y} = y$). Then the matrix (3.6) is reduced to

$$A(y) = \begin{bmatrix} -\dfrac{\partial \beta_2}{\partial y} & 1 \cdots 0 \\ \vdots & \ddots \\ -\dfrac{\partial \beta_{n-1}}{\partial y} & 0 \cdots 1 \\ -\dfrac{\partial \beta_n}{\partial y} & 0 \cdots 0 \end{bmatrix}.$$

The above matrix can be rendered constant and Hurwitz by selecting

$$\beta_i(y) = k_i y, \qquad i = 2, \dots, n$$

and choosing the constants k_i appropriately. Moreover, we can select $\gamma_{21} = \gamma_{23} = 0$. As a result, Assumption 3 is trivially satisfied for any linear function $\kappa_{21}(\cdot)$ by taking γ sufficiently large and condition (3.14) holds.

4.2. *Systems with ISS zero dynamics*

Consider the system (3.2) and suppose that Assumptions 1 and 2 hold for $y^* = 0$, *i.e.* the η-subsystem is ISS with respect to y. Then condition (3.10) reduces to

$$\dot{V}_1 \leq -\kappa_{11}(\eta) + \gamma_{12}(|y|),$$

i.e. $\gamma_{11} = 0$, hence condition (3.14) holds. Finally, Assumption 3 is simplified with $\gamma_{23} = 0$. Note that, in this case, we could define new perturbation functions

$$\Delta_i'(\eta, y) = \phi_i(y)^T \eta + \Delta_i(\eta, y), \qquad i = 1, \dots, n$$

and select the functions $\beta_i(y)$ as in Section 4.1 to yield a constant Hurwitz matrix A, thus recovering the design proposed in [8].

4.3. *Unperturbed systems*

Assumption 3 and condition (3.14) can be relaxed in the case of an unperturbed system, *i.e.* a system with $\Delta_i(\cdot) = 0$, as the following corollary shows.

Corollary 1: *Consider a system described by equations of the form (3.2) with $\Delta_i(\cdot) = 0$, $i = 0, 1, \ldots, n$, and such that Assumption 2 holds. Suppose that there exist functions $\beta_i(y)$, $i = 1, \ldots, n$ and a positive definite matrix P such that*

$$z^T \left(A(y)^T P + PA(y) \right) z \leq -\kappa_{21}(|z|)$$

for any y, where $A(y)$ is given by (3.6). Then there exists a dynamic output feedback control law, described by equations of the form (3.3), such that the closed-loop system (3.2)–(3.3) is GAS.

Proof: We simply verify that Theorem 1 applies. To begin with, note that Assumption 1 is trivially satisfied, and Assumptions 2 and 3 hold by hypothesis. Consider now conditions (3.13) and note that the function γ_{21} is zero, hence γ_2 can be arbitrarily selected. Hence, condition (3.14) holds. □

4.4. *Linear perturbed systems*

Consider a linear system described by equations of the form[b]

$$\begin{aligned} \dot{\eta} &= F_0\eta + Gy + \Delta_0(\eta, y) \\ \dot{x}_1 &= x_2 + F_1^T\eta + \Delta_1(\eta, y) \\ &\vdots \\ \dot{x}_i &= x_{i+1} + F_i^T\eta + \Delta_i(\eta, y) \\ &\vdots \\ \dot{x}_n &= u + F_n^T\eta + \Delta_n(\eta, y) \\ y &= x_1 \end{aligned} \tag{3.18}$$

with $\eta \in \mathbb{R}^m$ and suppose that Assumption 1 holds for quadratic functions ρ_{i1} and ρ_{i2} and Assumption 2 holds for a linear function $y^*(\eta + z_1)$. The

[b]The form (3.18) can be obtained, for instance, from any transfer function of relative degree n and order $n+m$, whose high-frequency gain is known and its coefficients belong to a known range.

system (3.18) can be written in matrix form as

$$\begin{bmatrix} \dot{\zeta} \\ \dot{x}_1 \end{bmatrix} = \left[\begin{array}{c|c} & G \\ A_0 & \vdots \\ & 0 \\ \hline C_0 & 0 \end{array}\right] \begin{bmatrix} \zeta \\ x_1 \end{bmatrix} + \left[\begin{array}{c} 0 \\ \vdots \\ 1 \\ \hline 0 \end{array}\right] u + \Delta(\eta, y)$$

$$y = \begin{bmatrix} 0 \; 1 \end{bmatrix} \begin{bmatrix} \zeta \\ x_1 \end{bmatrix},$$

where $\zeta = \left[\eta^T, x_2, \ldots, x_n\right]^T$, $\Delta = \left[\Delta_0^T, \Delta_2, \ldots, \Delta_n, \Delta_1\right]^T$,

$$A_0 = \begin{bmatrix} F_0 & 0\,0 \cdots & 0 \\ F_2^T & 0\,1 \cdots & 0 \\ \vdots & \vdots & \ddots \\ F_{n-1}^T & 0\,0 \cdots & 1 \\ F_n^T & 0\,0 \cdots & 0 \end{bmatrix}, \quad C_0 = \begin{bmatrix} F_1^T \; 1 \; 0 \cdots 0 \end{bmatrix}.$$

Define the function

$$\beta(y) = Ky,$$

where K is a constant vector, and note that the matrix (3.6) can be written as

$$A = A_0 - KC_0.$$

As a result, Assumption 3 can be replaced by the following:

Assumption 4: There exists a vector K and a constant $\kappa_{21} > 0$ such that

$$z^T \left(A^T P + PA + \frac{P\left(I + KK^T\right)P}{\gamma} \right) z + \gamma\gamma_{23}|z_1|^2 \leq -\kappa_{21}|z|^2.$$

Remark 6: If the system (3.18) is detectable for $\Delta = 0$, then the pair $\{A_0, C_0\}$ is also detectable, hence there exists a positive definite matrix P such that the matrix $A^T P + PA$ is negative definite. Note that, due to the presence of γ_{23}, this does not imply (in general) that Assumption 4 holds. However, if the system (3.18) is also minimum-phase, then $\gamma_{23} = 0$ and Assumption 4 is always satisfied for sufficiently large γ.

Finally, conditions (3.10) and (3.12) reduce respectively to

$$\dot{V}_1 \leq -\kappa_{11}|\eta|^2 + \gamma_{11}|z_1|^2 + \gamma_{12}|\tilde{y}|^2$$

and

$$\dot{V}_2 \le -\kappa_{21}|z|^2 + \gamma_{21}|\eta|^2 + \gamma_{22}|\tilde{y}|^2.$$

Hence, the small-gain condition (3.14) reduces to

$$\frac{\gamma_{11}}{(1-\varepsilon_1)\kappa_{11}} \frac{\gamma_{21}}{(1-\varepsilon_2)\kappa_{21}} < 1.$$

5. Examples

5.1. *A nonminimum-phase system*

In this section we apply the proposed method to a simple example, whose zero dynamics are linear and unstable, hence the result in [8] is not applicable. Consider the three-dimensional system

$$\begin{aligned} \dot{\eta} &= \eta + y + \delta(t)y \\ \dot{x}_1 &= x_2 + \left(1+y^2\right)\eta \\ \dot{x}_2 &= u + \left(2+y^2\right)\eta \\ y &= x_1, \end{aligned} \tag{3.19}$$

where $\delta(t)$ is an unknown disturbance such that $|\delta(t)| \le p$, for all t, with $p \in [0,1)$ a known constant. Hence, Assumption 1 is satisfied with $\rho_1(|\eta|) = 0$ and $\rho_2(|y|) = p^2y^2$.

Assumption 2 is also satisfied with the function

$$y^*(\eta) = -k_1\eta,$$

for some positive constant k_1. In fact, the time-derivative of the function $V_1(\eta) = \eta^2/2$ along the trajectories of the system

$$\dot{\eta} = \eta + (1+\delta(t))\left(y^*(\eta+z_1) + \tilde{y}\right) \tag{3.20}$$

is given by

$$\dot{V}_1 = -\left(k_1\left(1+\delta(t)\right) - 1\right)\eta^2 - k_1\left(1+\delta(t)\right)z_1\eta + \left(1+\delta(t)\right)\tilde{y}\eta,$$

which implies

$$\dot{V}_1 \le -\left(k_1\left(1-p\right) - 1 - \gamma_1\right)\eta^2 + \frac{k_1^2(1+p)^2}{2\gamma_1}z_1^2 + \frac{(1+p)^2}{2\gamma_1}\tilde{y}^2, \tag{3.21}$$

for some $\gamma_1 > 0$. Hence, for $k_1 > (1+\gamma_1)/(1-p)$, the system (3.20) is ISS with respect to z_1 and $\tilde{y}$. The error dynamics (3.5) are given by the system

$$\begin{bmatrix} \dot{z}_1 \\ \dot{z}_2 \end{bmatrix} = \left[\begin{array}{c|c} 1 - \dfrac{\partial\beta_1}{\partial y}\left(1+y^2\right) & -\dfrac{\partial\beta_1}{\partial y} \\ \hline 2 + y^2 - \dfrac{\partial\beta_2}{\partial y}\left(1+y^2\right) & -\dfrac{\partial\beta_2}{\partial y} \end{array}\right] \begin{bmatrix} z_1 \\ z_2 \end{bmatrix} - \begin{bmatrix} \delta(t)y \\ 0 \end{bmatrix}.$$

Consider the Lyapunov function $V_2(z) = z^T P z$ with

$$P = \begin{bmatrix} 1 & 0 \\ 0 & a \end{bmatrix},$$

where $a > 1$ is a constant. Assigning the functions $\beta_1(y)$ and $\beta_2(y)$ so that

$$\frac{\partial \beta_1}{\partial y} = \frac{a}{1+y^2}, \qquad \frac{\partial \beta_2}{\partial y} = \frac{2+y^2}{1+y^2} - \frac{1}{(1+y^2)^2}$$

yields

$$\dot{V}_2 \leq -c_1 z_1^2 - c_2 z_2^2 + \gamma_2 p^2 y^2 + \frac{1}{4\gamma_2} z_1^2,$$

for some $\gamma_2 > 0$, where $c_2 = 1 + c_1 = a$. Noting that

$$y^2 \leq (1+d)\,\tilde{y}^2 + k_1^2 \left(1 + \frac{1}{d}\right)^2 \eta^2 + k_1^2 \left(1 + \frac{1}{d}\right)(1+d)\, z_1^2$$

with $d > 0$ yields

$$\begin{aligned} \dot{V}_2 \leq & -\left(c_1 - \frac{1}{4\gamma_2} - \gamma_2 p^2 k_1^2 \left(1 + \frac{1}{d}\right)(1+d)\right) z_1^2 - c_2 z_2^2 \\ & + \gamma_2 p^2 k_1^2 \left(1 + \frac{1}{d}\right)^2 \eta^2 + \gamma_2 p^2 (1+d)\, \tilde{y}^2. \end{aligned}$$

Hence, for sufficiently large c_1, Assumption 3 is satisfied. The design is completed by choosing all the constants in the foregoing inequalities to satisfy the small-gain condition. Clearly, such a selection is always possible since the constants c_1 and c_2 can be chosen arbitrarily large.

Figure 3.2 shows the response of the closed-loop system to the initial conditions $\eta(0) = -1$, $x_1(0) = x_2(0) = 0$ for various disturbances $\delta(t)$.

Remark 7: Applying the change of co-ordinates $\xi_1 = x_1$, $\xi_2 = x_2 + (1 + x_1^2)\eta$, the system (3.19) can be transformed into the system

$$\begin{aligned} \dot{\eta} &= \eta + \xi_1 + \delta(t)\xi_1 \\ \dot{\xi}_1 &= \xi_2 \\ \dot{\xi}_2 &= u + \left(3 + 2\xi_1^2 + 2\xi_1\xi_2\right)\eta + \left(1 + \xi_1^2\right)\xi_1\,(1 + \delta(t)) \\ y &= \xi_1, \end{aligned}$$

for which the result in [7] is applicable. However, its application hinges upon the hypothesis (see Assumption 2 in [7]) that a robust global output

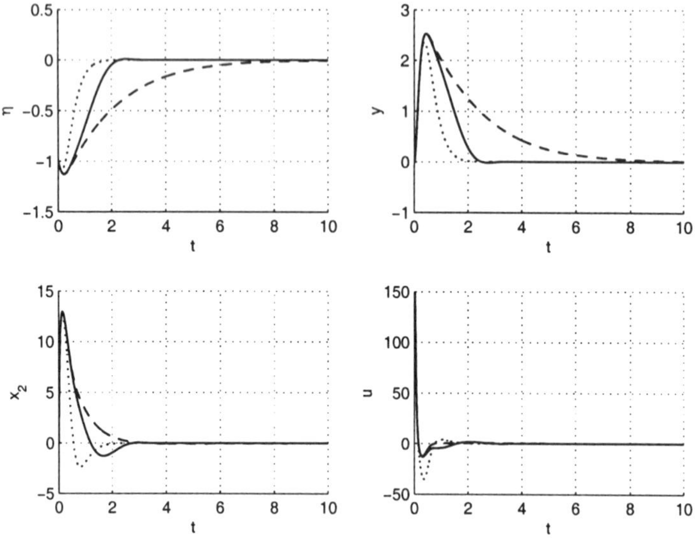

Figure 3.2. Initial response of the system (3.19) for various disturbances. Dotted line: $\delta(t) = 0$. Dashed line: $\delta(t) = -0.4$. Solid line: $\delta(t) = -0.4\cos(t)$.

feedback stabilizer is available for the auxiliary system

$$\begin{aligned}
\dot{\eta} &= \eta + \xi_1 + \delta(t)\xi_1 \\
\dot{\xi}_1 &= u_a \\
y_a &= \left(3 + 2\xi_1^2 + 2\xi_1 u_a\right)\eta + \left(1 + \xi_1^2\right)\xi_1\left(1 + \delta(t)\right)
\end{aligned}$$

with input u_a and output y_a. Although it may be possible in this case to find such a stabilizer, it is certainly not a trivial task.

5.2. *Output feedback stabilization of a nonlinear benchmark system*

In this section we propose a new globally stabilizing output feedback controller for a *translational oscillator with a rotational actuator* (TORA), which has been considered as a benchmark nonlinear system, see [2] for details. Output feedback controllers requiring measurement of the rotational and translational positions but not of the velocities have been proposed in [3] and [9], while controllers using measurements of the rotational position alone have appeared in [3, 5].

The TORA system, depicted in Figure 3.3, is described by the equations

$$(M+m)\,\ddot{x}_d + ml\left(\ddot{\theta}\cos\theta - \dot{\theta}^2\sin\theta\right) = -kx_d$$
$$\left(J+ml^2\right)\ddot{\theta} + m\ddot{x}_d l\cos\theta = \tau,$$

where θ is the angle of rotation, x_d is the translational displacement and τ is the control torque. The positive constants k, l, J, M and m denote the spring stiffness, the radius of rotation, the moment of inertia, the mass of the cart and the eccentric mass, respectively. Define the co-ordinates[c]

$$\begin{aligned}\eta_1 &= x_d + \frac{ml}{M+m}\sin\theta\\ \eta_2 &= \dot{x}_d + \frac{ml}{M+m}\dot{\theta}\cos\theta\\ x_1 &= \theta\\ x_2 &= \dot{\theta}\psi(\theta)\end{aligned}$$

with

$$\psi(\theta) = \sqrt{(J+ml^2)(M+m) - m^2l^2\cos^2\theta}$$

and the control input

$$u = (M+m)\tau.$$

Note that $\psi(\theta) > 0$ by definition, hence the above transformation is well-defined. In these co-ordinates the system is described by a set of equations of the form (3.2), namely

$$\begin{aligned}\dot{\eta} &= \begin{bmatrix}0 & 1\\ -\epsilon_2 & 0\end{bmatrix}\eta + \begin{bmatrix}0\\ \epsilon_3\sin y\end{bmatrix}\\ \dot{x}_1 &= \frac{1}{\psi(y)}x_2\\ \dot{x}_2 &= \frac{1}{\psi(y)}u + \frac{k\epsilon_1\cos y}{\psi(y)}\eta_1 - \frac{\epsilon_2\epsilon_1^2\sin y\cos y}{\psi(y)}\\ y &= x_1,\end{aligned} \tag{3.22}$$

where

$$\epsilon_1 = ml, \qquad \epsilon_2 = \frac{k}{M+m}, \qquad \epsilon_3 = \frac{kml}{(M+m)^2}.$$

[c]The co-ordinate transformation used here follows [2] and [9], but avoids the normalizations and time scaling.

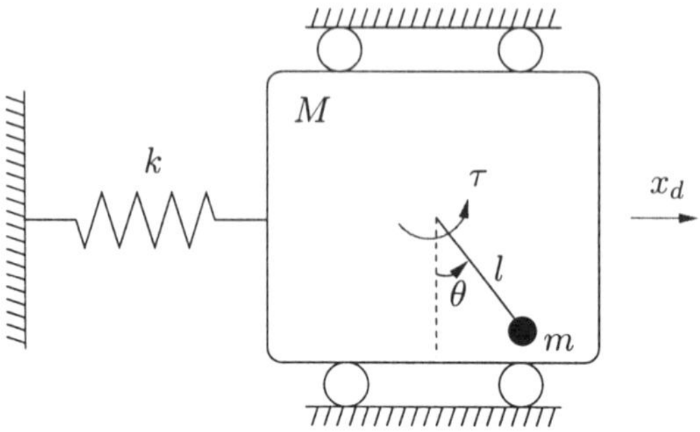

Figure 3.3. A translational oscillator with a rotational actuator (TORA).

We assume that only the output y is available for measurement, thus the system is only weakly minimum-phase. The control objective is to stabilize the system around the origin, so that both the translational displacement and the rotation angle converge to zero.

Following the construction of Section 3, we define the estimation errors

$$\begin{aligned} z_1 &= \hat{\eta} - \eta + \beta_1(y) \\ z_2 &= \hat{x}_2 - x_2 + \beta_2(y) \end{aligned}$$

and the update laws

$$\begin{aligned} \dot{\hat{\eta}} &= \begin{bmatrix} 0 & 1 \\ -\epsilon_2 & 0 \end{bmatrix} (\hat{\eta} + \beta_1(y)) + \begin{bmatrix} 0 \\ \epsilon_3 \sin y \end{bmatrix} - \frac{\partial \beta_1}{\partial y} \frac{1}{\psi(y)} (\hat{x}_2 + \beta_2) \\ \dot{\hat{x}}_2 &= \frac{1}{\psi(y)} u + \left[\frac{k\epsilon_1 \cos y}{\psi(y)}, 0 \right] (\hat{\eta} + \beta_1(y)) \\ &\quad - \frac{\epsilon_2 \epsilon_1^2 \sin y \cos y}{\psi(y)} - \frac{\partial \beta_2}{\partial y} \frac{1}{\psi(y)} (\hat{x}_2 + \beta_2(y)) . \end{aligned}$$

Selecting the functions

$$\beta_1(y) = \left[k_1 k \epsilon_1 \sin y,\, 0 \right]^T, \qquad \beta_2(y) = k_2 y,$$

where $k_1 > 0$, $k_2 > 0$ are arbitrary constants, yields the error dynamics

$$\dot{z} = \left[\begin{array}{cc|c} 0 & 1 & -\dfrac{k_1 k \epsilon_1 \cos y}{\psi(y)} \\ -\epsilon_2 & 0 & 0 \\ \hline \dfrac{k\epsilon_1 \cos y}{\psi(y)} & 0 & -\dfrac{k_2}{\psi(y)} \end{array} \right] z. \tag{3.23}$$

Note that the Lyapunov function $V_2(z) = z^T P z$ with $P = \mathrm{diag}\,(1, 1/\epsilon_2, k_1)$ is such that

$$\dot{V}_2(z) = -\frac{k_1 k_2}{\psi(y)} z_2^2 \leq -\frac{k_1 k_2}{\sqrt{(J + ml^2)(M + m)}} z_2^2,$$

hence z is bounded and $z_2 \in \mathcal{L}_2$. It follows from boundedness of $\dot{z}$ and Barbalat's lemma that $\lim_{t\to\infty} z_2 = 0$. Although this property is weaker than Assumption 3, it will be shown that it is sufficient to construct a globally asymptotically stabilizing control law.

Towards this end, consider the η-subsystem and the function

$$y^*(\eta + z_1) = -\tan^{-1}(\big[\,0,\ k_0\,\big]\,(\eta + z_1)),$$

where $k_0 > 0$ is a constant, and note that the function $V_1(\eta) = \left(\epsilon_2 \eta_1^2 + \eta_2^2\right)/2\epsilon_3$ is such that

$$\dot{V}_1(\eta) = -\frac{k_0 \eta_2 (\eta_2 + z_{12})}{\sqrt{1 + k_0^2 (\eta_2 + z_{12})^2}} \cos\tilde{y} + \frac{\eta_2}{\sqrt{1 + k_0^2(\eta_2 + z_{12})^2}} \sin\tilde{y}.$$

Using the identity $\cos\tilde{y} = 1 - 2\sin^2(\tilde{y}/2)$ and the inequalities $2ab \leq da^2 + b^2/d$ for any $d > 0$ and $|\sin(x)/x| \leq 1$, after some calculations, we obtain

$$\begin{aligned}\dot{V}_1(\eta) \leq &-\left(1 - \frac{1}{\gamma_{11}} - \frac{1}{\gamma_{12}}\right) \frac{k_0 \eta_2^2}{\sqrt{1 + k_0^2(\eta_2 + z_{12})^2}} \\ &+ \frac{\gamma_{11}}{4} \frac{k_0 z_{12}^2}{\sqrt{1 + k_0^2(\eta_2 + z_{12})^2}} \\ &+ \frac{\gamma_{12} k_0}{2} (\eta_2 + z_{12})^2 \tilde{y}^2 + \frac{\gamma_{12}}{2k_0} \tilde{y}^2\end{aligned}$$

for some constants $\gamma_{11} > 1$, $\gamma_{12} > 1$. Consider now the output error $\tilde{y} = y - y^*$. Defining the control law as in (3.15) yields the system

$$\begin{aligned}\dot{\tilde{x}}_1 &= \frac{1}{\psi(y)} \left(\lambda_1(y, \hat{\eta}) + \tilde{x}_2 - z_2\right) \\ \dot{\tilde{x}}_2 &= \frac{1}{\psi(y)} \left(\lambda_2(y, \hat{x}_2, \hat{\eta}) + \frac{\partial x_2^*}{\partial y} z_2\right).\end{aligned}$$

Selecting the functions $\lambda_i(\cdot)$ as

$$\begin{aligned}\lambda_1(y, \hat{\eta}) &= -a_1 \psi(y) \tilde{y} - \varepsilon \tilde{y} - \delta \psi(y) (\eta_2 + z_{12})^2 \tilde{y} \\ \lambda_2(y, \hat{x}_2, \hat{\eta}) &= -a_2 \psi(y) \tilde{x}_2 - \tilde{y} - \varepsilon \left(\frac{\partial x_2^*}{\partial y}\right)^2 \tilde{x}_2,\end{aligned}$$

where a_1, a_2, ε and δ are positive constants, renders $W(\tilde{x}) = \frac{1}{2}\left(\tilde{y}^2 + \tilde{x}_2^2\right)$ an ISS Lyapunov function for the $\tilde{x}$-subsystem. In particular, we have

$$\dot{W}(\tilde{x}) \leq -a_1\tilde{y}^2 - a_2\tilde{x}_2^2 - \delta(\eta_2 + z_{12})^2\tilde{y}^2 + \frac{1}{2\varepsilon\psi(y)}z_2^2.$$

As a result, since z_2 goes to zero asymptotically, so does $\tilde{y}$. This, from (3.23), implies that the entire vector z converges to zero. By combining $W(\tilde{x})$ with $V_1(\eta)$, we conclude that the $(\eta, \tilde{x})$-subsystem with input z and output η_2 is *input-to-output stable* [20]. Since it is also zero-state detectable, η converges to zero. Hence, the origin is globally asymptotically stable.

The closed-loop system has been simulated using the parameters $J = 0.0002175$ kg/m^2, $M = 1.3608$ kg, $m = 0.096$ kg, $l = 0.0592$ m and $k = 186.3$ N/m and the initial state $\eta_1(0) = 0.025$, $\eta_2(0) = x_1(0) = x_2(0) = 0$. The controller parameters have been set to $k_0 = 4$, $a_1 = a_2 = 1$. For the sake of comparison we have used the same parameters and initial conditions as in the paper [5], which also provides a globally asymptotically stabilizing controller requiring only measurement of the rotational position.

Figure 3.4 shows the response of the closed-loop system in the original co-ordinates x_d and θ, the control torque τ and the estimation error z. The convergence of x_d and θ to zero can be seen in Figure 3.5, where their norm has been plotted in logarithmic scale. Comparing with [5] we see that the response is considerably faster, while the control effort remains within the physical constraints given in [2], namely $|\tau| \leq 0.1$ Nm. The performance can be further improved (at the expense of the control effort) by increasing the parameter k_0.

6. Conclusions

The problem of output feedback stabilization of a class of nonlinear systems with dynamic uncertainties has been studied. It has been shown that, by using a novel observer design tool together with a standard backstepping construction and a small-gain condition, it is possible to obtain a globally stabilizing output feedback control law. The proposed method applies to systems with unstable zero dynamics, thus extending the result in [8]. It also allows for cross-terms between the output and the unmeasured states to appear in the system equations, hence it is more general than the observer backstepping method used in [13]. The method has been illustrated by means of a contrived example of a nonminimum-phase nonlinear system. The proposed approach has been used to design a globally asymptotically

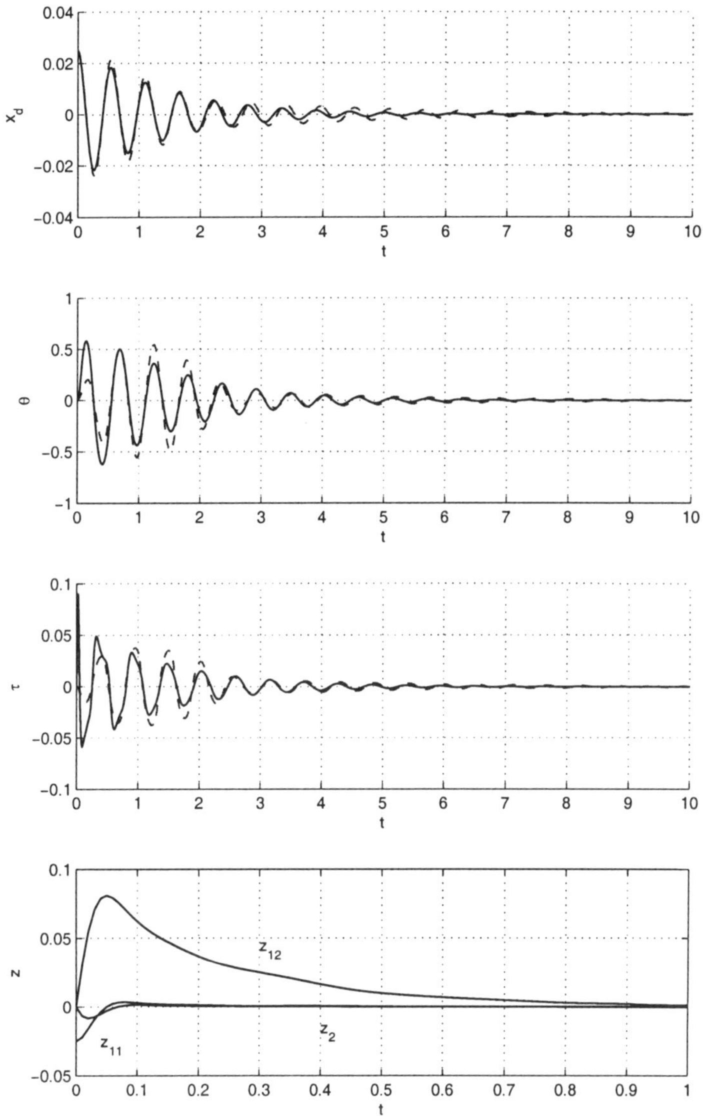

Figure 3.4. Initial response of the TORA system. Dashed line: Passivity-based controller. Solid line: Proposed controller.

stabilizing controller for the benchmark translational oscillator/rotational actuator (TORA) system, where only the rotation angle is measurable.

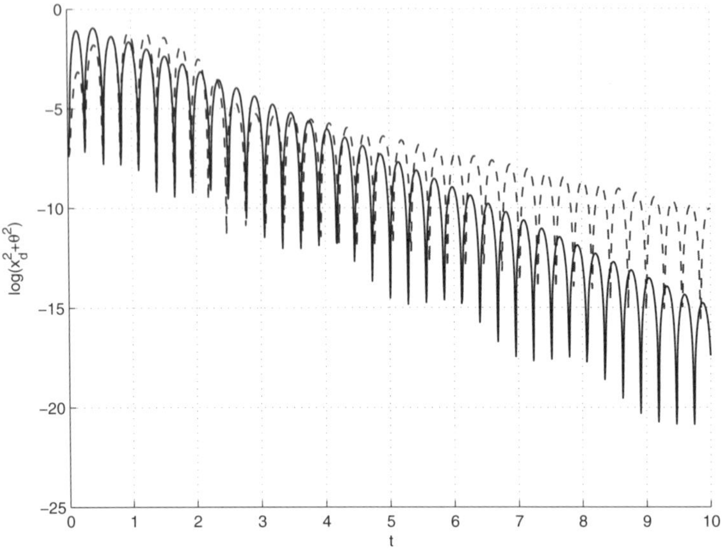

Figure 3.5. Convergence rate of the TORA system. Dashed line: Passivity-based controller. Solid line: Proposed controller.

Bibliography

1. A. Astolfi and R. Ortega. Immersion and invariance: a new tool for stabilization and adaptive control of nonlinear systems. *IEEE Trans. Automatic Control*, 48(4):590–606, 2003.
2. R.T. Bupp, D.S. Bernstein, and V.T. Coppola. A benchmark problem for nonlinear control design: problem statement, experimental testbed and passive nonlinear compensation. In *Proc. American Control Conference, Seattle, Washington*, pages 4363–4367, 1995.
3. T. Burg and D. Dawson. Additional notes on the TORA example: a filtering approach to eliminate velocity measurements. *IEEE Trans. Control Systems Technology*, 5(5):520–523, 1997.
4. Z. Ding. Adaptive stabilization of a class of nonlinear systems with unstable internal dynamics. *IEEE Trans. Automatic Control*, 48(10):1788–1792, 2003.
5. G. Escobar, R. Ortega, and H. Sira-Ramírez. Output-feedback global stabilization of a nonlinear benchmark system using a saturated passivity-based controller. *IEEE Trans. Control Systems Technology*, 7(2):289–293, 1999.
6. A. Isidori. *Nonlinear Control Systems II*. Springer-Verlag, London, 1999.
7. A. Isidori. A tool for semiglobal stabilization of uncertain nonminimum-phase nonlinear systems via output feedback. *IEEE Trans. Automatic Control*, 45(10):1817–1827, 2000.
8. Z.P. Jiang. A combined backstepping and small-gain approach to adaptive output feedback control. *Automatica*, 35(6):1131–1139, 1999.

9. Z.P. Jiang and I. Kanellakopoulos. Global output feedback tracking for a benchmark nonlinear system. *IEEE Trans. Automatic Control*, 45(5):1023–1027, 2000.
10. Z.P. Jiang, I. Mareels, and Y. Wang. A Lyapunov formulation of the nonlinear small-gain theorem for interconnected ISS systems. *Automatica*, 32(8):1211–1215, 1996.
11. D. Karagiannis, A. Astolfi, and R. Ortega. Two results for adaptive output feedback stabilization of nonlinear systems. *Automatica*, 39(5):857–866, 2003.
12. H.K. Khalil and A. Saberi. Adaptive stabilization of a class of nonlinear systems using high-gain feedback. *IEEE Trans. Automatic Control*, 32(11):1031–1035, 1987.
13. M. Krstić, I. Kanellakopoulos, and P. Kokotović. *Nonlinear and Adaptive Control Design*. John Wiley and Sons, New York, 1995.
14. R. Marino and P. Tomei. Global adaptive output-feedback control of nonlinear systems, part I: linear parameterization. *IEEE Trans. Automatic Control*, 38(1):17–32, 1993.
15. R. Marino and P. Tomei. Global adaptive output-feedback control of nonlinear systems, part II: nonlinear parameterization. *IEEE Trans. Automatic Control*, 38(1):33–48, 1993.
16. R. Marino and P. Tomei. *Nonlinear Control Design: Geometric, Adaptive and Robust*. Prentice-Hall, London, 1995.
17. L. Praly. Asymptotic stabilization via output feedback for lower triangular systems with output dependent incremental rate. *IEEE Trans. Automatic Control*, 48(6):1103–1108, 2003.
18. L. Praly and Z.P. Jiang. Linear output feedback with dynamic high gain for nonlinear systems. *Systems & Control Letters*, 53(2):107–116, 2004.
19. C. Qian and W. Lin. Output feedback control of a class of nonlinear systems: a nonseparation principle paradigm. *IEEE Trans. Automatic Control*, 47(10):1710–1715, 2002.
20. E.D. Sontag. On the input-to-state stability property. *European Journal of Control*, 1:24–36, 1995.
21. E.D. Sontag and Y. Wang. On characterizations of the input-to-state stability property. *Systems & Control Letters*, 24(5):351–359, 1995.

CHAPTER 4

Matching in the Method of Controlled Lagrangians and IDA-Passivity Based Control

G. Blankenstein[1], R. Ortega[2] and A. J. van der Schaft[3]

[1]*SYSTeMS, Ghent University, Technologiepark 914, 9052 Zwijnaarde, Belgium, E-mail: Guido.Blankenstein@UGent.be*

[2]*Laboratoire des Signaux et Systèmes, Supélec, 91192 Gif-sur-Yvette, France, E-mail: Romeo.Ortega@lss.supelec.fr*

[3]*Dept. of Applied Mathematics, University of Twente, P.O. Box 217, 7500 AE Enschede, The Netherlands, E-mail: a.j.vanderschaft@math.utwente.nl*

This chapter reviews the method of controlled Lagrangians and the interconnection and damping assignment passivity based control (IDA-PBC) method. Both methods have been presented recently in the literature as means to stabilize a desired equilibrium point of an Euler–Lagrange system, respectively Hamiltonian system, by searching for a stabilizing structure preserving feedback law. The conditions under which two Euler–Lagrange or Hamiltonian systems are equivalent under feedback are called the matching conditions (consisting of a set of nonlinear PDEs). Both methods are applied to the general class of underactuated mechanical systems and it is shown that the IDA-PBC method contains the controlled Lagrangians method as a special case by choosing an appropriate closed-loop interconnection structure. Moreover, explicit conditions are derived under which the closed-loop Hamiltonian system is integrable, leading to the introduction of gyroscopic terms. The λ-method as introduced in recent papers for the controlled Lagrangians method transforms the matching conditions into a set of linear PDEs. In this chapter the method is extended, transforming the matching conditions obtained in the IDA-PBC method into a set of quasi-linear and linear PDEs.

1. Introduction

Recently there has been a lot of interest in the stabilization of underactuated mechanical systems using methods that preserve the mathematical structure of the system. A mechanical system is called underactuated if the number of control inputs is strictly less than the number of degrees of freedom of the system. Such systems often occur for example in robotics, and are generally difficult to control. While fully actuated mechanical systems admit an arbitrary shaping of the potential energy by means of feedback, and therefore a stabilization to any desired equilibrium, such a strategy is in general not possible for underactuated systems. Indeed, underactuation puts a severe restriction on the possibilities to shape the potential energy. In certain cases this problem can be overcome by also modifying the kinetic energy of the system, thus leading to a new mechanical system with a *modified total energy*. A well known example is given by the inverted pendulum on a cart. This is an underactuated system since only the horizontal position of the cart can be controlled directly by a force in this direction, whereas by the absence of a torque the angle of the pendulum is uncontrolled. For this system it is not possible to stabilize the upright position of the pendulum by potential energy shaping only. However, allowing in addition the shaping of kinetic energy does stabilize the upright position of the pendulum, as well as the horizontal position of the cart. The closed-loop system is again described by a mechanical system, with a modified positive definite total energy function.

1.1. *Controlled Lagrangians*

The idea of kinetic energy shaping has led to a method for stabilizing underactuated mechanical systems, called the *method of controlled Lagrangians.* This method was introduced by [8, 9, 11] for the stabilization of relative equilibria of mechanical systems with symmetry.

Starting point is an underactuated mechanical control system described by the forced Euler–Lagrange equations with a Lagrangian being the difference of the kinetic and potential energy of the system. The system is assumed to admit a symmetry, in fact, the Lagrangian is assumed to be invariant under the action of an Abelian Lie group (in the case of a cart and pendulum this means that the Lagrangian is independent of the horizontal position of the cart). The idea now is to stabilize a relative equilibrium of the system (*i.e.* the upright position of the pendulum, irrespective of the horizontal position of the cart) by searching for a suitable (stabilizing) closed-loop system which is again in Euler–Lagrange format *and* preserves the symmetry of the system. This is done by proposing a class of Lagrangians, called *controlled Lagrangians*, which preserve the symmetry of the system, and investigating which of these Lagrangians can possibly be obtained as a closed-loop Lagrangian by choosing a suitable feedback law for the original system. The conditions under which such a feedback law exists are called *matching conditions*, and in case these conditions are satisfied the original control system and the closed-loop Euler–Lagrange system are said to *match*. The feedback law can be calculated by using the symmetry properties of the system. The class of controlled Lagrangians proposed by Bloch *et al.* [8, 9, 11] consists of Lagrangians being the difference of a *shaped kinetic energy* and the potential energy of the original system. That is, the kinetic energy is modified (in a certain restricted way), whereas the potential energy of the system remains unchanged. In general, the matching conditions for this class of controlled Lagrangians are described by a set of nonlinear partial differential equations to be solved for the closed-loop Lagrangian. In special cases, the so-called simplified matching assumptions [11], defining a restrictive but useful class of possible closed-loop controlled Lagrangians, these PDEs are automatically solved. The desired relative equilibrium is locally stabilized by finding a controlled Lagrangian, satisfying the matching assumptions, such that the total energy of the closed-loop system is (usually negative) definite around this equilibrium.

This method has proved to work well for the examples of stabilization of an inverted pendulum on a cart or an inverted spherical pendulum and the stabilization of a satellite with an internal rotor, see [8,9, 11] for details.

The method of Bloch *et al.* [8,9,11] concerning mechanical systems with symmetry, has been refined in the work of [1–3] to describe the stabilization of equilibria of general mechanical systems, see also the work of [16]. The idea is to stabilize a desired equilibrium by searching for a closed-loop Euler–Lagrange system with a *modified total energy*, *i.e.* in addition to the shaping of kinetic energy also the shaping of potential energy is allowed. Again, the matching conditions are described by a set of nonlinear PDEs. In [2,4] the so-called λ-method is presented to convert these nonlinear PDEs into a set of *linear* PDEs. The method is designed for general mechanical systems and does not require any symmetry of the system. In fact, in general the symmetries present in the original system will be destroyed by the shaping of the potential energy in order to stabilize a desired equilibrium *point*. For the cart and pendulum this means that besides stabilizing the upright position of the pendulum, as in the method of Bloch *et al.* [8, 9, 11], simultaneously the position of the cart is stabilized towards a desired horizontal position. We remark that the need for potential energy shaping to stabilize an equilibrium point has also been recognized in [10, 12], where the term *symmetry-breaking potential* has been used.

The method of controlled Lagrangians has been extended in the work of [17] to describe the matching of general Euler–Lagrange systems. These systems are not restricted to be of a mechanical nature, that is, the Lagrangian is not necessarily given by the difference of a kinetic and a potential energy. Under a regularity assumption on the Lagrangian the matching conditions define a set of nonlinear PDEs, generalizing the PDEs described previously for mechanical systems.

Finally, we would like to remark that recently some results have been obtained in [18, 28] extending the method of controlled Lagrangians to also include the matching and stabilization of Euler–Lagrange systems with (nonholonomic) constraints.

1.2. *Interconnection and damping assignment*

At the same time, on the Hamiltonian side a method has been developed to stabilize port-controlled Hamiltonian systems, [20, 21]. Port-controlled Hamiltonian systems have shown to be instrumental in the network modeling of energy conserving physical systems. They strictly contain the class of Euler–Lagrange systems. See [25] and the references therein for more information on the development and the use of port-controlled Hamiltonian systems. Analogously to the method of controlled Lagrangians, the idea is to stabilize a desired equilibrium point of the system by searching for a suitable closed-loop system which is again in port-controlled Hamiltonian format. The closed-loop system is defined by changing the internal interconnection structure (*i.e.* the skew-symmetric structure matrix corresponding to the Poisson bracket of the system) and the Hamiltonian (*i.e.* energy) function of the system. The conditions under which these changes lead to a system that can possibly be obtained as a closed-loop system of the original system, by choosing a suitable feedback law, constitute a new set of matching conditions. These are a set of nonlinear PDEs to be solved for the closed-loop Hamiltonian *and* the closed-loop interconnection structure. The principal (energy) concept used to stabilize the system is *passivity*, and since the closed-loop system is defined by shaping the internal interconnection structure of the system, the term *interconnection and damping assignment passivity based control (IDA-PBC)* has been coined to describe this method.[a] We refer to [20, 21] for more details on the method and on the

[a]The method described in [20, 21] additionally allows the shaping of the damping structure of the system. However, in this chapter we will not consider this possibility, see the remarks afterwards.

underlying passivity concept. It is important to notice that the possibility of also changing the interconnection structure, in addition to changing the Hamiltonian function, gives an extra degree of freedom to the IDA-PBC method with respect to the controlled Lagrangians method. Furthermore, since the class of port-controlled Hamiltonian systems strictly contains the class of forced Euler–Lagrange systems, the IDA-PBC method is more generally applicable than the controlled Lagrangians method. In [20, 21] it has been shown that the method can be used to stabilize electrical systems such as power converters, electromechanical systems, *e.g.* synchronous motors, and mass-balance systems. The application of IDA-PBC to mechanical systems has been described in [20, 22]. The method has been extended to systems with constraints in [5]. We refer to [23] for a recent survey on the IDA-PBC method.

1.3. *Contributions and outline of the chapter*

In Section 2 we discuss the matching of general Euler–Lagrange systems. Necessary and sufficient conditions are derived for two Euler–Lagrange systems to match, resulting in a set of nonlinear PDEs to be solved for the closed-loop Lagrangian. The method of [11] for mechanical systems with symmetry is reviewed, and the matching conditions obtained in that method are given an interpretation in terms of the matching of kinetic and potential energy. Section 3 recalls the matching of port-controlled Hamiltonian systems, as used in the IDA-PBC method. In Section 4 both methods, applied to the class of mechanical systems, are compared. It is shown that the controlled Lagrangians method is strictly included in the IDA-PBC method (see however Remark 9 for a novel extension of the controlled Lagrangians method, yielding equivalence of both methods). Furthermore, the λ-method as described in [2] for the controlled Lagrangians method is extended to the IDA-PBC method. It is shown that the matching conditions, consisting of a set of nonlinear PDEs, can be transformed into an equivalent set of one quasi-linear and two linear PDEs, to be solved recursively. In Section 5 the extra degree of freedom provided by the IDA-PBC method, *i.e.* the shaping of the internal interconnection structure, is used to discuss the integrability of the closed-loop Hamiltonian system. Necessary and sufficient conditions are given for the closed-loop system to be integrable, leading to the introduction of gyroscopic terms in the closed-loop system. Section 6 contains the conclusions.

Some further details and proofs can be found in the journal version of this chapter [6]. A brief survey was presented in [7].

Important remarks: Before continuing with the technical part of the chapter it is important to make the following two remarks. Firstly, notice that this chapter is not concerned with the actual stabilization of equilibrium points of Euler–Lagrange or Hamiltonian systems. The (asymptotic) stabilization of equilibria is the aim of the papers [11, 12, 20–22] where the controlled Lagrangians method and the IDA-PBC method are introduced. In this chapter we are merely interested in the matching of Euler–Lagrange, respectively Hamiltonian systems, which is the fundamental concept underlying both stabilization methods.

Secondly, for simplicity of exposition we do not consider any natural damping to be present in the control system, nor the introduction of energy dissipation by feedback in the closed-loop system. That is, we consider all systems to be energy conserving. The introduction of damping by feedback, called damping injection or damping assignment, is a very important issue in the methods as described in [11, 12, 20–22] to *asymptotically* stabilize an equilibrium which is made stable by shaping the Lagrangian, respectively the Hamiltonian and the internal interconnection structure, of the system. The inclusion of damping assignment in the results of this chapter should be straightforward. Indeed, for mechanical systems with no natural damping feeding back the passive output results (under some detectability condition) in an asymptotically stable system. In this case the damping does not appear in the matching conditions, see [22].

Notation: Let $L(q, \dot{q})$ be a smooth function, then $\partial_q L$ denotes the partial derivative of L with respect to q and $\partial_{\dot{q}} L$ denotes the partial derivative of L with respect to $\dot{q}$ (these are $n \times 1$ matrices). The second order derivatives of L (which are $n \times n$ matrices) are denoted by $\partial_{qq} L, \partial_{q\dot{q}} L$ *etc.*Furthermore, if $\Theta(q, \dot{q}) \in \mathbb{R}^n$ is a smooth vector-valued function of $(q, \dot{q})$, then $\partial_q \Theta$ denotes the $n \times n$ matrix with (i, j)-th entry being $\partial_{q_j} \Theta_i(q, \dot{q})$.

2. Matching of Euler–Lagrange Systems

In this section we describe the matching of Euler–Lagrange systems.

2.1. *General matching conditions*

Consider a forced Euler–Lagrange system with configuration space $\mathcal{Q}$, taken for simplicity to be equal to $\mathbb{R}^n$, and described by a Lagrangian $L : T\mathcal{Q} \to \mathbb{R}$,

$$\frac{d}{dt}\partial_{\dot{q}} L(q, \dot{q}) - \partial_q L(q, \dot{q}) = G(q)u. \tag{4.1}$$

The matrix $G(q) : \mathbb{R}^m \to T_q^*\mathcal{Q} \simeq \mathbb{R}^n$, with rank $G = m$, defines the force fields corresponding to the input $u \in \mathbb{R}^m$. Note that if $m = n$, then (4.1) describes a *fully actuated* Euler–Lagrange system, whereas the system is *underactuated* if (and only if) $m < n$. Consider a second, autonomous Euler–Lagrange system, defined by a Lagrangian $L_c : T\mathcal{Q} \to \mathbb{R}$ (the subscript c suggestively stands for closed-loop),

$$\frac{d}{dt}\partial_{\dot{q}} L_c(q,\dot{q}) - \partial_q L_c(q,\dot{q}) = 0. \tag{4.2}$$

The question we ask ourselves is whether the system (4.2) can be obtained as a possible closed-loop system corresponding to (4.1) by choosing a suitable control law u. If (4.2) is a possible closed-loop system of (4.1) then we say that the systems (4.1) and (4.2) *match.*

Now, consider the system (4.1), and let $G^\perp(q) : (\mathbb{R}^{n-m})^T \to (\mathbb{R}^n)^T$ denote a full rank left annihilator of $G(q)$, *i.e.* $G^\perp(q)G(q) = 0$, $\forall q \in \mathcal{Q}$. Note that from (4.1) it follows that

$$G^\perp(q)\left(\frac{d}{dt}\partial_{\dot{q}} L(q,\dot{q}) - \partial_q L(q,\dot{q})\right) = 0. \tag{4.3}$$

Consider the system (4.2). First notice that $\mathbb{R}^n = \operatorname{Im} G(q) \oplus \operatorname{Im} (G^\perp)^T(q)$. This implies that (4.2) is equivalent to the following two equations:

$$G^T(q)\left(\frac{d}{dt}\partial_{\dot{q}} L_c(q,\dot{q}) - \partial_q L_c(q,\dot{q})\right) = 0, \tag{4.4}$$

$$G^\perp(q)\left(\frac{d}{dt}\partial_{\dot{q}} L_c(q,\dot{q}) - \partial_q L_c(q,\dot{q})\right) = 0. \tag{4.5}$$

The first of these two equations can always be obtained from (4.1) by choosing the control

$$u = (G^T G)^{-1} G^T \left[\left(\frac{d}{dt}\partial_{\dot{q}} L - \partial_q L\right) - \left(\frac{d}{dt}\partial_{\dot{q}} L_c - \partial_q L_c\right)\right] \in \mathbb{R}^m, \tag{4.6}$$

where we left out the arguments $(q, \dot{q})$ for clarity (notice that indeed $G^T G$ is square and has full rank m). This leads to the following proposition.

Proposition 1: *The systems (4.1) and (4.2) match if and only if Eq. (4.5) holds along solutions of the system (4.1, 4.6) (equivalently (4.3, 4.4)).*

Remark 1: If rank $G = n$ then $G^\perp = 0$ and Eq. (4.5) is trivially satisfied, for any arbitrary closed-loop Lagrangian L_c. This corresponds to the well known fact that in case the system is fully actuated, its dynamics can be modified arbitrarily.

Equation (4.5) is referred to as the matching conditions. Following common terminology we call the closed-loop Lagrangian L_c the *controlled Lagrangian.*

Recall that the matching conditions (4.5) have to be satisfied *along solutions* of the system (4.1, 4.6), or equivalently (4.3, 4.4). Now take into account the regularity of the Lagrangians L and L_c, that is $\partial_{\dot{q}\dot{q}}L$ and $\partial_{\dot{q}\dot{q}}L_c$ are invertible. Then by eliminating the accelerations, the matching conditions (4.5) can be written as a set of nonlinear partial differential equations, to be satisfied for all $(q, \dot{q})$. Furthermore, the control law (4.6) is seen to be a state feedback control law. The construction is as follows:

Writing out the system (4.1) gives

$$(\partial_{\dot{q}\dot{q}}L)\ddot{q} + (\partial_{q\dot{q}}L)\dot{q} - \partial_q L = Gu. \tag{4.7}$$

Assuming that the Lagrangian is regular the system can be written as

$$\ddot{q} = -(\partial_{\dot{q}\dot{q}}L)^{-1}(\partial_{q\dot{q}}L)\dot{q} + (\partial_{\dot{q}\dot{q}}L)^{-1}\partial_q L + (\partial_{\dot{q}\dot{q}}L)^{-1}Gu. \tag{4.8}$$

Equivalently, the system (4.2) can be written as (assuming regularity)

$$\ddot{q} = -(\partial_{\dot{q}\dot{q}}L_c)^{-1}(\partial_{q\dot{q}}L_c)\dot{q} + (\partial_{\dot{q}\dot{q}}L_c)^{-1}\partial_q L_c. \tag{4.9}$$

The systems (4.1) and (4.2) match, for some suitably defined control law u, if the solutions of both systems are the same. That is, $(q(t), u(t)))$ is a solution of (4.1) if and only if $q(t)$ is a solution of (4.2), or equivalently, $(q(t), u(t)))$ satisfies (4.8) if and only if $q(t)$ satisfies (4.9). It follows that (4.1) and (4.2) match if and only if

$$\begin{aligned} &-(\partial_{\dot{q}\dot{q}}L)^{-1}(\partial_{q\dot{q}}L)\dot{q} + (\partial_{\dot{q}\dot{q}}L)^{-1}\partial_q L + (\partial_{\dot{q}\dot{q}}L)^{-1}Gu = \\ &-(\partial_{\dot{q}\dot{q}}L_c)^{-1}(\partial_{q\dot{q}}L_c)\dot{q} + (\partial_{\dot{q}\dot{q}}L_c)^{-1}\partial_q L_c, \end{aligned} \tag{4.10}$$

which can be written as

$$Gu = \{\partial_{q\dot{q}}L - (\partial_{\dot{q}\dot{q}}L)(\partial_{\dot{q}\dot{q}}L_c)^{-1}(\partial_{q\dot{q}}L_c)\}\dot{q} - \partial_q L + (\partial_{\dot{q}\dot{q}}L)(\partial_{\dot{q}\dot{q}}L_c)^{-1}\partial_q L_c. \tag{4.11}$$

Using the left annihilator $G^{\perp}$ of G, (4.11) can be equivalently written as

$$\begin{aligned} &G^{\perp}\Big(\{\partial_{q\dot{q}}L - (\partial_{\dot{q}\dot{q}}L)(\partial_{\dot{q}\dot{q}}L_c)^{-1}(\partial_{q\dot{q}}L_c)\}\dot{q} - \partial_q L + (\partial_{\dot{q}\dot{q}}L)(\partial_{\dot{q}\dot{q}}L_c)^{-1}\partial_q L_c\Big) \\ &= 0. \end{aligned} \tag{4.12}$$

Proposition 2: *The systems (4.1) and (4.2) match if and only if the matching conditions (4.12) hold. In that case, the state feedback control law is explicitly given by*

$$u = (G^T G)^{-1} G^T (\text{rhs of } (4.11)). \tag{4.13}$$

Remark 2: Writing out (4.6) and using (4.9) it is easy to show that the control laws defined in (4.6) and (4.13) are the same. Notice that the control law is a state feedback law, depending only on q and $\dot{q}$.

Equation (4.12) is equivalent to the matching conditions of [17], Eq. (5). Furthermore, notice that (4.12) defines a set of nonlinear PDEs, where L is given and L_c acts as the unknown variable. The set of solutions L_c of (4.12) describes all the possible Euler-Lagrangian closed-loop systems (4.2) that can be obtained from (4.1) by a suitable choice (*i.e.* (4.13)) of the control law.

2.2. *Mechanical systems*

In case the Euler–Lagrange systems (4.1) and (4.2) both describe a mechanical system, then the matching conditions (4.12) can be split into two parts. The first part describes the shaping of kinetic energy, whereas the second part describes the shaping of potential energy.

Assume that (4.1) describes an (under)actuated mechanical system, that is, L is the difference of kinetic and potential energy

$$L(q,\dot{q}) = \frac{1}{2}\dot{q}^T M(q)\dot{q} - V(q), \tag{4.14}$$

where $M = M^T$ describes the generalized mass matrix of the system. We assume that M is invertible, which is equivalent to L being regular (the usual assumption is that M is positive definite.) We consider control laws which render the closed-loop system to be a mechanical system, that is, of the form (4.2) with controlled Lagrangian being of the form

$$L_c(q,\dot{q}) = \frac{1}{2}\dot{q}^T M_c(q)\dot{q} - V_c(q), \tag{4.15}$$

for some *shaped* generalized mass matrix $M_c = M_c^T$ (assumed to be invertible) and potential energy function V_c. In this case, the matching conditions (4.12) become

$$\begin{aligned} G^{\perp}(q)\Big[\{\partial_q(M(q)\dot{q}) - M(q)M_c^{-1}(q)\partial_q(M_c(q)\dot{q})\}\dot{q} - \partial_q(\frac{1}{2}\dot{q}^T M(q)\dot{q}) \\ + \partial_q V(q) + M(q)M_c^{-1}(q)[\partial_q(\frac{1}{2}\dot{q}^T M_c(q)\dot{q}) - \partial_q V_c(q)]\Big] = 0\,. \end{aligned} \tag{4.16}$$

Collecting the terms dependent, respectively independent, on $\dot{q}$ we see that (4.16) can be equivalently written as a set of two nonlinear PDEs in $M_c(q)$

and $V_c(q)$,

$$G^{\perp}(q)\Big[\{\partial_q(M(q)\dot{q}) - M(q)M_c^{-1}(q)\partial_q(M_c(q)\dot{q})\}\dot{q} \\ - \partial_q(\frac{1}{2}\dot{q}^T M(q)\dot{q}) + M(q)M_c^{-1}(q)[\partial_q(\frac{1}{2}\dot{q}^T M_c(q)\dot{q})]\Big] = 0 \quad (4.17)$$

and

$$G^{\perp}(q)\big[\,\partial_q V(q) - M(q)M_c^{-1}(q)\partial_q V_c(q)\,\big] = 0. \quad (4.18)$$

Equation (4.17) matches the *kinetic* energy and is independent of the potential energy, whereas Eq. (4.18) matches the *potential* energy of the closed-loop system and depends on the shaped generalized mass matrix M_c. Notice that (4.17) defines a homogeneous polynomial in $\dot{q}$, whereas (4.18) is independent of $\dot{q}$.

The λ-method of Auckly *et al.* [4] Equations (4.17) and (4.18) constitute a set of two *nonlinear* PDEs in M_c and V_c. In [1–4] a method has been presented to solve (4.17, 4.18) by recursively solving a set of three *linear* PDEs, thereby greatly reducing the complexity of finding solutions. Let us translate this method into our notation.

Consider Eq. (4.17) and notice that this equation has to hold for all points $(q,\dot{q}) \in T\mathcal{Q}$, whereby q and $\dot{q}$ should be seen as independent variables (*i.e.* the state of the system). This means that (4.17) can be equivalently written as (at a point $q_0 \in \mathcal{Q}$)

$$G^{\perp}(q_0)M(q_0)\Big[M^{-1}(q_0)\partial_q(M(q)v)|_{q_0}v - M^{-1}(q_0)\partial_q(\frac{1}{2}v^T M(q)v)|_{q_0} \\ + (\partial_q X)|_{q_0}v - M_c^{-1}(q_0)\partial_q(M_c(q)v)|_{q_0}v + M_c^{-1}(q_0)\partial_q(\frac{1}{2}v^T M_c(q)v)|_{q_0} \\ - (\partial_q X)|_{q_0}v\Big] = 0, \quad (4.19)$$

for all vector fields $X \in T\mathcal{Q}$ with $X(q_0) = v \in T_{q_0}\mathcal{Q}$. In (4.19) we recognize the expression for the covariant derivative, see [19]. The covariant derivative, denoted by ∇, assigns to two vector fields $X, Y \in T\mathcal{Q}$ a third one denoted by $\nabla_X Y \in T\mathcal{Q}$, called the covariant derivative of Y with respect to X. It is uniquely defined by the kinetic energy metric $g(X,Y)(q) = X(q)^T M(q)Y(q)$, $X, Y \in T\mathcal{Q}$.[b] (The symbol ∇ is also called the Levi-Civita connection corresponding to the metric g.) Let $\hat{\nabla}$ denote

[b] In our notation: $2\nabla_X Y = M^{-1}\partial_q(MX)Y + M^{-1}\partial_q(MY)X - M^{-1}\partial_q(X^T MY) + [X,Y] + M^{-1}(\partial_q X)^T Y + M^{-1}(\partial_q Y)^T X$, with $X, Y \in T\mathcal{Q}$.

the covariant derivative corresponding to the metric defined by the matrix M_c. Then (4.19) can be written as (suppressing the argument q_0)

$$G^{\perp}M\left[\nabla_X X - \hat{\nabla}_X X\right] = 0, \quad \forall X \in TQ. \tag{4.20}$$

This is exactly the matching condition as given in [2], Eq. (1.4), where $G^{\perp}M$ is denoted by P, see also [1,3]. Writing out the expression for the covariant derivative in the coefficients of X using the Christoffel symbols results in the matching conditions as given in [16], Theorem 1. Furthermore, the control law given in [16], Theorem 1, equals the control law defined by (4.13).

We can polarize (4.20) to get

$$\begin{aligned}
0 &= \frac{1}{2}G^{\perp}M\Big[\nabla_{X+Y}(X+Y) - \hat{\nabla}_{X+Y}(X+Y) \\
&\qquad\qquad - (\nabla_X X - \hat{\nabla}_X X) - (\nabla_Y Y - \hat{\nabla}_Y Y)\Big] \\
&= \frac{1}{2}G^{\perp}M\left[\nabla_X Y + \nabla_Y X - \hat{\nabla}_X Y - \hat{\nabla}_Y X\right] \\
&= G^{\perp}M\left[\nabla_X Y - \hat{\nabla}_X Y\right], \quad \forall X, Y \in TQ,
\end{aligned} \tag{4.21}$$

where we used that $\nabla_X Y - \nabla_Y X = [X, Y] = \hat{\nabla}_X Y - \hat{\nabla}_Y X$, which follows easily from the formula for the covariant derivative. Recall that $G^{\perp}$ denotes a full rank left annihilator of G (*i.e.* normalizing G to $[0 \quad I]^T$ this means that $G^{\perp} = [I \quad 0]$). Instead, let $\bar{G}^{\perp}$ denote an orthogonal projection matrix, *i.e.* $(\bar{G}^{\perp})^T = \bar{G}^{\perp}$ and $(\bar{G}^{\perp})^2 = \bar{G}^{\perp}$, such that $\bar{G}^{\perp}G = 0$. Normalizing G to $[0 \quad I]^T$ this means that

$$\bar{G}^{\perp} = \begin{bmatrix} I & 0 \\ 0 & 0 \end{bmatrix}. \tag{4.22}$$

Then (4.21) still holds when one writes $\bar{G}^{\perp}$ instead of $G^{\perp}$. Now introduce a 'new' matrix variable by $\lambda = M_c^{-1}M$. Then a linear PDE in λ is obtained by taking $X = \lambda\bar{G}^{\perp}MX'$ and $Y = Y'$ and premultiplying (4.21) by $(X')^T M$. After some algebra, eliminating Y', this results in the following equation (suppressing the prime and writing X for X'):

$$\begin{aligned}
0 &= X^T M\bar{G}^{\perp}\lambda^T\Big\{[\partial_q(M\bar{G}^{\perp}MX)]^T - [\partial_q(\bar{G}^{\perp}MX)]^T M - M\partial_q(\bar{G}^{\perp}MX)\Big\} \\
&\quad + X^T M\bar{G}^{\perp}\Big\{[\partial_q(\lambda\bar{G}^{\perp}MX)]^T M + M\partial_q(\lambda\bar{G}^{\perp}MX) - [\partial_q(M\lambda\bar{G}^{\perp}MX)]^T\Big\}, \\
&\forall X \in TQ.
\end{aligned} \tag{4.23}$$

Observe that (4.23) is a *linear* PDE in λ. However, notice that a solution is only defined with respect to the image of $\bar{G}^{\perp}$, *i.e.* a solution is only defined for $\lambda\bar{G}^{\perp}M$. Equation (4.23) is called the λ-*equation* and corresponds to Eq. (1.11) in [2].

The complete solution λ (or, equivalently, M_c) of the kinetic energy matching condition (4.17) can be found by solving another linear PDE. Indeed, premultiply (4.17) by M to get

$$0 = M\bar{G}^{\perp}\lambda^T\{\partial_q(\frac{1}{2}\dot{q}^T M_c\dot{q}) - \partial_q(M_c\dot{q})\dot{q}\} + M\bar{G}^{\perp}\{\partial_q(M\dot{q})\dot{q} - \partial_q(\frac{1}{2}\dot{q}^T M\dot{q})\}, \tag{4.24}$$

$\forall(q,\dot{q}) \in TQ$. Given a solution $\lambda\bar{G}^{\perp}M$ of (4.23) this is a *linear* PDE in M_c. Equation (4.24) corresponds to Eq. (1.12) in [2] (with $Z = \dot{q}$ and eliminating X from 1.12).

Finally, given M_c, the potential energy matching condition (4.18) is a *linear* PDE in V_c. It can also be written in terms of a solution $\lambda\bar{G}^{\perp}M$ of (4.23) by premultiplying (4.18) by M to obtain:

$$0 = M\bar{G}^{\perp}\partial_q V - M\bar{G}^{\perp}\lambda^T\partial_q V_c. \tag{4.25}$$

This equation corresponds to Eq. (1.13) in [2].

In [2, 3] it is shown that the matching conditions (4.17, 4.18) can be solved by solving the equivalent set of three linear PDEs (4.23, 4.24, 4.25). That is, first solving (4.23) for $\lambda\bar{G}^{\perp}M$, then (4.24) for M_c, and finally (4.25) for V_c.

2.3. *Mechanical systems with symmetry*

In this section we review the controlled Lagrangians method as introduced by [8, 9, 11] for mechanical systems with symmetry. In particular, we interpret the matching conditions obtained in those papers in terms of the matching of kinetic and potential energy as described by the PDEs (4.17, 4.18).

Consider a mechanical system with configuration space an n-dimensional manifold $\mathcal{Q} \simeq \mathbb{R}^n$. Let the configuration coordinates be denoted by $q = (x, \theta) \in \mathbb{R}^n$. Here $x \in \mathbb{R}^{n-m}$ are called the *shape variables* and $\theta \in \mathbb{R}^m$ are called the *group variables*. We assume that the group variables are fully actuated, whereas the shape variables are unactuated, this corresponds to $G = [0 \quad I_m]^T$. Furthermore, we assume that the Lagrangian of the system does not depend on the variables θ (we call θ *cyclic variables*).

Remark 3: The mathematical construction used in [11] is to consider a principal fiber bundle $\mathcal{Q} \to \mathcal{Q}/\mathcal{G}$ corresponding to the regular action of an Abelian (*i.e.* commutative) Lie group $\mathcal{G}$ on $\mathcal{Q}$. Then $x \in \mathcal{Q}/\mathcal{G}$ and $\theta \in \mathcal{G}$, and the Lagrangian L being cyclic in θ is equivalent to assuming that L is invariant under the action of the group G.

The forced Euler–Lagrange equations become

$$\frac{d}{dt}\partial_{\dot{x}}L - \partial_x L = 0, \tag{4.26}$$

$$\frac{d}{dt}\partial_{\dot{\theta}}L = u, \tag{4.27}$$

with

$$L(x, \dot{x}, \dot{\theta}) = \frac{1}{2}\dot{q}^T M(x)\dot{q} - V(x), \qquad \dot{q} = (\dot{x}, \dot{\theta}). \tag{4.28}$$

As explained in [11] quite a large class of mechanical systems fall within this description. The goal of the controlled Lagrangians method described in [11] is to stabilize a *relative equilibrium*[c] ($x = x_e$, $\dot{x} = 0$, $\dot{\theta} = 0$) of the system. This is done by searching for a stabilizing closed-loop Euler-Lagrangian system which preserves the symmetry of the system. In [11] a class of controlled Lagrangians is proposed which have the property that θ is a cyclic variable for L_c. This class can be described as follows: First, decompose the generalized mass matrix M as follows

$$M = \begin{bmatrix} M^{xx} & M^{x\theta} \\ M^{\theta x} & M^{\theta\theta} \end{bmatrix}, \tag{4.29}$$

according to the decomposition $q = (x, \theta)$. Define the shaped generalized mass matrix as follows

$$M_c = \begin{bmatrix} M^{xx} + M^{x\theta}\tau + \tau^T M^{\theta x} + \tau^T(M^{\theta\theta} + \sigma)\tau & M^{x\theta} + \tau^T M^{\theta\theta} \\ M^{\theta x} + M^{\theta\theta}\tau & M^{\theta\theta} \end{bmatrix}. \tag{4.30}$$

Here, $\tau(x) \in \mathbb{R}^{m\times n}$ and $\sigma(x) \in \mathbb{R}^{m\times m}$ are matrices only depending on the shape variables. In [11] τ is called a 'Lie algebra valued horizontal one-form', which means that it works only on vectors in the shape space $\mathbb{R}^{n-m}$ and takes values in $\mathbb{R}^m$. The matrix σ is called the 'changed metric acting on horizontal vectors', which means that it changes the mass matrix in the direction of the shape variables. The controlled Lagrangian is then defined by, corresponding to formula (2.11) in [11],

$$L_c(x, \dot{x}, \dot{\theta}) = \frac{1}{2}\dot{q}^T M_c(x)\dot{q} - V(x), \qquad \dot{q} = (\dot{x}, \dot{\theta}). \tag{4.31}$$

[c]The term relative equilibrium is used in reduction theory. It denotes an equilibrium in the shape variables, whereas motion with constant velocity (or better, momentum) in the group variables is allowed. In our case the relative equilibrium has zero velocity in the group variables. The configuration θ of the group variables however is unspecified.

It is important to notice that only the kinetic energy is changed whereas the potential energy of the system is left unchanged. Since the controlled Lagrangian preserves symmetry, *i.e.* L_c does not depend on θ, the corresponding Euler–Lagrange system looks like

$$\frac{d}{dt}\partial_{\dot{x}} L_c - \partial_x L_c = 0, \tag{4.32}$$

$$\frac{d}{dt}\partial_{\dot{\theta}} L_c = 0. \tag{4.33}$$

The idea of the method of [11] is to shape the kinetic energy, by choosing suitable matrices τ and σ, in order to obtain a closed-loop Euler-Lagrangian system (4.31, 4.32, 4.33) for which the desired relative equilibrium is stable. The conditions under which L_c can be obtained as a possible closed-loop Lagrangian by choosing a suitable control law for the system (4.26, 4.27, 4.28) are the matching conditions of [11]. In general, they consist of a set of nonlinear PDEs in the components of the matrices τ and σ. In the next paragraph the derivation of these matching conditions is described.

The matching conditions of Bloch *et al.* [11] In [11] the result of proposition 1 is used to deduce conditions under which the systems (4.26, 4.27, 4.28) and (4.31, 4.32, 4.33) match. That is, they give conditions under which (4.32) holds along solutions of (4.26, 4.33). Towards this objective denote the x-component of the Euler–Lagrange equations as:

$$\mathcal{E}_x(L_c) = G^{\perp}\left(\frac{d}{dt}\partial_{\dot{q}} L_c - \partial_q L_c\right) = \frac{d}{dt}\partial_{\dot{x}} L_c - \partial_x L_c. \tag{4.34}$$

Subtracting (4.3), equivalently (4.26), this becomes

$$\begin{aligned}\mathcal{E}_x(L_c) &= G^{\perp}\left(\frac{d}{dt}\partial_{\dot{q}} L_c - \partial_q L_c - \frac{d}{dt}\partial_{\dot{q}} L + \partial_q L\right)\\ &= G^{\perp}\Big((I - MM_c^{-1})M_c\ddot{q} + \partial_q(M_c\dot{q})\dot{q} - \partial_q(M\dot{q})\dot{q}\\ &\qquad - \partial_q(\frac{1}{2}\dot{q}^T M_c\dot{q}) + \partial_q(\frac{1}{2}\dot{q}^T M\dot{q})\Big),\end{aligned} \tag{4.35}$$

assuming M_c is invertible.

Now notice that (4.33) defines the first integral $\partial_{\dot{\theta}} L_c$ of the controlled Lagrangian system. Decompose M_c, defined in (4.30), according to the decomposition $q = (x, \theta)$ and write

$$M_c = \begin{bmatrix} M_c^{xx} & M_c^{x\theta} \\ M_c^{\theta x} & M_c^{\theta\theta} \end{bmatrix}. \tag{4.36}$$

Then

$$\partial_{\dot{\theta}} L_c = M_c^{\theta x} \dot{x} + M_c^{\theta\theta} \dot{\theta}, \tag{4.37}$$

which gives by (4.33), taking into account that θ is a cyclic variable,

$$M_c^{\theta x} \ddot{x} + M_c^{\theta\theta} \ddot{\theta} + \partial_x (M_c^{\theta x} \dot{x}) \dot{x} + \partial_x (M_c^{\theta\theta} \dot{\theta}) \dot{x} = 0. \tag{4.38}$$

Assuming that $M_c^{\theta\theta}$ is invertible (notice that a sufficient condition for $M_c^{\theta\theta}$ to be invertible is that M_c is definite) this results in

$$\ddot{\theta} = -(M_c^{\theta\theta})^{-1} M_c^{\theta x} \ddot{x} - (M_c^{\theta\theta})^{-1} \left(\partial_x (M_c^{\theta x} \dot{x}) \dot{x} + \partial_x (M_c^{\theta\theta} \dot{\theta}) \dot{x} \right). \tag{4.39}$$

Using (4.39) we can calculate

$$\begin{aligned} M_c \ddot{q} &= \begin{bmatrix} M_c^{xx} \ddot{x} + M_c^{x\theta} \ddot{\theta} \\ M_c^{\theta x} \ddot{x} + M_c^{\theta\theta} \ddot{\theta} \end{bmatrix} \\ &= \begin{bmatrix} S_c \ddot{x} - M_c^{x\theta} (M_c^{\theta\theta})^{-1} \left(\partial_x (M_c^{\theta x} \dot{x}) \dot{x} + \partial_x (M_c^{\theta\theta} \dot{\theta}) \dot{x} \right) \\ - \left(\partial_x (M_c^{\theta x} \dot{x}) \dot{x} + \partial_x (M_c^{\theta\theta} \dot{\theta}) \dot{x} \right) \end{bmatrix}, \end{aligned} \tag{4.40}$$

where $S_c := M_c^{xx} - M_c^{x\theta} (M_c^{\theta\theta})^{-1} M_c^{\theta x}$ is precisely the Schur-complement of the matrix M_c. Since we assume that M_c is invertible, it follows that S_c is invertible, see [14], p. 46.

Now substitute (4.40) into (4.35). The only terms of $\mathcal{E}_x(L_c)$ involving accelerations are given by

$$G^{\perp} (I - M M_c^{-1}) \begin{bmatrix} I \\ 0 \end{bmatrix} S_c \ddot{x}. \tag{4.41}$$

Bloch *et al.* [11] define their first matching condition, Assumption M-1, in such a way as to cancel all the terms in $\mathcal{E}_x(L_c)$ that involve the accelerations $\ddot{x}$. Since S_c is invertible, we have the following proposition, valid with respect to the class of controlled Lagrangians (4.30, 4.31) considered in [11]. (Recall that $G^{\perp} = [I_{n-m} \quad 0]$.)

Proposition 3: *The matching condition M-1 of [11] is equivalent to the condition*

$$[I_{n-m} \quad 0] (I_n - M M_c^{-1}) \begin{bmatrix} I_{n-m} \\ 0 \end{bmatrix} = 0. \tag{4.42}$$

Condition (4.42) is an algebraic condition on the kinetic energy metric defined by M_c. Assuming (4.42) holds, let us calculate $\mathcal{E}_x(L_c)$. First calculate that

$$-\partial_q(M_c\dot{q})\dot{q} = \begin{bmatrix} * \\ -\left(\partial_x(M_c^{\theta x}\dot{x})\dot{x} + \partial_x(M_c^{\theta\theta}\dot{\theta})\dot{x}\right) \end{bmatrix}. \tag{4.43}$$

Then after substitution of (4.40) into (4.35) and using (4.42) and (4.43), Eq. (4.35) becomes

$$\begin{aligned} \mathcal{E}_x(L_c) &= G^{\perp}\big(-(I - MM_c^{-1})\partial_q(M_c\dot{q})\dot{q} + \partial_q(M_c\dot{q})\dot{q} - \partial_q(M\dot{q})\dot{q} \\ &\quad - \partial_q(\frac{1}{2}\dot{q}^T M_c\dot{q}) + \partial_q(\frac{1}{2}\dot{q}^T M\dot{q})\big) \\ &= G^{\perp}\big(MM_c^{-1}\partial_q(M_c\dot{q})\dot{q} - \partial_q(M\dot{q})\dot{q} - \partial_q(\frac{1}{2}\dot{q}^T M_c\dot{q}) + \partial_q(\frac{1}{2}\dot{q}^T M\dot{q})\big). \end{aligned} \tag{4.44}$$

From the fact that θ is a cyclic variable for L_c it follows using (4.42) that

$$G^{\perp}\partial_q(\frac{1}{2}\dot{q}^T M_c\dot{q}) = G^{\perp}\begin{bmatrix} I \\ 0 \end{bmatrix}\partial_x(\frac{1}{2}\dot{q}^T M_c\dot{q}) = G^{\perp}MM_c^{-1}\partial_q(\frac{1}{2}\dot{q}^T M_c\dot{q}). \tag{4.45}$$

Finally, this results in the following equation for $\mathcal{E}_x(L_c)$:

$$\begin{aligned} \mathcal{E}_x(L_c) &= G^{\perp}\Big(\{MM_c^{-1}\partial_q(M_c\dot{q}) - \partial_q(M\dot{q})\}\dot{q} \\ &\qquad - MM_c^{-1}\partial_q(\frac{1}{2}\dot{q}^T M_c\dot{q}) + \partial_q(\frac{1}{2}\dot{q}^T M\dot{q})\Big). \end{aligned} \tag{4.46}$$

This corresponds to Eq. (2.25) in [11]. Bloch *et al.* [11] proceed by giving two conditions, *i.e.* Assumption M-2 and Assumption M-3, under which $\mathcal{E}_x(L_c)$ is identically zero, thereby accomplishing matching.

Interpretation of the matching conditions According to Section 2.2 the systems (4.26, 4.27, 4.28) and (4.31, 4.32, 4.33) match if and only if the two PDEs (4.17, 4.18) hold. Notice that (4.18), describing the matching of the potential energy, in this case becomes the *algebraic equation*

$$G^{\perp}\left[\left(I - M(x)M_c^{-1}(x)\right)\partial_q V(x)\right] = 0. \tag{4.47}$$

In the sequel we will interpret the matching conditions obtained by [11] in terms of the conditions (4.17) and (4.47).

As described above the assumptions M-1, M-2 and M-3 accomplish matching for the class of controlled Lagrangians (4.30, 4.31) considered in [11]. According to proposition 3, condition M-1 is equivalent to (4.42). Now consider the matching condition (4.47) for the potential energy. Since

θ is a cyclic variable for V, we have that

$$\partial_q V(x) = \begin{bmatrix} I \\ 0 \end{bmatrix} \partial_x V(x). \tag{4.48}$$

However, this means that (4.42) implies (4.47). Actually, this holds for any function V which is independent of the variables θ.

Proposition 4: *Assumption M-1 of [11] implies that the unchanged potential energy V matches.*

In other words, assumption M-1 takes care of the matching of potential energy. Notice that similarly to (4.47), assumption M-1 describes an *algebraic equation* on the kinetic energy matrix M_c.

Secondly, assuming that condition M-1 holds, we calculated $\mathcal{E}_x(L_c)$ to be as in (4.46). The condition that $\mathcal{E}_x(L_c)$ is equal to zero is precisely the matching condition (4.17) for the kinetic energy.

Proposition 5: *Assume that condition M-1 holds. Then assumptions M-2 and M-3 are equivalent to the matching condition (4.17) on the kinetic energy.*

In other words, assumptions M-2 and M-3 take care of the matching of kinetic energy. Notice that similar to (4.17), assumptions M-2 and M-3 define a set of nonlinear PDEs, to be solved for the kinetic energy matrix M_c (or its determining components τ and σ).

The above two propositions give an interpretation of the matching conditions as defined in [11] in terms of the matching of kinetic and potential energy.

Observe that to conclude if a certain controlled Lagrangian can be obtained as a closed-loop Lagrangian (*i.e.* matches) one needs to check the nonlinear PDEs (4.17, 4.18). In case one considers the class of systems and controlled Lagrangians as defined in [11] this comes down to checking the algebraic condition (4.42) and the nonlinear PDE (4.17) (or equivalently, checking assumptions M-1, M-2, M-3). In [11] a set of conditions, called the *simplified matching assumptions*, is given under which (4.42) and (4.17) automatically hold. Let us translate these conditions into the notation used in this chapter. Recall the decomposition of the matrix M as in (4.29) and denote $\Delta := M^{x\theta}(M^{\theta\theta})^{-1}M^{\theta x}$. The second and fourth of the simplified matching assumptions [11] can be translated as follows:

[SM-1] $M^{\theta\theta}(x) = M^{\theta\theta}$ is a constant (invertible) matrix,

$$\text{[SM-2]} \quad \partial_{x_j} M^{x_i\theta_k} = \partial_{x_i} M^{x_j\theta_k}, \quad i,j = 1,\ldots,n-m, \; k = 1,\ldots,m.$$

As remarked in [11], these conditions imply that the mechanical connection corresponding to the system is flat, that is, the system lacks gyroscopic forces. The first and third of the simplified matching assumptions [11] can be translated into taking[d]

$$\tau = \kappa (M^{\theta\theta})^{-1} M^{\theta x}, \quad \sigma = -\frac{1}{\kappa} M^{\theta\theta}, \tag{4.49}$$

for some arbitrary nonzero constant $\kappa \in \mathbb{R}$, which can be seen as a design parameter. This results in the shaped kinetic energy matrix M_c,

$$\text{[SM-3]} \quad M_c = \begin{bmatrix} M^{xx} + \kappa(\kappa+1)\Delta & (\kappa+1)M^{x\theta} \\ (\kappa+1)M^{\theta x} & M^{\theta\theta} \end{bmatrix}.$$

Now we can state the following proposition [11].

Proposition 6: *Assume that the Lagrangian (4.28) satisfies assumptions SM-1 and SM-2. Take the controlled Lagrangian L_c to be of the form (4.31), with M_c as in SM-3 (for arbitrary κ). Then L_c is a matching Lagrangian, that is, the systems (4.26, 4.27, 4.28) and (4.31, 4.32, 4.33) match.*

Although the assumptions SM-1, SM-2 and SM-3 are quite restrictive[e], they seem to work well for the matching and stabilization of a number of interesting systems like the inverted pendulum on a cart and the spherical inverted pendulum. See [11] for worked examples.

2.4. *The cart and pendulum*

In this section we want to make a few remarks on the matching methods we have described so far, taking as a guideline the example of an inverted pendulum on a cart. This system was first stabilized using the method of controlled Lagrangians by [8,11]. We described this method in the previous section. The method has two key features:

(I) The method stabilizes a *relative* equilibrium.

In the case of the cart and pendulum this means that the upright position of the pendulum is stabilized, irrespective of the horizontal position of the cart.

[d] For $\kappa = 0$: take $\tau = 0$ and σ any matrix. Then $M_c = M$.

[e] However, in the case $n = 2$, $m = 1$ (*e.g.* inverted pendulum on a cart) assumptions M-1, M-2, M-3 and assumptions SM-1, SM-2, SM-3 are equivalent, as can easily be seen.

(II) The kinetic energy of the closed-loop system is *negative* definite.

This means that the closed-loop system simulates a mechanical system with *negative* masses and inertias, which is physically not very appealing.[f]

The first problem can easily be overcome by allowing also the shaping of potential energy (recall that in the method of [11] the potential energy was unchanged). This destroys the symmetry present in the system but in return stabilizes the group variables (*i.e.* the position of the cart) at a desired equilibrium point. Extending the above method by also including potential energy shaping was described in [10,12]. In those papers, the kinetic energy is still shaped according to assumptions SM-1, SM-2 and SM-3, and in addition the potential energy is also shaped (by introducing a new matching assumption). This solves the first problem, however, it cannot solve the second problem. In fact, for the cart and pendulum example, it can easily be checked that taking the shaped kinetic energy according to assumptions SM-1, SM-2 and SM-3, the potential energy can never be shaped in such a way that the stabilizing closed-loop kinetic energy is positive definite at the desired equilibrium (*i.e.* upright position of the pendulum, cart at a desired horizontal position). This seems to be a structural property of the method as described in [11, 12].

On the other hand, if we consider the more general matching conditions as described in Section 2.2, then problems (I) and (II) are absent. Indeed, as shown in [2, 16], it is possible to stabilize the cart and pendulum system at the desired equilibrium point, such that the total energy of the closed-loop system is *positive* definite. This means that the closed-loop system corresponds to a physically existing mechanical system, with positive masses and inertias. Remark that indeed the corresponding shaped kinetic energy matrix does not have the form as in SM-3.

We conclude that although the controlled Lagrangians method, and the corresponding (simplified) matching assumptions, described in [11, 12] and Section 2.3, can be very helpful in solving the matching conditions and stabilizing a mechanical system, for a large class of examples it leads to closed-loop systems having a negative definite total energy, something which is physically not very appealing and can become problematic in the presence of damping. This problem does not occur when one shapes the energy

[f]The problem of a negative definite kinetic energy becomes serious in the presence of physical damping. Indeed physical damping *dissipates* energy, pushing the state towards a *minimum* of the energy. This means that in order for the controlled Lagrangians method to work the (usually unknown) damping has to be compensated, see also [26, 27].

according to the more general matching conditions described in Section 2.2, see [2, 16] for examples.

3. Matching of Port-Controlled Hamiltonian Systems

In [20, 21] a method has been developed to stabilize a desired equilibrium point of a port-controlled Hamiltonian system. The class of port-controlled Hamiltonian systems strictly contains the class of regular Euler–Lagrange systems. The method is called the *interconnection and damping assignment passivity based control (IDA-PBC)* method. Analogously to the method of controlled Lagrangians the basic idea is to search for a closed-loop system with stable desired equilibrium point which is again in port-controlled Hamiltonian format. As in the previously described method this leads to a set of matching conditions, described by a set of nonlinear PDEs.

In this section we recall the method developed in [20, 21] and its application to mechanical systems.

3.1. *General matching conditions*

Consider a port-controlled Hamiltonian system of the form

$$\dot{z} = J(z)\partial_z H(z) + g(z)u, \tag{4.50}$$

where $z \in \mathcal{M}$ (a manifold), $J(z) = -J^T(z) : T_z^*\mathcal{M} \to T_z\mathcal{M}$ is a skew-symmetric matrix (or better, vector bundle map) describing the internal interconnection structure of the system, $g(z) : \mathbb{R}^m \to T_z\mathcal{M}$ describes the input vector fields corresponding to the input $u \in \mathbb{R}^m$ and $H(z)$ is the Hamiltonian (or energy) function of the system. The objective of IDA-PBC is to stabilize a desired equilibrium point of the system. Analogously to the method of controlled Lagrangians this goal is pursued by considering static state feedback laws which render the closed-loop system in port-controlled Hamiltonian format. That is, the closed-loop system is described by the equations

$$\dot{z} = J_d(z)\partial_z H_d(z). \tag{4.51}$$

Here, $J_d(z) = -J_d^T(z)$ denotes the closed-loop interconnection matrix and $H_d(z)$ the closed-loop Hamiltonian function. The system (4.51) can be obtained from (4.50) by state feedback $u = u(z)$ if and only if

$$J_d(z)\partial_z H_d(z) = J(z)\partial_z H(z) + g(z)u(z). \tag{4.52}$$

Let $g^{\perp}(z)$ denote a full rank left annihilator of $g(z)$, then (4.52) can be equivalently written as

$$g^{\perp}(z)\big[J_d(z)\partial_z H_d(z) - J(z)\partial_z H(z)\big] = 0, \tag{4.53}$$

which are the matching conditions of the IDA-PBC method [20,21]. Notice that the matching conditions (4.53) define a set of nonlinear PDEs, to be solved for the shaped Hamiltonian H_d *and* the shaped interconnection matrix J_d. If the matching conditions are satisfied, *i.e.* the systems (4.50) and (4.51) match, then the corresponding state feedback law is explicitly given by

$$u(z) = (g^T(z)g(z))^{-1}g^T(z)\big\{J_d(z)\partial_z H_d(z) - J(z)\partial_z H(z)\big\}. \tag{4.54}$$

Remark 4: In [20,21] the following equivalent form of the matching conditions can be found: Write $J_a = J_d - J$ and $H_a = H_d - H$, then Eq. (4.52) becomes

$$(J(z) + J_a(z))\partial_z H_a(z) = -J_a(z)\partial_z H(z) + g(z)u(z), \tag{4.55}$$

and the matching conditions (4.53) get the form

$$g^{\perp}(z)\big[(J(z) + J_a(z))\partial_z H_a(z) + J_a(z)\partial_z H(z)\big] = 0, \tag{4.56}$$

which is a set of nonlinear PDEs to be solved for H_a and J_a.

Remark 5: Suppose (4.50) represents a *linear* port-controlled Hamiltonian system, *i.e.* $\dot{z} = JQz + gu$ for constant matrices $J = -J^T$, g, and Hamiltonian function $H(z) = \frac{1}{2}z^TQz$, $Q = Q^T$, and suppose that also the closed-loop system (4.51) is a linear system. It has been shown in [24] that in this case the matching conditions (4.53), as well as the conditions for stability of the closed-loop system, can be transformed into a set of linear matrix inequalities (LMIs). Powerful algorithms for solving these LMIs are available in several software packages.

Remark 6: *Equivalence under state feedback.* The closed-loop system (4.51) does not include the description of external inputs. This stems from the fact that the IDA-PBC method is designed to construct feedback controllers $u = u(z)$ which stabilize an assigned equilibrium point z^*, that is, the closed-loop system (4.51) has a stable equilibrium point at z^*. The addition of external inputs to the closed-loop system, yielding

$$\dot{z} = J_d(z)\partial_z H_d(z) + g(z)v, \quad v \in \mathbb{R}^m, \tag{4.57}$$

can be of importance in reaching additional control objectives. For instance, feeding back the passive output $y = g^T\partial_z H_d$ by $v = -Ky$, $K > 0$, yields under suitable assumptions *asymptotic* stability, see [22]. However, the addition of external inputs to the closed-loop system does not change the matching conditions (4.53). The systems (4.50) and (4.51) are *equivalent* under state feedback $u(z,v) = \alpha(z) + v$ if and only if (4.53) holds. The corresponding control law $\alpha(z)$ is defined by (4.54). Of course, an analogous remark can be made for the controlled Lagrangians method.

3.2. *Mechanical systems*

In this section we apply the method described above to mechanical systems, see [22]. A mechanical system can be described by a port-controlled Hamiltonian system of the form (4.50),

$$\begin{bmatrix} \dot{q} \\ \dot{p} \end{bmatrix} = \begin{bmatrix} 0 & I_n \\ -I_n & 0 \end{bmatrix} \begin{bmatrix} \partial_q H \\ \partial_p H \end{bmatrix} + \begin{bmatrix} 0 \\ G(q) \end{bmatrix} u, \tag{4.58}$$

where (q,p) (consisting of configuration coordinates q and momenta p) denote coordinates for the state space $\mathcal{M} = T^*\mathcal{Q} \simeq \mathbb{R}^{2n}$, with $\mathcal{Q} \simeq \mathbb{R}^n$ denoting the configuration space of the mechanical system. The matrix $G(q) : \mathbb{R}^m \to T_q^*\mathcal{Q} \simeq \mathbb{R}^n$ defines the force fields corresponding to the input $u \in \mathbb{R}^m$. The Hamiltonian function $H(q,p)$ is given by the total, *i.e.* kinetic plus potential, energy in the system

$$H(q,p) = \frac{1}{2} p^T M^{-1}(q) p + V(q), \tag{4.59}$$

where $M = M^T$ describes the generalized mass matrix of the system, and is assumed to be invertible (for most physical systems M will be positive definite). Note that from (4.58) and (4.59) it follows that the momenta are defined as usual by $p = M(q)\dot{q}$. Following [22] we propose the shaped Hamiltonian function $H_d(q,p)$ to be again of the form (4.59),

$$H_d(q,p) = \frac{1}{2} p^T M_d^{-1}(q) p + V_d(q), \tag{4.60}$$

for some shaped generalized mass matrix $M_d = M_d^T$ (assumed to be invertible) and potential energy function $V_d(q)$. The shaped interconnection matrix is taken to be in the most general form

$$J_d(q,p) = \begin{bmatrix} 0 & M^{-1}(q) M_d(q) \\ -M_d(q) M^{-1}(q) & J_2(q,p) \end{bmatrix}, \tag{4.61}$$

for some skew-symmetric matrix $J_2(q,p)$. Then, system (4.51) becomes

$$\begin{bmatrix} \dot{q} \\ \dot{p} \end{bmatrix} = \begin{bmatrix} 0 & M^{-1} M_d \\ -M_d M^{-1} & J_2 \end{bmatrix} \begin{bmatrix} \partial_q H_d \\ \partial_p H_d \end{bmatrix}. \tag{4.62}$$

Remark 7: Since $\dot{q}$ is a nonactuated coordinate, it follows that the relationship $\dot{q} = M^{-1}(q)p$ should also hold in closed-loop. Fixing (4.51) and (4.60) this explains the first row of the matrix J_d.

In this case the matching conditions (4.53) become

$$G^{\perp}\left[\partial_q H - M_d M^{-1}\partial_q H_d + J_2 M_d^{-1} p\right] = 0. \tag{4.63}$$

Using (4.59) and (4.60) and collecting terms dependent, respectively independent, of p we see that (4.63) can be equivalently written as a set of two nonlinear PDEs

$$\begin{aligned} G^{\perp}(q)\Big[\partial_q(\frac{1}{2}p^T M^{-1}(q)p) - M_d(q)M^{-1}(q)\partial_q(\frac{1}{2}p^T M_d^{-1}(q)p) \\ + J_2(q,p)M_d^{-1}(q)p\Big] = 0, \end{aligned} \tag{4.64}$$

and

$$G^{\perp}(q)\left[\partial_q V(q) - M_d(q)M^{-1}(q)\partial_q V_d(q)\right] = 0. \tag{4.65}$$

Like in the Lagrangian case, Eq. (4.64) matches the kinetic energy and is independent of the potential energy, whereas Eq. (4.65) matches the potential energy of the closed-loop system (and depends on M_d). The PDEs contain the unknown variables M_d and V_d, whereas the matrix J_2 acts as a free parameter which can be suitably chosen to allow the PDEs to be solvable for specific choices of M_d and V_d (directed by the stabilization objective). In case of matching the corresponding feedback law is given by (4.54)

$$u = (G^T G)^{-1} G^T\left\{\partial_q H - M_d M^{-1}\partial_q H_d + J_2 M_d^{-1} p\right\}. \tag{4.66}$$

Again remark that (4.64) and (4.65) define a set of nonlinear PDEs, which are in general not easy to solve. However, for a special class of systems these PDEs can be transformed into a set of nonlinear ODEs which are much easier to solve. This is described in [15]. The class of systems for which this transformation is possible is defined by the following assumptions: 1) the system is assumed to have n degrees of freedom and $n-1$ actuators (*i.e.* there is only one unactuated coordinate), and 2) the kinetic energy matrix M is assumed only to depend on the unactuated coordinate. This class of systems is quite common in underactuated mechanical systems and includes for instance the cart and pendulum example. By choosing the shaped kinetic energy matrix M_c to only depend on the unactuated coordinate, it can be shown that the set of PDEs (4.64, 4.65) can be transformed into an equivalent set of ODEs. In [15] the method is applied to the examples of a cart and pendulum system and a ball and beam system. For general systems we will show in Section 4.2 that the λ-method as described in Section 2.2 can also be used to simplify the process of solving the matching conditions (4.64) and (4.65), by transforming them into a set of quasi-linear and linear PDEs.

4. Comparison Between the Two Methods

In Sections 2 and 3 we described the matching of Euler–Lagrange systems, respectively of port-controlled Hamiltonian systems. Since the class of regular Euler–Lagrange systems is strictly contained in the class of port-controlled Hamiltonian systems, the method of Section 2 should be a special case of the more general method described in Section 3. In this section we consider both methods as applied to mechanical systems, see Sections 2.2 and 3.2, and show that Euler–Lagrange matching is a special case of port-controlled Hamiltonian matching. Notice that the IDA-PBC method has an extra degree of freedom with respect to the controlled Lagrangians method, in the sense that, in addition to shaping the total energy of the system, it is also possible to shape the internal interconnection structure of the system. This extra freedom means that the IDA-PBC method results in a larger class of matching closed-loop systems than the controlled Lagrangians method described in Section 2.2. This can be an important point in finding suitable stabilizing feedback controllers. Furthermore, the λ-method described in Section 2.2 is shown to be useful in solving the matching conditions obtained in the IDA-PBC method.

4.1. *The controlled Lagrangians case of IDA-PBC*

Consider a mechanical system described by the Euler–Lagrange system (4.1, 4.14). This system is equivalent via the Legendre transformation to the Hamiltonian system (4.58, 4.59). In Section 2.2 we gave conditions under which the autonomous Euler–Lagrange system (4.2, 4.15) matches with the system (4.1, 4.14). The system (4.2, 4.15) is equivalent to a canonical Hamiltonian system in the following way. Define the momenta to be

$$p_c = \partial_{\dot{q}} L_c = M_c(q)\dot{q}, \tag{4.67}$$

and the Hamiltonian by the Legendre transformation,

$$H_c(q, p_c) = \frac{1}{2} p_c^T M_c^{-1}(q) p_c + V_c(q). \tag{4.68}$$

Then the Euler–Lagrange system (4.2, 4.15) can be equivalently written as the Hamiltonian system

$$\begin{bmatrix} \dot{q} \\ \dot{p}_c \end{bmatrix} = \underbrace{\begin{bmatrix} 0 & I_n \\ -I_n & 0 \end{bmatrix}}_{J_c} \begin{bmatrix} \partial_q H_c \\ \partial_{p_c} H_c \end{bmatrix}. \tag{4.69}$$

It follows that in the particular case that we choose M_d and J_d such that the closed-loop Hamiltonian system (4.60, 4.62) is equivalent (by a coordinate transformation) to the Hamiltonian system (4.68, 4.69), then the IDA-PBC method effectively results in the controlled Lagrangians method. Indeed, we will show that for a certain choice of M_c (or equivalently, for M_d) and J_2 the systems (4.68, 4.69) and (4.60, 4.62) are equivalent, as well as the corresponding matching conditions (4.17, 4.18) and (4.64, 4.65). This means that for this particular choice of J_2 (and therefore of the shaped interconnection structure J_d) the IDA-PBC and the controlled Lagrangians method are equivalent.

The systems (4.68, 4.69) and (4.60, 4.62) are equivalent (by a coordinate transformation) if and only if the Hamiltonians H_c and H_d are equivalent and in addition the structure matrices J_c and J_d are equivalent. Notice that $p_c = M_c M^{-1} p$, and calculate H_c in the coordinates (q, p) to obtain

$$H_c(q,p) = \frac{1}{2} p^T M^{-1}(q) M_c(q) M^{-1}(q) p + V_c(q). \tag{4.70}$$

The Hamiltonians H_c and H_d are equivalent if and only if

$$M_c(q) = M(q) M_d^{-1}(q) M(q) \quad \text{and} \quad V_c(q) = V_d(q). \tag{4.71}$$

Notice that there is a one-to-one relation between M_c and M_d. (4.71) implies

$$p_c = M(q) M_d^{-1}(q) p. \tag{4.72}$$

The structure matrices J_c and J_d are the same if and only if J_d becomes in the coordinates (q, p_c) the canonical matrix J_c (in that case we call (q, p_c) *canonical coordinates* for the matrix J_d). This means that the Poisson brackets of the coordinates (q, p_c) should satisfy

$$\{q, q\}_d = 0, \quad \{q, p_c\}_d = I_n \quad \text{and} \quad \{p_c, p_c\}_d = 0, \tag{4.73}$$

where $\{\cdot,\cdot\}_d$ denotes the Poisson bracket corresponding to the structure matrix J_d. It is easy to check that the first two conditions in (4.73) are satisfied, while for the last one:

$$\begin{aligned} \{p_c, p_c\}_d &= \{M M_d^{-1} p, M M_d^{-1} p\}_d \\ &= -[\partial_q (M M_d^{-1} p)]^T + \partial_q (M M_d^{-1} p) + M M_d^{-1} J_2 M_d^{-1} M. \end{aligned} \tag{4.74}$$

Thus $\{p_c, p_c\}_d$ is equal to zero if and only if

$$J_2(q,p) = M_d M^{-1}\big[[\partial_q(MM_d^{-1}p)]^T - \partial_q(MM_d^{-1}p)\big]M^{-1}M_d. \tag{4.75}$$

(For clarity we left out the argument q of the matrices M and M_d.) Note that J_2 is clearly skew-symmetric. In conclusion, the Hamiltonian systems (4.68, 4.69) and (4.60, 4.62) are equivalent if and only if conditions (4.71) and (4.75) hold.

Remark 8: The entries of the matrix J_2 in (4.75) can equivalently be written as

$$(J_2)_{ij}(q,p) = -p^T M_d^{-1} M\,[(M^{-1}M_d)_i, (M^{-1}M_d)_j], \quad i,j = 1,\ldots,n. \tag{4.76}$$

($[\cdot,\cdot]$ denotes the Lie bracket of vector fields.) This formulation was suggested in [20], although with swapped indices due to an unfortunate typo.

Since under conditions (4.71, 4.75) the Euler–Lagrange system (4.2, 4.15) and the Hamiltonian system (4.60, 4.62) are equivalent, the corresponding matching conditions (4.17, 4.18) and (4.64, 4.65) should also be equivalent. Indeed, it is easy to see that (4.71) implies that the matching conditions (4.18) and (4.65), describing the matching of potential energy, are the same. Furthermore, after some lengthy computations it can be shown that (4.64) is equal to (4.17) if J_2 is defined as in (4.75). Since under conditions (4.71, 4.75) the matching conditions (4.17, 4.18) (or equivalently (4.12)) and (4.64, 4.65) (or equivalently (4.63)) are equal, it follows immediately that also the corresponding feedback laws (4.13) and (4.66) are equal. In conclusion, we have the following proposition:

Proposition 7: *Consider the controlled Lagrangians method described in Section 2 and the IDA-PBC method described in Section 3, both applied to the class of mechanical systems (see Sections 2.2, 3.2 respectively). The IDA-PBC method is equivalent to the controlled Lagrangians method if and only if the shaped interconnection structure is chosen as in (4.75). The controlled Lagrangian L_c and the shaped Hamiltonian H_d are related by (4.71).*

Remark 9: Proposition 7 states that the controlled Lagrangians method as described in Section 2.2 is a special case of the more general IDA-PBC method (namely, with J_2 chosen equal to (4.75)). Independently from the present work, the controlled Lagrangians method has been extended in [13] in such a way that for mechanical systems it becomes *equivalent* with the IDA-PBC method. Essentially, instead of restricting to systems of the form (4.2), they also allow to include some external forces into the closed-loop Euler–Lagrange system (*i.e.* the right hand side of (4.2) is not necessarily equal to zero, but can be any external force). In this way, it is possible to write any mechanical Hamiltonian system in Euler–Lagrange format by including the nonintegrable part of the Hamiltonian system (corresponding to the failure of the Jacobi identity by the Poisson bracket) as an external (gyroscopic) force into the Euler–Lagrange system. Notice that this method only works for the class of simple mechanical systems (*i.e.* with total energy consisting of kinetic plus potential energy). Considering this larger class of closed-loop Euler–Lagrange systems in [13] it is shown that for simple mechanical systems the controlled Lagrangians method is equivalent to the IDA-PBC method.

4.2. *The λ-method for Hamiltonian matching*

In Section 2.2 we described the λ-method of [2]. This method describes a way to solve the matching condition (4.17), a nonlinear PDE in M_c, by recursively solving the two linear PDEs (4.23) and (4.24). In this section we will show that the method can also be used to solve the matching condition (4.64) obtained in the IDA-PBC procedure. However, instead of recursively solving two linear PDEs, we now have to solve one *quasi-linear* PDE and afterwards a linear PDE. Solving the quasi-linear PDE might be simplified by using the freedom in J_2.

Without loss of generality we may write the skew-symmetric matrix J_2 as

$$J_2(q,p) = M_d M^{-1}\big[[\partial_q(MM_d^{-1}p)]^T - \partial_q(MM_d^{-1}p)\big]M^{-1}M_d + U(q,p), \tag{4.77}$$

where $U(q,p)$ is a skew-symmetric matrix, free to choose by the designer. According to the results in the previous section, Eq. (4.64) then results in

$$G^{\perp}\Big[\{\partial_q(M\dot{q}) - MM_c^{-1}\partial_q(M_c\dot{q})\}\dot{q} - \partial_q(\tfrac{1}{2}\dot{q}^T M\dot{q}) + MM_c^{-1}[\partial_q(\tfrac{1}{2}\dot{q}^T M_c\dot{q})] + U(q,M\dot{q})M^{-1}M_c\dot{q}\Big] = 0, \quad \forall (q,\dot{q}) \in TQ. \tag{4.78}$$

As explained in Section 2.2 this can be equivalently written as

$$G^{\perp}M\left[\nabla_X X - \hat{\nabla}_X X + M^{-1}U(q,MX)M^{-1}M_c X\right] = 0, \quad \forall X \in TQ. \tag{4.79}$$

Equations (4.78) and (4.79) clearly show the extra freedom, represented by U, obtained in the IDA-PBC method with respect to the controlled Lagrangians method (Eqs. (4.17) and (4.20) respectively). Consider (4.78) and notice that in order to satisfy the matching condition the term $G^{\perp}U(q, M\dot{q})M^{-1}M_c\dot{q}$ has to be quadratic in $\dot{q}$. Therefore we take $U(q,p)$ to be linear in its second component. In that case we can write

$$U(q,p) = \sum_{k=1}^{n} p_k U_k(q), \quad U_k^T = -U_k, \tag{4.80}$$

where p_k denotes the k-th component of the vector p.

Remark 10: In general U can also be chosen to include terms independent of p. These terms however will not be present in the quadratic (in $\dot{q}$) part of matching condition. Indeed, they should satisfy a matching condition of their own (see Section 5.3). Terms in U independent of p come up in the matching of *integrable* Hamiltonian systems, see Section 5.

Next we will show that the nonlinear PDE (4.78), or equivalently (4.79), can be solved by first solving a quasi-linear PDE in $\lambda = M_c^{-1}M$ and afterwards a linear PDE in M_c. First, define the skew-symmetric matrices W_k by

$$U_k = 2\lambda^T W_k \lambda, \quad \text{i.e. } U(q,p) = 2\sum_{k=1}^{n} p_k \lambda^T W_k(q)\lambda. \tag{4.81}$$

Then (4.79) becomes

$$G^{\perp}M\left[\nabla_X X - \hat{\nabla}_X X\right] + 2\sum_{k=1}^{n} G^{\perp}(MX)_k \lambda^T W_k X = 0, \quad \forall X \in TQ, \tag{4.82}$$

where $(MX)_k$ denotes the k-th component of the vector MX. We can polarize this equation to obtain the equivalent condition

$$G^{\perp}M\left[\nabla_X Y - \hat{\nabla}_X Y\right] + \sum_{k=1}^{n} G^{\perp}\left[(MX)_k \lambda^T W_k Y + (MY)_k \lambda^T W_k X\right] = 0,$$
$$\forall X, Y \in TQ. \tag{4.83}$$

As in the original method of [2], see Section 2.2, consider (4.83) with the orthogonal projection matrix $\bar{G}^{\perp}$ instead of $G^{\perp}$. Furthermore, take $X = \lambda\bar{G}^{\perp}MX'$ and $Y = Y'$ and premultiply (4.83) by $(X')^T M$. Then the summation on the left hand side of (4.83) becomes

$$\sum_{k=1}^{n}\Big(\underbrace{(M\lambda\bar{G}^{\perp}MX')_k}_{\in\mathbb{R}}(X')^T M\bar{G}^{\perp}\lambda^T W_k Y' + \underbrace{((X')^T M\bar{G}^{\perp}\lambda^T W_k\lambda\bar{G}^{\perp}MX')}_{\in\mathbb{R}} M_{k*}Y'\Big), \tag{4.84}$$

where M_{k*} denotes the k-th row of the matrix M. As described in Section 2.2 the first term of the left hand side of (4.83) will result in the right hand side of the λ-equation (4.23). Then by eliminating Y' the nonlinear PDE (4.83) becomes (suppressing the prime and writing X for X'):

$$\begin{aligned} 0 &= X^T M\bar{G}^{\perp}\lambda^T\Big\{[\partial_q(M\bar{G}^{\perp}MX)]^T - [\partial_q(\bar{G}^{\perp}MX)]^T M - M\partial_q(\bar{G}^{\perp}MX)\Big\} \\ &\quad + X^T M\bar{G}^{\perp}\Big\{[\partial_q(\lambda\bar{G}^{\perp}MX)]^T M + M\partial_q(\lambda\bar{G}^{\perp}MX) - [\partial_q(M\lambda\bar{G}^{\perp}MX)]^T\Big\} \\ &\quad + \sum_{k=1}^{n}\Big((M\lambda\bar{G}^{\perp}MX)_k\, X^T M\bar{G}^{\perp}\lambda^T W_k + (X^T M\bar{G}^{\perp}\lambda^T W_k\lambda\bar{G}^{\perp}MX)\, M_{k*}\Big) \\ &\forall X \in TQ. \end{aligned} \tag{4.85}$$

This is a quasi-linear PDE in the sense that the derivatives of λ appear linear in the equation but the summation contains terms quadratic in the components of λ. Equation (4.85) can be regarded as the λ-equation for the matching of port-controlled Hamiltonian systems. Analogously to (4.23) it can be solved for $\lambda\bar{G}^{\perp}M$.

Remark 11: Remember that the skew-symmetric matrices W_k are designer chosen matrices. Exploiting the freedom in W_k might simplify the search for solutions of (4.85). Furthermore, notice that by taking $W_k = 0$, *i.e.* $U(q,p) = 0$, Eq. (4.85) results in the original λ-equation (4.23) (a *linear* PDE in λ), and the method reduces to the method of [2].

Once we have found a solution $\lambda\bar{G}^{\perp}M$ (together with some suitably chosen matrices W_k) of (4.85), the complete solution λ (or, equivalently, M_c) of the kinetic energy matching condition (4.78) can be found by solving a linear PDE. Indeed, premultiply (4.78) by M to obtain:

$$\begin{aligned} 0 &= M\bar{G}^{\perp}\lambda^T\{\partial_q(\tfrac{1}{2}\dot{q}^T M_c\dot{q}) - \partial_q(M_c\dot{q})\dot{q}\} + M\bar{G}^{\perp}\{\partial_q(M\dot{q})\dot{q} - \partial_q(\tfrac{1}{2}\dot{q}^T M\dot{q})\} \\ &\quad + 2\sum_{k=1}^{n}(M\dot{q})_k\, M\bar{G}^{\perp}\lambda^T W_k\dot{q}, \quad \forall (q,\dot{q}) \in TQ. \end{aligned} \tag{4.86}$$

Given a solution $\lambda\bar{G}^{\perp}M$ of (4.85), this is a linear PDE in M_c.

In conclusion, this suggests the following approach for solving the nonlinear matching PDE (4.64): First solve the λ-equation (4.85) for $\lambda\bar{G}^{\perp}M$, thereby choosing suitable matrices W_k. Afterwards solve (4.86) for M_c. Then the solution of (4.64) is given by $M_d = MM_c^{-1}M = M\lambda$ and J_2 as in (4.77), where $U(q,p)$ is defined in (4.81).

5. Integrability

In the previous section we showed that if we choose J_2 to be equal to (4.75), or equivalently (4.76), then there exist canonical coordinates (q, p_c) such that in these coordinates the structure matrix J_d (4.61) becomes the canonical matrix J_c. By Darboux's Theorem the existence of canonical coordinates is equivalent to the Poisson bracket satisfying the Jacobi identity. In this case we call the Poisson bracket, or equivalently J_d, *integrable.*

5.1. *Integrability of the structure matrix*

In this section we give necessary and sufficient conditions for the structure matrix J_d to be integrable. Recall the structure matrix J_d (4.61):

$$J_d(q,p) = \begin{bmatrix} 0 & M^{-1}(q)M_d(q) \\ -M_d(q)M^{-1}(q) & J_2(q,p) \end{bmatrix}. \tag{4.87}$$

Assume the matrix J_d is integrable and let the canonical coordinates be denoted by $(q_c, p_c) = (q_c(q,p), p_c(q,p))$. Without loss of generality we can assume that $q_c = q$. (See [6] for a precise statement and a proof of this.) Thus, let $(q_c, p_c) = (q, p_c(q,p))$ be canonical coordinates for J_d. This means that the relations (4.73) must be satisfied. Calculate

$$\{q, p_c\}_d = [I \quad 0] \begin{bmatrix} 0 & M^{-1}M_d \\ -M_dM^{-1} & J_2 \end{bmatrix} \begin{bmatrix} [\partial_q p_c]^T \\ [\partial_p p_c]^T \end{bmatrix} = M^{-1}M_d[\partial_p p_c]^T, \tag{4.88}$$

which is equal to I_n if and only if

$$p_c(q,p) = M(q)M_d^{-1}(q)p + Q(q), \tag{4.89}$$

with $Q(q)$ any smooth vector-valued function of the coordinates q. Secondly, use (4.89) to calculate

$$\begin{aligned} \{p_c, p_c\}_d = & -[\partial_q(MM_d^{-1}p)]^T - [\partial_q Q]^T + \partial_q(MM_d^{-1}p) + \partial_q Q \\ & + MM_d^{-1}J_2M_d^{-1}M. \end{aligned} \tag{4.90}$$

This is equal to zero if and only if

$$J_2 = M_d M^{-1}\left[[\partial_q(MM_d^{-1}p)]^T - \partial_q(MM_d^{-1}p)\right]M^{-1}M_d + \\ M_d M^{-1}\left[[\partial_q Q]^T - \partial_q Q\right]M^{-1}M_d. \tag{4.91}$$

We find it convenient to write $J_2(q,p) = J_2^\circ(q,p) + \hat{J}(q)$, with J_2° equal to (4.75) and

$$\hat{J}(q) = M_d M^{-1}\left[[\partial_q Q]^T - \partial_q Q\right]M^{-1}M_d. \tag{4.92}$$

So, if J_d is integrable then J_2 necessarily has the form (4.91). Conversely, if J_2 has the form (4.91), then clearly $q_c = q$ and p_c (4.89) are canonical coordinates for J_2. Notice that $Q(q) = 0$ yields $\hat{J} = 0$ and consequently $J_2 = J_2^\circ$, for which the canonical coordinates are $(q, p_c) = (q, MM_d^{-1}p)$ as we have seen in the previous section.

Proposition 8: *The structure matrix J_d defined in (4.87) is integrable if and only if J_2 has the form (4.91), for some smooth vector-valued function $Q(q)$.*

5.2. *Gyroscopic terms*

Consider the Hamiltonian H_d expressed in the canonical coordinates (q, p_c). For $(q, p_c) = (q, MM_d^{-1}p)$, corresponding to J_2°, the Hamiltonian H_d (4.60) becomes the canonical Hamiltonian H_c (4.68) with M_c and V_c defined by (4.71). Similar to H_d the canonical Hamiltonian H_c has the form of the sum of kinetic and potential energy. However, this is not the case anymore for $\hat{J} \neq 0$. Indeed, take $\hat{J}$ as in (4.92), then in the canonical coordinates the Hamiltonian H_d becomes the canonical Hamiltonian H_c defined by (substituting $p = M_d M^{-1}(p_c - Q)$ into (4.60)):

$$H_c(q_c, p_c) = \frac{1}{2}p_c^T M^{-1}M_d M^{-1}p_c - p_c^T M^{-1}M_d M^{-1}Q \\ + \frac{1}{2}Q^T M^{-1}M_d M^{-1}Q + V_d. \tag{4.93}$$

The canonical Hamiltonian includes the *gyroscopic terms*

$$-p_c^T M^{-1}M_d M^{-1}Q, \tag{4.94}$$

which are terms linear in the p-variables (the momenta). In addition the potential energy is augmented to be

$$V_c = \frac{1}{2}Q^T M^{-1}M_d M^{-1}Q + V_d. \tag{4.95}$$

Thus in case $\hat{J}$ is defined as in (4.92), then the system (4.60, 4.61, 4.62) becomes in the canonical coordinates $q_c = q$ and p_c (4.89) the canonical Hamiltonian system (4.69, 4.93). If $Q(q)$ is chosen to be nonzero then gyroscopic terms are introduced into the system and in addition the potential energy is augmented.

Remark 12: The canonical Hamiltonian system (4.69, 4.93) corresponds via the inverse Legendre transformation to the Euler–Lagrange system (4.2) with Lagrangian defined by

$$L_c(q, \dot{q}) = \frac{1}{2}\dot{q}^T M(q) M_d^{-1}(q) M(q)\dot{q} + \dot{q}^T Q(q) - V_d(q). \tag{4.96}$$

An interesting question is if the gyroscopic terms introduced by $\hat{J}$ are intrinsic or not, defined in the following way:

Definition 1: The gyroscopic terms are called *intrinsic* if there does not exist a canonical transformation $(q_c, p_c) \mapsto (\bar{q}_c, \bar{p}_c)$ such that in the new coordinates $(\bar{q}_c, \bar{p}_c)$ the Hamiltonian (4.93) becomes the quadratic Hamiltonian

$$\bar{H}_c(\bar{q}_c, \bar{p}_c) = \frac{1}{2}\bar{p}_c^T \Lambda^{-1}(\bar{q}_c)\bar{p}_c + \bar{V}(\bar{q}_c), \tag{4.97}$$

for some Λ and $\bar{V}$.

That is, the gyroscopic terms are intrinsic if they cannot be removed by a canonical coordinate transformation (and therefore the Hamiltonian cannot be transformed into the form of kinetic plus potential energy). The following proposition gives an answer to the above question [6].

Proposition 9: *The gyroscopic terms are intrinsic to the closed-loop system if and only if* $[\partial_q Q]^T \neq \partial_q Q$ *(which is equivalent to* $\hat{J} \neq 0$*).*

5.3. *Integrability and matching*

Consider the matching condition (4.64) for the kinetic energy and plug in J_2 as defined in (4.91) to get

$$\begin{aligned} G^{\perp}(q)\Big[\partial_q(\frac{1}{2}p^T M^{-1}(q)p) - M_d(q)M^{-1}(q)\partial_q(\frac{1}{2}p^T M_d^{-1}(q)p) \\ + J_2^{\circ}(q,p)M_d^{-1}(q)p\Big] + G^{\perp}(q)\left[\hat{J}(q)M_d^{-1}(q)p\right] = 0. \end{aligned} \tag{4.98}$$

This equation has to hold for all $(q,p) \in TQ$. Since the first part of (4.98) is quadratic in p (recall that J_2° is linear in p) and the second part is linear in p, it follows that (4.98) holds for all (q,p) if and only if the following two conditions hold:

$$G^\perp(q)\Big[\partial_q(\frac{1}{2}p^T M^{-1}(q)p) - M_d(q)M^{-1}(q)\partial_q(\frac{1}{2}p^T M_d^{-1}(q)p) \\ + J_2^\circ(q,p)M_d^{-1}(q)p\Big] = 0, \tag{4.99}$$

and

$$G^\perp(q)\hat{J}(q)M_d^{-1}(q) = G^\perp M_d M^{-1}\left[[\partial_q Q]^T - \partial_q Q\right]M^{-1} = 0, \tag{4.100}$$

for all $(q,p) \in TQ$. Equation (4.99) is nothing but the matching condition (4.64) with $J_2 = J_2^\circ$. Since it is equivalent to the matching condition (4.17), see Section 4, it can be solved by the λ-method. Equation (4.100) defines a matching condition for $\hat{J}$. Given a solution M_d of (4.99), it is a linear PDE in Q. It can also be written in terms of a solution $\lambda\bar{G}^\perp M$ of the λ-equation (4.23) by premultiplying (4.100) with M to obtain (notice that $\lambda = M_c^{-1}M = M^{-1}M_d$):

$$M\bar{G}^\perp\lambda^T\left[[\partial_q Q]^T - \partial_q Q\right]M^{-1} = 0. \tag{4.101}$$

This result leads to the following parameterization of matching integrable Hamiltonian systems:

Proposition 10: *Assume that the Hamiltonian system (4.60, 4.61, 4.62) with $J_2 = J_2^\circ$ (4.75) satisfies the matching conditions (4.64, 4.65), i.e. matches with the port-controlled Hamiltonian system (4.58, 4.59). Then every Hamiltonian system (4.60, 4.61, 4.62, 4.91), with $\hat{J}$ satisfying condition (4.100), is integrable and matches with the port-controlled Hamiltonian system (4.58, 4.59). Furthermore, this class of systems (parameterized by $\hat{J}$) describes exactly all the possible integrable Hamiltonian systems with Hamiltonian (4.60) that match with (4.58, 4.59).*

We remark that the Hamiltonian matching described in Proposition 10 can also be interpreted as Lagrangian matching with the closed-loop Lagrangian given by (4.96).

6. Conclusions

In this chapter we reviewed two recently developed methods for the stabilization of underactuated mechanical systems. The first is the controlled

Lagrangians method, defined for Euler–Lagrange systems. The second is the interconnection and damping assignment passivity based control (IDA-PBC) method, which considers port-controlled Hamiltonian systems. The fundamental idea underlying both methods is that of matching, that is, finding a suitable closed-loop Euler–Lagrange, respectively port-controlled Hamiltonian, system which stabilizes the desired equilibrium point (the conditions under which the corresponding control law exists are called matching conditions).

The controlled Lagrangians method as originally introduced in [11] for mechanical systems with symmetry is reviewed and the matching conditions obtained in that paper are interpreted in terms of kinetic and potential energy matching. Since the class of Euler–Lagrange systems is contained in the class of port-controlled Hamiltonian systems, the IDA-PBC method includes the controlled Lagrangians method as a special case. In fact, the possibility of shaping not only the energy function but also the interconnection structure of the system gives an extra degree of freedom to the IDA-PBC method. It is shown that for a particular choice of this interconnection structure the IDA-PBC method results in the controlled Lagrangians method. Furthermore the integrability of the closed-loop Hamiltonian system is investigated. Explicit (necessary and sufficient) conditions on the interconnection structure are given under which the closed-loop Hamiltonian system is integrable (*i.e.* corresponds to an Euler–Lagrange system). In general, this includes the introduction of intrinsic gyroscopic terms in the closed-loop system.

The matching conditions generally consist of a set of nonlinear PDEs, to be solved either for the closed-loop Lagrangian function (in the controlled Lagrangians method) or for the closed-loop Hamiltonian function and the interconnection structure (in case of the IDA-PBC method). The λ-method described in [2] for the controlled Lagrangians method converts these nonlinear PDEs into a set of linear PDEs, to be solved recursively. It is shown that the λ-method can also be applied to the PDEs obtained in the IDA-PBC method, leading to set of quasi-linear and linear PDEs to be solved recursively.

Acknowledgments

G. Blankenstein would like to thank Dr. Johan Hamberg of the Swedish Defense Research Establishment for helpful discussions and remarks.

Bibliography

1. F. Andreev, D. Auckly, L. Kapitanski, S. Gosavi, A. Kelkar, and W. White. Matching, linear systems, and the ball and beam. *Automatica*, 38(12):2147–2152, 2002.
2. D. Auckly, L. Kapitanski, and W. White. Control of nonlinear underactuated systems. *Comm. Pure and Applied Math.*, 53:354–369, 2000.
3. D. Auckly and L. Kapitanski. Mathematical problems in the control of underactuated systems. *CRM Proceedings and Lecture Notes*, 27:41–52, 2000.
4. D. Auckly and L. Kapitanski. On the lambda-equations for matching control laws. *SIAM J. Control and Optimization*, 41(5):1372–1388, 2002.
5. G. Blankenstein. Matching and stabilization of constrained systems. *Int. Symp. Math. Theory of Networks and Systems, Notre Dame, Indiana*, 2002.
6. G. Blankenstein, R. Ortega, and A.J. van der Schaft. The matching conditions of controlled Lagrangians and IDA-passivity based control. *Int. J. Control*, 75(9):645–665, 2002.
7. G. Blankenstein, R. Ortega, and A.J. van der Schaft. Matching of Euler-Lagrange and Hamiltonian systems. *Proc. 15th IFAC World Congress, Barcelona, Spain*, 2002.
8. A. Bloch, N. Leonard, and J.E. Marsden. Stabilization of mechanical systems using controlled Lagrangians. *Proc. 36th IEEE Conf. Decision and Control, San Diego, California*, pages 2356–2361, 1997.
9. A.M. Bloch, N.E. Leonard, and J.E. Marsden. Matching and stabilization by the method of controlled Lagrangians. *Proc. 37th IEEE Conf. Decision and Control, Tampa, Florida*, pages 1446–1451, 1998.
10. A.M. Bloch, N.E. Leonard, and J.E. Marsden. Potential shaping and the method of controlled Lagrangians. *Proc. 38th IEEE Conf. Decision and Control, Phoenix, Arizona*, pages 1653–1658, 1999.
11. A.M. Bloch, N.E. Leonard, and J.E. Marsden. Controlled Lagrangians and the stabilization of mechanical systems I: the first matching theorem. *IEEE Trans. Automatic Control*, 45(12):2253–2269, 2000.
12. A.M. Bloch, D.E. Chang, N.E. Leonard, and J.E. Marsden. Controlled Lagrangians and the stabilization of mechanical systems II: potential shaping. *IEEE Trans. Automatic Control*, 46(10):1556–1571, 2001.
13. D.E. Chang, A.M. Bloch, N.E. Leonard, J.E. Marsden, and C.A. Woolsey. The equivalence of controlled Lagrangian and controlled Hamiltonian systems. *ESAIM: Control, Optimisation and Calculus of Variations*, 8:393–422, 2002.
14. F.R. Gantmacher. *Théorie des Matrices.* Vol. 1, Dunod, Paris, 1966.
15. F. Gómez-Estern, R. Ortega, F.R. Rubio, and J. Aracil. Stabilization of a class of underactuated mechanical systems via total energy shaping. *Proc. 40th IEEE Conf. Decision and Control, Orlando, Florida*, 2001.
16. J. Hamberg. General matching conditions in the theory of controlled Lagrangians. *Proc. 38th IEEE Conf. Decision and Control, Phoenix, Arizona*, pages 2519–2523, 1999.
17. J. Hamberg. Controlled Lagrangians, symmetries and conditions for strong

matching. In *Preprints IFAC Workshop on Lagrangian and Hamiltonian Methods for Nonlinear Control, N.E. Leonard and R. Ortega (Eds.)*, Princeton, 2000.

18. J. Hamberg. Simplified conditions for matching and for generalized matching in the theory of controlled Lagrangians. *Proc. American Control Conference, Chicago, Illinois*, 2000.
19. J.E. Marsden and T. Ratiu. *Introduction to Mechanics and Symmetry.* Springer-Verlag, 2nd ed., 1999.
20. R. Ortega, A.J. van der Schaft, I. Mareels, and B. Maschke. Putting energy back in control. *IEEE Control Systems Magazine*, 21(2):18–33, 2001.
21. R. Ortega, A. van der Schaft, B. Maschke, and G. Escobar. Interconnection and damping assignment passivity-based control of port-controlled Hamiltonian systems. *Automatica*, 38:585–596, 2002.
22. R. Ortega, M.W. Spong, F. Gómez-Estern, and G. Blankenstein. Stabilization of underactuated mechanical systems via interconnection and damping assignment. *IEEE Trans. Automatic Control*, 47(8):1218–1233, 2002.
23. R. Ortega and E. Garcia-Canseco. Interconnection and damping assignment passivity-based control: a survey. *European Journal of Control*, 10:432–450, 2004.
24. S. Prajna, A. van der Schaft, and G. Meinsma. An LMI approach to stabilization of linear port-controlled Hamiltonian systems. *Systems & Control Letters*, 45:371–385, 2002.
25. A.J. van der Schaft. *L_2-Gain and Passivity Techniques in Nonlinear Control.* Springer-Verlag, 2nd ed., 2000.
26. C.A. Woolsey, A.M. Bloch, N.E. Leonard, and J.E. Marsden. Physical dissipation and the method of controlled Lagrangians. *Proc. European Control Conference, Porto, Portugal*, pages 2570–2575, 2001.
27. C. Woolsey, C.K. Reddy, A.M. Bloch, D.E. Chang, N.E. Leonard, and J.E. Marsden. Controlled Lagrangian systems with gyroscopic forcing and dissipation. *European Journal of Control*, 10, 2004.
28. D.V. Zenkov, A.M. Bloch, N.E. Leonard, and J.E. Marsden. Matching and stabilization of low-dimensional nonholonomic systems. *Proc. 39th IEEE Conf. Decision and Control, Sydney, Australia*, pages 1289–1295, 2000.

CHAPTER 5

Virtual Constraints for the Orbital Stabilization of the Pendubot

F. Grognard[1] and C. Canudas de Wit[1]

[1] *INRIA Sophia-Antipolis Projet COMORE, 2004 route des Lucioles, BP 93, 06902 Sophia-Antipolis Cedex, France, E-mail: frederic.grognard@sophia.inria.fr*

[2] *Laboratoire d'Automatique de Grenoble, ENSIEG, BP 46, 38402 Saint-Martin d'Hères Cedex, France, E-mail: canudas@lag.ensieg.inpg.fr*

A method for the generation of attractive and neutrally stable limit cycles for nonlinear systems is presented. It consists in designing an output that, when regulated through a suitable feedback, forces the existence of a limit cycle or neutral oscillations in the zero dynamics. Conditions are then given to ensure that those characteristics of the zero dynamics translate to the whole system. A particular focus is placed on the generation of neutrally stable oscillations through that method, because it is not always easy to build an output that results in the existence of a limit-cycle in the zero dynamics. A special case where such a difficulty arises is given in the analysis of oscillations generation around the upper vertical for the Pendubot. The regulation of the output results in neutrally stable oscillations, and we present a method for ensuring that those oscillations converge towards the desired ones.

1. Introduction

In many applications the natural operating mode of a control system is an oscillating one. However, the oscillations are not always present in the open-loop dynamics. Therefore, it is relevant to study new control design methods forcing the internal system dynamics (or a given output) to present a pre-specified limit cycle. Examples of such systems are: walking mechanisms (the full system state should behave periodically), rotating machines (the internal states, *i.e.* current and flux, are oscillatory if the torque output is kept constant), the synchronization of a vertically landing aircraft with the oscillation of a platform (*e.g.* an aircraft carrier), *etc.*

There exist some published works addressing problems in this category.

Under the hypothesis of the existence of a limit cycle, Hauser and Chung [8] present a setup for the computation of Lyapunov functions, allowing to determine if the given limit cycle is exponentially stable. Nevertheless no procedure is presented for the generation of the limit cycle itself. This latter problem has recently been addressed by the works of Marconi *et al.* [12], Aracil *et al.* [1], Sepulchre and Stan [18], Westervelt *et al.* [19] and Canudas de Wit *et al.* [4]. In Marconi *et al.* [12], the authors deal with the problem of tracking an oscillatory signal. The addressed problem is the motion synchronization of a vertically landing aircraft with the oscillation of a platform. In Aracil *et al.* [1], oscillations of the Furuta pendulum are stabilized through an energy-shaping approach, and passivity is used as a mean to generate oscillations in Sepulchre and Stan [18]. Westervelt *et al.* [19] and Canudas de Wit *et al.* [4] present approaches that have been motivated by the walking mechanism, for which the natural operating mode is a periodic one, with Westervelt studying the zero dynamics of a controlled biped (which are hybrid due to the impacts) and Canudas de Wit having designed a feedback law that generates globally stable orbits for an underactuated inverted pendulum.

In certain cases, the oscillatory internal behavior is a by-product of an output regulation problem. An example is the torque and flux norm regulation problem for the induction motor. The *linearization* of these two outputs (as originally proposed by De Luca and Ulivi [5]), leads to an oscillating behavior in the internal dynamics. Indeed, under this particular frame, the flux and current vector asymptotically converge to a linear stable oscillation [2]. In Canudas de Wit *et al.* [4], an output is designed and regulated such that the resulting zero dynamics of the system present a limit cycle. Furthermore, the family of outputs that is proposed allows for enough freedom such that the solutions of the zero dynamics can converge towards any prespecified closed curve. The formalization of this approach, with constructive methods for the design of outputs whose regulation generates oscillating behaviors were then given, along with the corresponding stability analyses [6,16]. The output, once regulated, is called a virtual constraint because it imposes a fixed relation between the states of the model, so that it reduces the number of degrees of freedom of the model.

In this chapter, we will complement the existing results [6, 16] and introduce their application on the Pendubot, a two-link planar robot whose only actuated joint is the "shoulder". We wish to generate oscillations for the Pendubot around the upward position: the oscillations of the first arm should have a given angular amplitude ($= 2\alpha_s$) and a prespecified period.

This chapter is structured as follows. In Section 1, we present results about the stability of periodic orbits in cascade systems. We then analyze two-dimensional oscillating behaviors in Section 2. A model of the Pendubot is presented in Section 3, followed with the presentation of conditions on the regulated output for the appearance of periodic orbits around the upper vertical. The control problem is then treated, for the Pendubot, in Section 4, which is concluded by some simulations. Finally, we give a conclusion.

2. Oscillations in Cascade Systems

2.1. *Attractive limit sets in cascade systems*

In this section we will show that a globally attractive limit-cycle in the zero dynamics of a system could result in a globally attractive limit cycle for the system itself. In order for that result to be valid, some conditions are to be satisfied for the interconnection between the zero dynamics and the rest of the system, and on the stability of the limit cycle in the zero dynamics. The affine system that we consider is analyzed in the normal form

$$\begin{cases} \dot{z} = f(z) + \psi(z,\xi) \\ \dot{\xi} = A\xi + Bu \end{cases} \tag{5.1}$$

where $\xi \in \mathbb{R}^r$, $z \in \mathbb{R}^{n-r}$, $u \in \mathbb{R}^m$ $(m \leq n)$, and the functions f and ψ are locally Lipschitz continuous on their domains of definition. The function ψ is such that $\psi(z,0) = 0$ for all z (without loss of generality; indeed, if it were not the case, f is redefined as $f(z)+\psi(z,0)$ and ψ as $\psi(z,\xi)-\psi(z,0)$). The zero dynamics, associated to the output $y = \xi$, are represented by $\dot{z} = f(z)$.

Many stabilization designs are known for this particular normal form (*e.g.* global asymptotic stabilization of the origin of the ξ dynamics through a feedback of ξ yields boundedness of the solutions of (5.1) and global asymptotic stability of the origin of (5.1) [17]). In this section, we impose the same kind of conditions as in Sepulchre *et al.* [17]: the interconnecting term $\psi(z,\xi)$ is linearly bounded in z when $\|z\| > M$ for some $M > 0$, so that finite escape time cannot occur. Also, there exists a positive semidefinite radially unbounded function $W(z)$, that decreases along the solutions of the system when $\|z\|$ is large, namely the following hold.

Assumption 1: There exist $M \geq 0$, and class $\mathcal{K}$ functions η_1 and η_2 such that

$$\|\psi(z,\xi)\| \leq \eta_1(\|\xi\|)\|z\| + \eta_2(\|\xi\|)$$

for $\|z\| \geq M$.

Assumption 2: There exists a positive semidefinite radially unbounded function $W(z)$ and positive constants C and M such that, for all $\|z\| > M$, the following holds:

(i) $L_f W(z) \leq 0$;
(ii) $\|\frac{\partial W}{\partial z}\| \; \|z\| \leq CW(z)$.

Point (ii) of Assumption 2 is classically satisfied by Lyapunov functions $W(z)$ produced by converse theorems for exponentially stable systems [11].

Nonlinear systems can present many different kinds of ω-limit sets other than an equilibrium point [7]. In this section we are interested in the case where one ω-limit set is a limit cycle. An important property of the limit cycles is that they are compact sets; therefore, we will analyze the situation where one ω-limit set is a compact set γ.

Definition 1: A compact set γ is "almost globally attractive" for the dynamics

$$\dot{x} = f(x) \tag{5.2}$$

with $x \in \mathbb{R}^n$, if it is attractive with basin of attraction containing the whole state space minus a set of Lebesgue measure zero.

In dimension 2, the simple situation where a limit cycle γ attracts every solution except those starting at the equilibrium inside the area circumscribed by γ fits into Definition 1: γ is almost globally attractive because the equilibrium is of Lebesgue measure zero.

We now consider the case where the union of the ω-limit sets of (5.2) is made of an almost globally attractive compact set γ and an equilibrium $\bar{x}$ such that γ and $\bar{x}$ are disjoint.

Remark 1: Note that $\bar{x}$ is unstable. Indeed, if $\bar{x}$ is stable without being asymptotically stable, there are solutions whose ω-limit set is neither $\bar{x}$ nor γ. On the other hand if $\bar{x}$ is asymptotically stable, the regions of attraction of $\bar{x}$ and γ are open, nonempty, connected sets [11]. Therefore, $\mathbb{R}^n$ needs to be covered by two disjoint, nonempty, open sets, which is a contradiction. Therefore $\bar{x}$ is unstable.

If we consider the existence of an almost globally attractive compact set γ in the z-dynamics, the next theorem gives conditions for the almost globally attractiveness of γ to translate into almost global attractiveness of $\Gamma \equiv \{(z, \xi) \in \mathbb{R}^n \mid z \in \gamma \text{ and } \xi = 0\}$ in the interconnected system.

Theorem 1: Suppose that Assumptions 1 and 2 are satisfied for system (5.1) for which $f : \mathbb{R}^{n-r} \to \mathbb{R}^{n-r}$ and $\psi : \mathbb{R}^{n-r} \times \mathbb{R}^r \to \mathbb{R}^{n-r}$ are locally Lipschitz continuous functions. Let the only invariant sets of $\dot{z} = f(z)$ be $z = 0$ and γ, respectively an equilibrium and an almost globally attractive compact set ($0 \notin \gamma$). If (A, B) is controllable, then any feedback of the form $u = k(\xi)$ guaranteeing that the origin of $\dot{\xi} = A\xi + Bk(\xi)$ is Globally Asymptotically Stable and Locally Exponentially Stable (GAS-LES) yields the following properties.

(i) Convergence of the solutions of

$$\begin{cases} \dot{z} = f(z) + \psi(z, \xi) \\ \dot{\xi} = A\xi + Bk(\xi) \end{cases} \tag{5.3}$$

to the unstable equilibrium $(z, \xi) = (0, 0)$ or to the compact set Γ.

(ii) If the set of initial conditions of solutions converging to $(0, 0)$ is of Lebesgue measure zero, the compact set Γ is almost globally attractive.

(iii) If the equilibrium of $\dot{z} = f(z)$ is hyperbolic, and if the global stable manifold of the origin $(0, 0)$ is a manifold whose dimension is globally defined and constant, the compact set Γ is almost globally attractive.

(iv) If γ is an exponentially stable periodic orbit for $\dot{z} = f(z)$, then Γ is an exponentially stable periodic orbit for (5.3).

Proof: Boundedness of the state along the solutions is a direct consequence of the boundedness of W along the solutions and the radial unboundedness of $W(z)$ [17].

Every solution of (5.3) converges to the set $E = \{(z, \xi) \in \mathbb{R}^n \,|\, \xi = 0\}$. LaSalle's invariance principle then asserts that every bounded solution converges to the largest invariant set in E. This set is defined by the largest invariant set of $\dot{z} = f(z)$: the origin $z = 0$ and the compact set γ. Because every solution of (5.3) is bounded, every solution either converges towards the origin $(0, 0)$ or towards the compact set Γ. Moreover, Remark 1 implies that $z = 0$ is an unstable equilibrium in the z-dynamics. The fact that $(z, \xi) = (0, 0)$ is unstable directly follows, which shows (i).

The basin of attraction of the origin $(0, 0)$ is the set of initial conditions of solutions not converging to Γ. Because it is of Lebesgue measure zero, the compact set Γ is almost globally attractive, which shows (ii).

If the origin $z = 0$ is a hyperbolic fixed point for $\dot{z} = f(z)$, the set of eigenvalues of the Jacobian linearization $\frac{\partial f}{\partial z}(0)$ contains $n_u \geq 1$ ($n_s \leq n -$

$r-1$) eigenvalues with positive (negative respectively) real parts ($n_s+n_u = n-r$).

Since the origin $\xi = 0$ is locally exponentially stable for $\dot{\xi} = A\xi + Bk(\xi)$, the Jacobian linearization $A + B\frac{\partial k}{\partial \xi}(0)$ has r eigenvalues with negative real parts. The Jacobian linearization of the complete system (5.3) is then

$$\mathcal{J} = \begin{pmatrix} \frac{\partial f}{\partial z}(0) + \frac{\partial \psi}{\partial z}(0,0) & \frac{\partial \psi}{\partial \xi}(0,0) \\ 0 & A + B\frac{\partial k}{\partial \xi}(0) \end{pmatrix}.$$

Because $\psi(z,0) = 0$ for all z, $\frac{\partial \psi}{\partial z}(0,0) = 0$. Therefore, the eigenvalues of $\mathcal{J}$ are those of $\frac{\partial f}{\partial z}(0)$ and of $A + B\frac{\partial k}{\partial \xi}(0)$. $\mathcal{J}$ has then $n_s + r \leq n-1$ eigenvalues with negative real parts and $n_u \geq 1$ eigenvalues with positive real parts. The stable manifold theorem for a fixed point [7] then states that there exists a local stable manifold $\mathcal{M}_s$ of dimension $n_s + r$ and a local unstable manifold $\mathcal{M}_u$ of dimension n_u at the origin. If the dimension of the global stable manifold is globally defined and constant, this global stable manifold also has a dimension $n_s + r$. Therefore the set of initial conditions of solutions that converge to the origin $(0,0)$ (and not to Γ) lies in a manifold of dimension smaller or equal to $n-1$. Because a manifold of dimension smaller or equal to $n-1$ is of Lebesgue measure zero in $\mathbb{R}^n$, the compact set Γ is almost globally attractive, which proves (iii).

Exponential stability of γ implies the existence of a Lyapunov function $V_1(z)$ that is decreasing along the solutions of $\dot{z} = f(z)$. This function satisfies

$$L_f V_1 \leq -k_1 \|z\|_\gamma^2$$

where $\|z\|_\gamma = \inf_{y\in\gamma} \|z-y\|$ [8]. On the other hand, a similar Lyapunov function $V_2(\xi)$ can be found for the ξ subsystem ($m_1\|\xi\|^2 \leq V_2(\xi) \leq m_2\|\xi\|^2$, $L_{A\xi+Bk(\xi)}V_2 \leq -m_3\|\xi\|^2$). As in Khalil [11], we then define the Lyapunov function:

$$V(z,\xi) = V_1(z) + 2k\sqrt{V_2(\xi)}$$

with $k > 0$, whose derivative is

$$\begin{aligned} \dot{V} &= L_f V_1(z) + L_\psi V_1(z) + k\frac{L_{A\xi+Bk(\xi)}V_2}{\sqrt{V_2}} \\ &\leq -k_1\|z\|_\gamma^2 + \frac{dV_1}{dz}\psi(z,\xi) - km_3\frac{\|\xi\|^2}{\sqrt{V_2}} \\ &\leq -k_1\|z\|_\gamma^2 - \frac{km_3}{m_1}\|\xi\| + \frac{dV_1}{dz}\psi(z,\xi) \\ &\leq -k_1\|z\|_\gamma^2 - \frac{km_3}{m_1}\|\xi\| + M\|\xi\| \end{aligned}$$

where the last inequality is valid in a small neighborhood of Γ because $\frac{dV_1}{dz}\psi(z,\xi)$ is continuously differentiable. Taking $k > \frac{m_1 M}{m_3}$ ensures negative definiteness of the derivative of V. The closed orbit Γ is therefore exponentially stable. This proves (iv). □

The origin of the complete closed-loop system behaves like a saddle point: it is attractive in some directions and repulsive in others. This is illustrated in Figure 5.1, where γ is a limit cycle, $n = 3$, and $r = 1$. In this figure, the z system is of dimension 2 with an almost globally attractive limit cycle γ. In order to clarify the figure, we suppose that the stable manifold $\mathcal{M}_s$ of the origin is the ξ axis. Therefore, every solution starting on that axis converge to the origin $(0,0,0)^T$. All other solutions converge to the limit cycle because the origin is only attractive in the ξ direction.

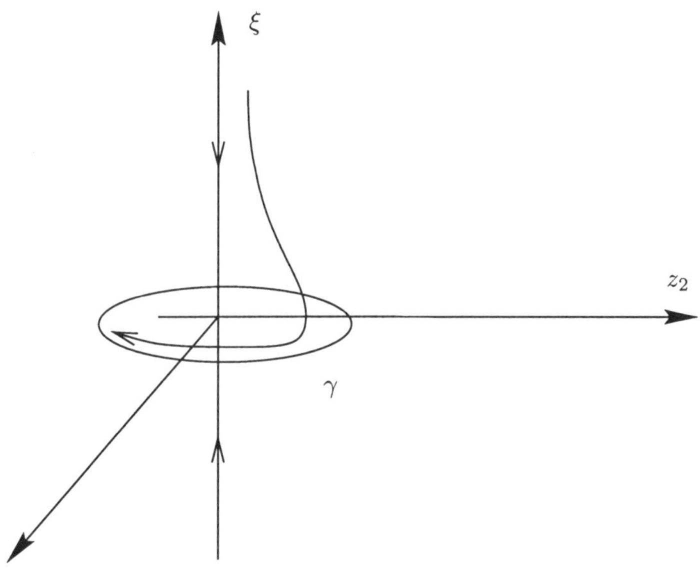

Figure 5.1. Almost global attractivity of a limit cycle. Solutions starting on the ξ axis converge to the origin. All others converge to γ.

The method that is then used for the generation of oscillations in an affine system in the form

$$\dot{x} = F(x) + G(x)u$$

consists in finding an output y that will be such that this system put into the normal form with $\xi_1 = y$ has the form (5.1) with an almost globally attractive limit-cycle in its zero dynamics. Such a method has been presented in Grognard and Canudas de Wit [6], but its application is difficult.

2.2. *Neutrally stable oscillations in cascade systems*

As we will see in the case of the Pendubot, the design of an output whose regulation generates an attractive limit cycle (as in the previous section) is not always an easy task. It is sometimes easier to find an output whose zero dynamics are neutrally stable, with an equilibrium surrounded by a continuum of periodic orbits. We will generalize this behavior to invariance of the level sets of a radially unbounded positive semidefinite Lyapunov function and to the following cascade system:

$$\begin{cases} \dot{z} = f(z) + \psi_z(z, \xi, u) \\ \dot{\xi} = \psi_\xi(\xi, u). \end{cases} \tag{5.4}$$

We therefore need two new assumptions that are closely related to the previous ones.

Assumption 3: There exist $M \geq 0$, and class $\mathcal{K}$ functions η_1 and η_2 such that

$$\|\psi_z(z, \xi, u)\| \leq \eta_1(\|(\xi, u)\|)\|z\| + \eta_2(\|(\xi, u)\|)$$

for $\|z\| \geq M$.

Assumption 4: There exists a positive semidefinite radially unbounded function $W(z)$ such that, for all $\|z\|$:

$$L_f W(z) = 0.$$

Then, we obtain a weaker result than Theorem 1.

Theorem 2: Suppose that Assumptions 3 and 4 are satisfied for system (5.4) for which $f : \mathbb{R}^{n-r} \to \mathbb{R}^{n-r}$, $\psi_z : \mathbb{R}^{n-r} \times \mathbb{R}^r \times \mathbb{R}^m \to \mathbb{R}^{n-r}$ and $\psi_\xi : \mathbb{R}^r \times \mathbb{R}^m \to \mathbb{R}^r$ are locally Lipschitz continuous functions. If the origin of the ξ subsystem is finite-time stabilizable, then any feedback in the form $u = k(\xi)$ guaranteeing that the origin of $\dot{\xi} = \psi\xi(\xi, k(\xi))$ is finite-time stabilized ensures that each solution of

$$\begin{cases} \dot{z} = f(z) + \psi_z(z, \xi, k(\xi)) \\ \dot{\xi} = \psi_\xi(\xi, k(\xi)) \end{cases}$$

reaches, in finite time, a region where $W(z)$ is constant and $\xi = 0$.

Proof: For any initial condition $(z(0), \xi(0))$, the controller $k(\xi)$ ensures that there exists a time $T(z(0), \xi(0))$ and some $Z(z(0), \xi(0)) \in \mathbb{R}^{n-r}$ such that $(z(T(z(0), \xi(0))), \xi(T(z(0), \xi(0)))) = (Z, 0)$ (Assumption 3 ensures that no escape of z in finite time can take place during that time-span so that Z is finite). For all $t \geq T$, we then have $\xi(t) = 0$, so that the remaining dynamics are

$$\dot{z} = f(z)$$

and the solution is such that $W(z)$ remains constant (equal to $W(Z)$). □

It is interesting to see that, if (5.4) can be rewritten in the following form

$$\begin{cases} \dot{z} = f(z) + \psi_z(z,\xi)\xi + \psi_{uz}(z,\xi,u)u \\ \dot{\xi} = \psi_\xi(\xi) + \psi_{u\xi}(z,\xi,u)u \end{cases}$$

with, among other conditions, $\dot{z} = f(z)$ neutrally stable, $\dot{\xi} = \psi_\xi(\xi)$ globally exponentially stable and $\psi_z(z,0) = 0$, the classical forwarding technique [10,13] can be applied to achieve stabilization of the solutions of the closed-loop system to a prespecified level-set. We do not develop this method because, in the Pendubot case, we do not have $\psi_z(z,0) = 0$.

Also note that, if (5.4) is in normal form [9], with u scalar and the ξ subsystem a chain of integrators, it is easy to build a finite-time controller (the time-optimal controller for the ξ subsystem, for example). However, the resulting controller is not very satisfying, because the level of $W(z)$ that is reached cannot be tuned. For the Pendubot, we will present heuristics that ensure convergence of the solutions to the desired level set (and desired oscillations) after having designed an output that ensures neutral stability of the oscillations in dimension 2. In the following section, we will be interested in analyzing those neutrally stable oscillations in planar systems.

3. Oscillations in the Plane

As stated earlier, the oscillations that we will generate will come from the zero dynamics, which is very interesting because the analysis of the cycles can then be made in the dimension of the zero dynamics (which is smaller than the dimension of the original system). By construction, this dimension will often be two. Therefore, we give two results for the analysis of cycles in two-dimensional systems.

Lemma 1: *Let $f : \mathbb{R} \times \mathbb{R} \to \mathbb{R}$ be a Lipschitz continuous function such that for all $(z_1, z_2) \in \mathbb{R}^2$ the function f is such that $f(z_1, z_2) = f(z_1, -z_2)$.*

Consider the system

$$\ddot{z} + f(z, \dot{z}) = 0. \tag{5.5}$$

If there exists $\bar{z} \in \mathbb{R}$ such that $f(\bar{z}, 0) = 0$ (($\bar{z}, 0$) is an equilibrium of (5.5)) and there exist $z_{min} < \bar{z} < z_{max}$ such that $f(z, 0) < 0$ in $[z_{min}, \bar{z})$ and $f(z, 0) > 0$ in $(\bar{z}, z_{max}]$, then there exists a neighborhood of $(z, \dot{z}) = (\bar{z}, 0)$ such that all solutions in that neighborhood are periodic orbits (the hypotheses are illustrated in Figure 5.2).

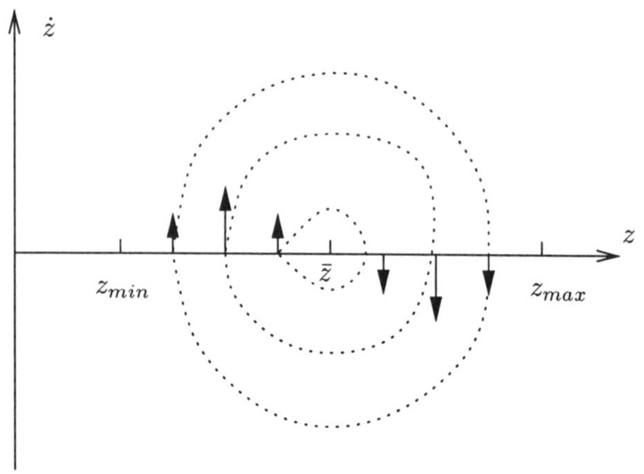

Figure 5.2. Generic form of behavior for systems satisfying the hypotheses of Lemma 1. The arrows indicate the direction of the field when $\dot{z} = 0$.

Proof: Let us first show that the phase plane is symmetric with respect to the z axis. We first define $(z_1, z_2) = (z, \dot{z})$. System (5.5) can then be rewritten as

$$\begin{cases} \dot{z}_1 = z_2 \\ \dot{z}_2 = -f(z_1, z_2). \end{cases} \tag{5.6}$$

The symmetry has to be shown with respect to the z_1 axis. Let us now reverse the time (replace t by $\tau = -t$) and replace z_2 by $-z_2$. The dynamics (5.6) then become

$$\begin{cases} \frac{dz_1}{d\tau} = z_2 \\ \frac{dz_2}{d\tau} = -f(z_1, -z_2) = -f(z_1, z_2). \end{cases}$$

The dynamics are unchanged. This means that, if $(z_1(t), z_2(t))$ is solution of (5.6), then $(z_1(t), -z_2(t))$ also is. The main difference is that, if $(z_1(t), z_2(t))$ is a solution with t increasing, then $(z_1(t), -z_2(t))$ is a solution with t decreasing: they *run* in opposite directions.

We will now constructively show the existence of a cycle for (5.6) around the equilibrium $(z_1, z_2) = (\bar{z}, 0)$.

Consider $(z_1(0), z_2(0)) = (z_{max}, 0)$. From (5.6), we see that $\dot{z}_2(0) < 0$; this means that z_2 starts decreasing as well as z_1. We will now show that $(z_1(t), z_2(t))$ reaches the axis $z_1 = \bar{z}$ in finite time.

First, we see that, as long as $z_1(t)$ is inside the interval $(\bar{z}, z_{max}]$, z_1 decreases. Indeed, if it were to increase at some time, this would mean that z_2 has become positive, so that it is gone through 0 with $z_1(t)$ inside the interval. However, $\dot{z}_2 < 0$ for $z_2 = 0$ and z_1 in the interval, which prevents z_2 from becoming positive and thus z_1 from increasing.

We now show that $(z_1(t), z_2(t))$ has to reach the axis $z_1 = \bar{z}$ at time T with $z_2(T) < 0$. We will show this by contradiction.

Suppose that $z_1(t)$ converges to $z_1^* > \bar{z}$ and never converges to $\bar{z}$. To make sure that z_1 does not keep decreasing beyond z_1^* (so that z_1^* is the limit in finite or infinite time of $z_1(t)$), there must exist $z_2^* \leq 0$ such that (z_1^*, z_2^*) is an equilibrium, which is not the case. Therefore, $z_1(t)$ has to converge to $\bar{z}$ in finite or infinite time.

Now suppose that $(z_1(t), z_2(t))$ converges to $(\bar{z}, 0)$ in finite or infinite time. This would create a solution with initial condition at $(z_{max}, 0)$ and going to $(\bar{z}, 0)$ through the region where z_2 is negative. By symmetry, there would exist a reverse-time solution with initial condition in $(z_{max}, 0)$ and going to $(\bar{z}, 0)$ through the region where z_2 is positive. The concatenation of both solutions in positive time creates a homoclinic curve. For the existence of a homoclinic curve, an equilibrium is required in the interior of the region defined by the curve, which is not the case. Therefore a solution starting at $(z_{max}, 0)$ cannot converge to $(\bar{z}, 0)$.

Therefore, there exists $z_2^+ < 0$ such that $(z_1(t), z_2(t))$ converges to $(\bar{z}, z_2^+)$ (See Figure 5.3). This takes place in finite time. Indeed, for a convergence to take place in infinite time, $(\bar{z}, z_2^+)$ must be an equilibrium, which is not the case.

Consider now an initial condition $(z_1(t), z_2(t)) = (z_{min}, 0)$ for system (5.6) in reverse time:

$$\begin{cases} \frac{dz_1}{d\tau} = -z_2 \\ \frac{dz_2}{d\tau} = f(z_1, z_2). \end{cases}$$

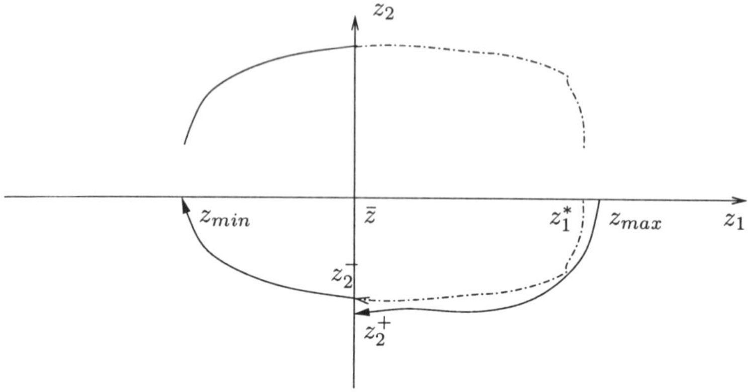

Figure 5.3. Construction of a cycle for Eq. (5.5).

The same reasoning can be held to show that there exists $z_2^- < 0$ such that $(z_1(\tau), z_2(\tau))$ reaches $(\bar{z}, z_2^-)$ in finite (reverse) time.

Suppose now, without loss of generality, that $z_2^+ < z_2^- < 0$. We will now consider all solutions starting at $(z_1(0), 0)$ with $z_1(0) \in [\bar{z}, z_{max}]$. All those solutions cross the axis $z_1 = \bar{z}$ in finite time. If $z_1(0) = \bar{z}$, this crossing takes place at $z_2 = 0$; if $z_1(0) = z_{max}$, this crossing takes place at $z_2 = z_2^+$. By continuity, for all $z_2^* \in [z_2^+, 0]$, there exists $z_1(0)$ such that the solution crosses the axis $z_1 = \bar{z}$ with $z_2 = z_2^*$. Pick $z_2^* = z_2^-$, and rename the corresponding $z_1(0)$ "z_1^*" (see the dash-dotted line in Figure 5.3).

We now have a solution going from $(z_1^*, 0)$ to $(\bar{z}, z_2^-)$ in finite time with $z_2 < 0$. We can then concatenate this solution with the solution going from $(\bar{z}, z_2^-)$ to $(z_{min}, 0)$ in finite time (which we had discovered in reverse time). We now have a solution going from $(z_1^*, 0)$ to $(z_{min}, 0)$ in finite time with $z_2 < 0$. By symmetry of the phase plane, we have a solution linking $(z_{min}, 0)$ to $(z_1^*, 0)$ in finite time with $z_2 > 0$. This creates a cycle Γ.

A similar reasoning can be held for any initial condition inside the region circumscribed by Γ (except the equilibrium). Γ defines the border of a neighborhood of the equilibrium inside which all solutions are cycles. □

Remark 2: The condition of Lemma 1 concerning the sign of $f(z, 0)$ for z belonging to an interval $[z_{min}, z_{max}]$ is satisfied if f is differentiable at $(\bar{z}, 0)$ and $\frac{\partial f}{\partial z}(\bar{z}, 0) > 0$.

The simplest example of this kind of system is the harmonic oscillator

$$\ddot{z} + \omega^2 z = 0. \tag{5.7}$$

The evenness of the function f with respect of $\dot{z}$ is obvious. The *isolated* equilibrium is $z = 0$, and $\frac{df}{dz} = \omega^2 > 0$.

A stronger result can be given when f has a particular structure, which results in the following form for (5.5):

$$\alpha(z)\ddot{z} + \beta(z)\dot{z}^2 + \gamma(z) = 0, \tag{5.8}$$

where we suppose that $\frac{\beta}{\alpha}$ and $\frac{\gamma}{\alpha}$ are Lipschitz continuous. This form is central in this chapter as it will be the one that the zero dynamics will take when generating oscillations for the Pendubot through the regulation of a linear output.

For (5.8), we first have a direct consequence of Lemma 1.

Corollary 1: *If $\frac{\beta}{\alpha}$ and $\frac{\gamma}{\alpha}$ are Lipschitz continuous functions defined on $\mathbb{R}$ then, for any root $\bar{z}$ of $\gamma(z)$, there exists a neighborhood of $(z, \dot{z}) = (\bar{z}, 0)$ such that all solutions in that neighborhood are cycles if γ is differentiable at $\bar{z}$ and*

$$\frac{d\gamma}{dz}(\bar{z})\alpha(\bar{z}) > 0.$$

Note that, if the opposite condition ($\frac{d\gamma}{dz}(\bar{z})\alpha(\bar{z}) < 0$) is satisfied at an equilibrium, this equilibrium is unstable (it suffices to analyze the linearization of the system around this equilibrium).

We have also shown in Perram *et al.* [14] that a general integral of system (5.8) could be built. It is described in the following result.

Theorem 3: Let $(z(t), \dot{z}(t))$ be the solution of system (5.8) with given initial conditions $(z_0, \dot{z}_0)$. If the function

$$\begin{aligned} I(z, \dot{z}, z_0, \dot{z}_0) = \dot{z}^2 - \exp\left\{-2\int_{z_0}^{z} \frac{\beta(\tau)}{\alpha(\tau)} d\tau\right\} \dot{z}_0^2 \\ + \exp\left\{-2\int_{z_0}^{z} \frac{\beta(\tau)}{\alpha(\tau)} d\tau\right\} \int_{z_0}^{z} \exp\left\{2\int_{z_0}^{s} \frac{\beta(\tau)}{\alpha(\tau)} d\tau\right\} \frac{2\gamma(s)}{\alpha(s)}\, ds \end{aligned} \tag{5.9}$$

exists, then it is finite and preserves its value ($= 0$) along the solution $(z(t), \dot{z}(t))$.

Using this full integral, and as hinted in Shiriaev and Canudas de Wit [16], we will almost always be able to build a first integral of system (5.8) (independent of the initial condition) as shown in the following result.

Lemma 2: *Given any solution $(z(t), \dot{z}(t))$ of (5.8) (with initial condition $(z_0, \dot{z}_0)$) and any $z_1^*, z_2^* \in \mathbb{R}$ such that $\alpha(s) \neq 0$ for s belonging to the intervals $[z_0, z_1^*]$ (or $[z_1^*, z_0]$) and $[z_0, z_2^*]$ (or $[z_2^*, z_0]$), the function*

$$\begin{aligned} V_{(z_1^*,z_2^*)}(z,\dot{z}) = \exp\left\{2\int_{z_1^*}^{z}\frac{\beta(\tau)}{\alpha(\tau)}d\tau\right\}\dot{z}^2 \\ +\int_{z_2^*}^{z}\exp\left\{2\int_{z_1^*}^{s}\frac{\beta(\tau)}{\alpha(\tau)}d\tau\right\}\frac{2\gamma(s)}{\alpha(s)}ds \end{aligned} \tag{5.10}$$

is constant along the solution $(z(t), \dot{z}(t))$ that are such that $\alpha(z(t)) \neq 0$ for all $t \geq 0$.

Proof: This result is a direct consequence of (5.9). Using the equality

$$\exp\left\{-2\int_{z_0}^{z}\frac{\beta(\tau)}{\alpha(\tau)}d\tau\right\} = \exp\left\{-2\int_{z_1^*}^{z}\frac{\beta(\tau)}{\alpha(\tau)}d\tau\right\}\exp\left\{2\int_{z_1^*}^{z_0}\frac{\beta(\tau)}{\alpha(\tau)}d\tau\right\}$$

which is valid for any $z_1^* \in \mathbb{R}$ satisfying the condition given in the lemma, we can multiply (5.9) by $\exp\left\{2\int_{z_1^*}^{z}\frac{\beta(\tau)}{\alpha(\tau)}d\tau\right\}$ and obtain:

$$\begin{aligned} \dot{z}^2\exp\left\{2\int_{z_1^*}^{z}\frac{\beta(\tau)}{\alpha(\tau)}d\tau\right\} - \exp\left\{2\int_{z_1^*}^{z_0}\frac{\beta(\tau)}{\alpha(\tau)}d\tau\right\}\dot{z}_0^2 \\ +\exp\left\{2\int_{z_1^*}^{z_0}\frac{\beta(\tau)}{\alpha(\tau)}d\tau\right\}\int_{z_0}^{z}\exp\left\{2\int_{z_1^*}^{s}\frac{\beta(\tau)}{\alpha(\tau)}d\tau\right\}\exp\left\{-2\int_{z_1^*}^{z_0}\frac{\beta(\tau)}{\alpha(\tau)}d\tau\right\}\frac{2\gamma(s)}{\alpha(s)}ds = 0 \end{aligned}$$

along the solution starting at $(z_0, \dot{z}_0)$.

The second term is constant, so that it can be put on the right hand side of the equality. The third term can be simplified (the first and third exponentials in that term are inverse of each other) so that we now have:

$$\begin{aligned} \dot{z}^2\exp\left\{2\int_{z_1^*}^{z}\frac{\beta(\tau)}{\alpha(\tau)}d\tau\right\} + \int_{z_0}^{z}\exp\left\{2\int_{z_1^*}^{s}\frac{\beta(\tau)}{\alpha(\tau)}d\tau\right\}\frac{2\gamma(s)}{\alpha(s)}ds = \\ \exp\left\{2\int_{z_1^*}^{z_0}\frac{\beta(\tau)}{\alpha(\tau)}d\tau\right\}\dot{z}_0^2. \end{aligned} \tag{5.11}$$

The second term of the left hand side can now be split into the sum of two integrals:

$$\begin{aligned} \int_{z_0}^{z}\exp\left\{2\int_{z_1^*}^{s}\frac{\beta(\tau)}{\alpha(\tau)}d\tau\right\}\frac{2\gamma(s)}{\alpha(s)}ds = \int_{z_0}^{z_2^*}\exp\left\{2\int_{z_1^*}^{s}\frac{\beta(\tau)}{\alpha(\tau)}d\tau\right\}\frac{2\gamma(s)}{\alpha(s)}ds \\ +\int_{z_2^*}^{z}\exp\left\{2\int_{z_1^*}^{s}\frac{\beta(\tau)}{\alpha(\tau)}d\tau\right\}\frac{2\gamma(s)}{\alpha(s)}ds \end{aligned}$$

with $\int_{z_0}^{z_2^*}\exp\left\{2\int_{z_1^*}^{s}\frac{\beta(\tau)}{\alpha(\tau)}d\tau\right\}\frac{2\gamma(s)}{\alpha(s)}ds$ going to the right hand side of (5.11), so that (5.11) becomes:

$$\begin{aligned} \exp\left\{2\int_{z_1^*}^{z}\frac{\beta(\tau)}{\alpha(\tau)}d\tau\right\}\dot{z}^2 + \int_{z_2^*}^{z}\exp\left\{2\int_{z_1^*}^{s}\frac{\beta(\tau)}{\alpha(\tau)}d\tau\right\}\frac{2\gamma(s)}{\alpha(s)}ds = \\ \exp\left\{2\int_{z_1^*}^{z_0}\frac{\beta(\tau)}{\alpha(\tau)}d\tau\right\}\dot{z}_0^2 + \int_{z_2^*}^{z_0}\exp\left\{2\int_{z_1^*}^{s}\frac{\beta(\tau)}{\alpha(\tau)}d\tau\right\}\frac{2\gamma(s)}{\alpha(s)}ds \end{aligned}$$

which shows the proposition: the function

$$V_{(z_1^*,z_2^*)}(z,\dot{z}) = \exp\left\{2\int_{z_1^*}^{z} \frac{\beta(\tau)}{\alpha(\tau)}d\tau\right\}\dot{z}^2 + \int_{z_2^*}^{z} \exp\left\{2\int_{z_1^*}^{s} \frac{\beta(\tau)}{\alpha(\tau)}d\tau\right\}\frac{2\gamma(s)}{\alpha(s)}ds$$

stays constant along the solution of (5.8) with initial condition $(z_0, \dot{z}_0)$. □

In this proposition we present a family of functions which stay constant along any solution of the system: the parameters z_1^* and z_2^* can be chosen almost freely independently of the initial conditions (with the restriction that the condition given in Lemma 2 is satisfied). The main difference with the function I is that the function V is independent of the initial condition, but that the constant value at which the function stays is not zero: it depends on the initial condition. In order to confirm this result, it can easily be computed that $\dot{V} = 0$ along the solutions.

Finally, it is interesting to notice that, behind the infinite number of functions that are integral functions for system (5.8) (defined by the different values that z_1^* and z_2^* can take), lies a single function. In fact, for any two pairs of parameters (z_1^*, z_2^*) and (z_a^*, z_b^*), there exists real constants A and B such that $V_{(z_1^*,z_2^*)}(z,\dot{z}) = AV_{(z_a^*,z_b^*)}(z,\dot{z}) + B$ for all $(z, \dot{z})$.

Considering the harmonic oscillator (5.7), we see that $\alpha(z) = 1$, $\beta(z) = 0$ and $\gamma(z) = \omega^2 z$. The construction (5.10) results in the function

$$V(z,\dot{z}) = \dot{z}^2 + \omega^2 z^2 + \omega^2 z_2^{*2}.$$

As z_2^* can be chosen freely, we take $z_2^* = 0$, so that V is the classical Lyapunov function for the harmonic oscillator.

The existence of a function V along which the solutions are constant is not sufficient to ensure the presence of cycles. Only if the constant levels of this function represent closed curve does it result into cycles. This function will later be useful to create attractive limit cycles for the Pendubot.

In Section 1, we have shown that an efficient method to generate an oscillating behavior in a nonlinear control system was the construction and regulation of an output that ensures the presence of oscillations in the zero dynamics of the system. After regulation of this output, the remaining dynamics are oscillating, so that the whole system presents oscillations. We have first exposed a case where the zero dynamics result in an attractive limit cycle, and have then shown that we could obtain a (weaker) result when those zero dynamics are simply neutrally stable: the resulting closed-loop dynamics present oscillations of unknown amplitude (depending on the initial condition). In Section 2, we have then presented tools for the analysis of those neutral oscillations in a special case of zero dynamics:

planar oscillations in the form (5.5) or (5.8); sufficient conditions for the existence of neutrally stable cycles and a first integral were given. We will now use these tools for the control and analysis of a particular mechanical system, the Pendubot, for which we will show how to go from neutrally stable to exponentially stable oscillations.

4. Control of the Pendubot

The following section will be devoted to the description of the Pendubot: a two-links planar robot with a motor at the shoulder and no motor at the elbow (see Figure 5.4).

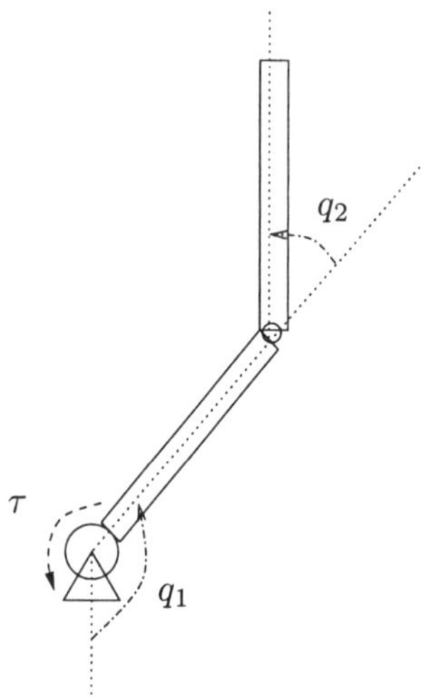

Figure 5.4. Coordinates position on the Pendubot.

A classical mechanical model for this robot is:

$$M(q)\ddot{q} + C(q, \dot{q})\dot{q} + g(q) = B\tau \tag{5.12}$$

where $q = [q_1\ q_2]^T \in \mathbb{R}^2$, and q_1 represents the angle of the first link with the lower vertical axis and q_2 the angle of the second link with the first link. If a motor is available at both joints, the robot is fully actuated and the solution of the problem is trivial. For the Pendubot, a torque can only be applied at the first joint: therefore, $\tau \in \mathbb{R}$, which means that the Pendubot is underactuated. The matrices that define the model are the following:

$$M(q) = \begin{pmatrix} m_1 l_1^2 + m_2(l_2^2 + L_1^2 + 2L_1 l_2 cos(q_2)) & m_2(l_2^2 + L_1 l_2 \cos(q_2)) \\ m_2(l_2^2 + L_1 l_2 \cos(q_2)) & m_2 l_2^2 \end{pmatrix}$$

$$C(q, \dot{q}) = \begin{pmatrix} -m_2 L_1 l_2 \sin(q_2)\dot{q}_2 & -m_2 L_1 l_2 \sin(q_2)(\dot{q}_1 + \dot{q}_2) \\ m_2 L_1 l_2 \sin(q_2)\dot{q}_1 & 0 \end{pmatrix}.$$

$$g(q) = \begin{pmatrix} g\left((m_1 l_1 + m_2 l_2)\sin(q_1) + m_2 l_2 \sin(q_1 + q_2)\right) \\ g m_2 l_2 \sin(q_1 + q_2) \end{pmatrix}$$

$$B = \begin{pmatrix} 1 \\ 0 \end{pmatrix}.$$

We wish to generate oscillations for the Pendubot around the upward position. It is desired for the oscillations of the first arm (linked to the shoulder) to have a given angular amplitude ($= 2\alpha_s$) and a prespecified period (T_s). Also, we wish that the oscillations take place with the second arm close to the vertical. A simple description of the oscillatory behavior would be to say that the robot goes back and forth between a rest position at the left of the vertical and a rest position at the right of the vertical axis. In terms of coordinates, this means that q_1 oscillates around π and q_2 oscillates around 0. If we take the desired behavior to the limit, we would like to have the first link oscillate around the upper vertical while the second one stays vertical. For the model, this translates into oscillations of q_1 around π, while $q_1 + q_2$ stays constant at the value π (this also results in $\dot{q}_1 = -\dot{q}_2$). If a control law that achieves this objective is built, the behavior of the complete system is given by the evolution of q_1 or q_2 once $q_1 + q_2 = \pi$: it represents the zero dynamics. For the Pendubot, we will concentrate on the evolution of q_2 based on the second equation of the model

$$\begin{aligned} 0 &= -m_2(l_2^2 + L_1 l_2 \cos(q_2))\ddot{q}_2 + m_2 l_2^2 \ddot{q}_2 + m_2 L_1 l_2 \sin(q_2)\dot{q}_2^2 \\ &= -\cos(q_2)\ddot{q}_2 + \sin(q_2)\dot{q}_2^2 = 0 \end{aligned}$$

where we have introduced the constraint $q_1 + q_2 = \pi$. This system is not in the appropriate form to apply Lemma 1. If oscillations are present, there is a rest position on the right of the vertical axis ($q_2 < 0$ and $\dot{q}_2 = 0$). At that point, the derivative of the angle satisfies $\dot{q}_2 = 0$, so that $\ddot{q}_2$ is forced to be zero by the zero dynamics: the whole system is at rest and cannot move afterwards. This is in contradiction with the fact that we were considering an oscillating solution. Therefore, no oscillation can be generated with the second link staying vertical.

4.1. *Sufficient conditions for oscillations*

Because an upward oscillatory behavior is not achievable when keeping the second link vertical, we will not impose such a strong constraint and the second link will also have to oscillate. Instead of having $q_1 + q_2 = \pi$, we will impose a more general constraint

$$q_1 + \varphi(q_2) = 0$$

by building an output $y = q_1 + \varphi(q_2)$ that we will regulate so that the system presents the desired oscillations. In other words, the zero dynamics must present oscillations.

The zero dynamics then take the following form (we study the second equation of the model when $y = 0$):

$$\begin{aligned}
0 &= -m_2(l_2^2 + L_1 l_2 \cos(q_2))(\frac{d^2\varphi}{dq_2^2}\dot{q}_2^2 + \frac{d\varphi}{dq_2}\ddot{q}_2) + m_2 l_2^2 \ddot{q}_2 \\
&\quad + m_2 L_1 l_2 \sin(q_2)(\frac{d\varphi}{dq_2})^2 \dot{q}_2^2 + g m_2 l_2 \sin(q_2 - \varphi(q_2)) \\
&= \left[(1 - \frac{d\varphi}{dq_2}) m_2 l_2^2 - \frac{d\varphi}{dq_2} m_2 L_1 l_2 \cos(q_2)\right] \ddot{q}_2 \\
&\quad + \left[m_2 L_1 l_2 \sin(q_2)(\frac{d\varphi}{dq_2})^2 - m_2(l_2^2 + L_1 l_2 \cos(q_2))\frac{\partial^2\varphi}{dq_2^2}\right] \dot{q}_2^2 \qquad (5.13) \\
&\quad + g m_2 l_2 \sin(q_2 - \varphi(q_2)) \\
&= \left[(1 - \frac{d\varphi}{dq_2}) l_2 - \frac{d\varphi}{dq_2} L_1 \cos(q_2)\right] \ddot{q}_2 \\
&\quad + \left[L_1 \sin(q_2)(\frac{d\varphi}{dq_2})^2 - (l_2 + L_1 \cos(q_2))\frac{\partial^2\varphi}{dq_2^2}\right] \dot{q}_2^2 + g \sin(q_2 - \varphi(q_2)).
\end{aligned}$$

These zero dynamics are exactly in the form (5.8)

$$\alpha(q_2)\ddot{q}_2 + \beta(q_2)\dot{q}_2^2 + \gamma(q_2) = 0$$

that was considered in Corollary 1. For any system in the form (5.14), we can then perform the analysis based on this result, and an integral function can be built based on Lemma 2. If this integral function is radially unbounded and there is a single equilibrium, then every solution is a cycle.

Analyzing (5.14) in the light of Corollary 1 will impose conditions on the function $\varphi(q_2)$ that will make sure that there are oscillating solutions around the upper vertical where $q_1 = \pi$ and $q_2 = 0$. In order to have an equilibrium in $q_2 = 0$, we must have $\gamma(0) = 0$, which translates into

$$\varphi(0) = k\pi$$

for some integer k. This equilibrium must correspond to $q_1 = \pi$, so that we must pick $k = -1$. The second condition that should be satisfied for the application of Corollary 1 is

$$\frac{d\gamma(0)}{dq_2}\alpha(0) > 0$$

which here becomes:

$$g(1 - \frac{d\varphi}{dq_2}(0)) \left[(1 - \frac{d\varphi}{dq_2}(0))l_2 - \frac{d\varphi}{dq_2}(0)L_1\right] > 0$$

that is

$$(l_2 + L_1)\frac{d\varphi}{dq_2}(0)^2 - (2l_2 + L_1)\frac{d\varphi}{dq_2}(0) + l_2 > 0$$

so that, cycles occur around the upper vertical if

$$\varphi(0) = -\pi \text{ and } \frac{l_2}{L_1 + l_2} < \frac{d\varphi}{dq_2}(0) < 1. \tag{5.14}$$

4.2. *Oscillations shaping*

Instead of looking at the problem of analyzing Eq. (5.14) for a given function φ, we can shape φ so that (5.14) has a desired form:

$$\alpha_d(q_2)\ddot{q}_2 + \beta_d(q_2)\dot{q}_2^2 + \gamma_d(q_2) = 0 \tag{5.15}$$

so that the oscillations of the zero dynamics follow a prespecified behavior. In order to have equivalence of the behaviors, we should first notice that we can multiply (5.15) by an arbitrary function $\phi(q_2) \neq 0$ without changing the behavior of the desired system. We can then match (5.14) and (5.15):

$$\begin{cases} (1 - \frac{d\varphi}{dq_2})l_2 - \frac{d\varphi}{dq_2}L_1\cos(q_2) & = \alpha_d(q_2)\phi(q_2) \\ L_1\sin(q_2)(\frac{d\varphi}{dq_2})^2 - (l_2 + L_1\cos(q_2))\frac{\partial^2\varphi}{dq_2^2} & = \beta_d(q_2)\phi(q_2) \\ g\sin(q_2 - \varphi(q_2)) & = \gamma_d(q_2)\phi(q_2) \end{cases}$$

and, with $\phi(q_2)$ fixed, we obtain a set of ordinary differential equations with q_2 as independent variable and φ the unknown solution. If this set of equations is solvable, it is very difficult to find an analytic expression for these solutions for given $\alpha_d, \beta_d, \gamma_d$. In order to avoid the use of a numerical solution of this equation, we simply consider the case where the output is linear.

4.3. *Linear output*

A particular case of the previous constraint is the case where φ is affine: $\varphi(q_2) = aq_2 - b$, so that the regulation of the output y yields $q_1 + aq_2 - b = 0$. It directly comes from the condition for oscillations (5.14) that we must have

$$b = \pi \text{ and } \frac{l_2}{l_2 + L_1} < a < 1.$$

In order to stay close to the case where $q_1 + q_2 = \pi$, we rewrite a as $a = 1 - \epsilon$ (with ϵ small), so that the constraint is rewritten as $q_1 + q_2 = \epsilon q_2 + \pi$ and the condition for oscillation becomes

$$0 < \epsilon < \frac{L_1}{L_1 + l_2}. \tag{5.16}$$

From the constraint $q_1 + q_2 = \epsilon q_2 + \pi$, it results that, when $q_1 < \pi$, we have $q_2 > 0$ and $q_1 + q_2 > \pi$, so that, when the robot is on the right of the vertical axis, the second link leans slightly to the left (and conversely, when $q_1 > \pi$). The evolution of the Pendubot during a half-cycle follows the illustration of Figure 5.5.

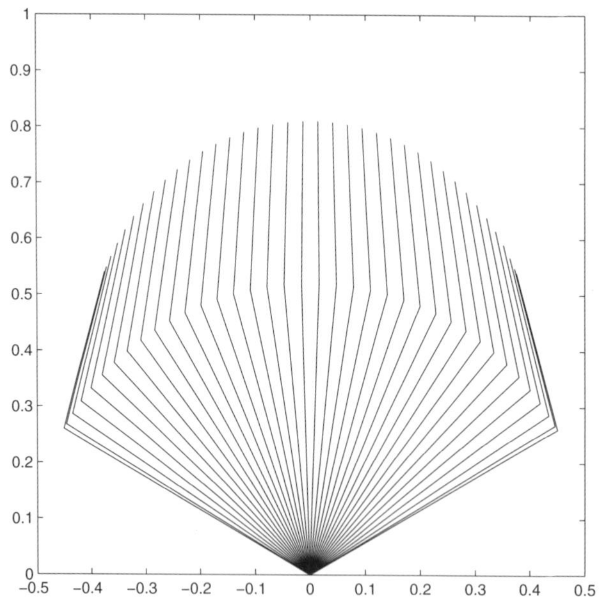

Figure 5.5. Evolution of the Pendubot during a half-cycle for $\epsilon = 0.2$ and $\alpha_s = \frac{\pi}{3}$.

The notations of (5.14) then simplify and the zero dynamics become:

$$(\epsilon l_2 - (1-\epsilon)L_1 \cos(q_2))\,\ddot{q}_2 + L_1 \sin(q_2)(1-\epsilon)^2 \dot{q}_2^2 + g\sin(\epsilon q_2 + \pi) = 0$$

so that the dynamics are rewritten as:

$$((1-\epsilon)L_1 \cos(q_2) - \epsilon l_2)\,\ddot{q}_2 - L_1 \sin(q_2)(1-\epsilon)^2 \dot{q}_2^2 + g\sin(\epsilon q_2) = 0. \quad (5.17)$$

A first consequence of this constraint (5.16) on ϵ, is that $\alpha(q_2) = (1-\epsilon)L_1 cos(q_2) - \epsilon l_2$ is not sign definite, so that the phase plane of (5.17) will be made of vertical strips separated by vertical lines corresponding to the roots of $\alpha(q_2)$. In the case where ϵ does not satisfy (5.16), we have the behavior of the zero dynamics that is illustrated in Figure 5.6. Numerous cycles are observed around different equilibria. However, no cycle is observed around the equilibrium $q_2 = 0$ (which corresponds to the equilibrium $q_1 = \pi$), so that the desired behavior is not achieved.

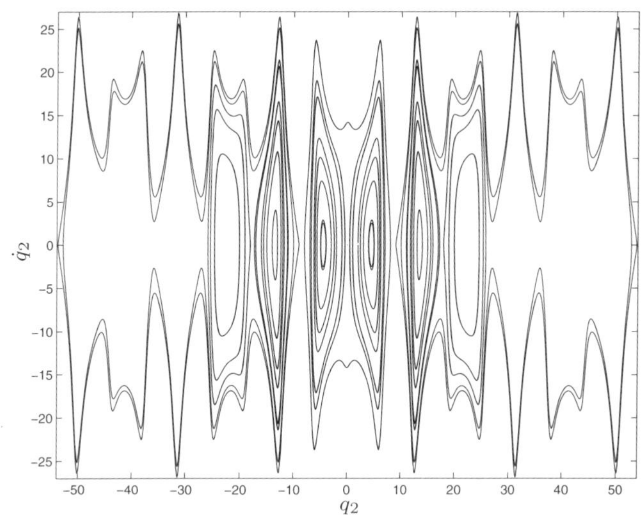

Figure 5.6. $(q_2, \dot{q}_2)$ phase plane when the constraint (5.16) is not satisfied by ϵ.

We will now concentrate on the oscillations of (5.17) in the region surrounding $(q_2, \dot{q}_2) = (0,0)$ where $\alpha(q_2) \neq 0$. This constraint on α imposes $\epsilon l_2 < (1-\epsilon)L_1 \cos(q_2)$ for all q_2 in the region. For any ϵ, this constraint can only be satisfied if q_2 does not reach $\frac{\pi}{2}$: for a given ϵ, q_2 must belong to an interval:

$$-\arccos\left(\frac{\epsilon l_2}{(1-\epsilon)L_1}\right) < q_2 < \arccos\left(\frac{\epsilon l_2}{(1-\epsilon)L_1}\right). \quad (5.18)$$

From this, we see that, the smaller ϵ is, the larger the angle α can be: as ϵ tends to zero, the interval tends to $[-\frac{\pi}{2}, \frac{\pi}{2}]$. The conditions of Lemma 1 and 2 are then well satisfied for the equilibrium $\bar{q}_2 = 0$, so that oscillations take place around the upper vertical axis.

In Figure 5.7, the symmetries of the phase planes are illustrated through the picture of the cycles. The vertical dotted lines represent the limit induced by (5.18).

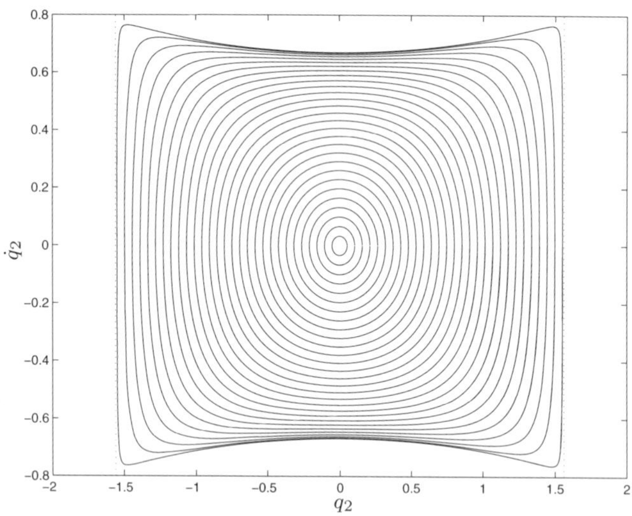

Figure 5.7. Phase plane of the zero dynamics (5.17) for $\epsilon = 0.02$.

Integral function

As (5.17) fits in the format (5.8), we can deduce from Lemma 2 the form of the integral function that is kept constant along the solutions of (5.17):

$$\begin{aligned} V_{(q_2^{*a}, q_2^{*b})}(q_2, \dot{q}_2) &= \exp\left\{2\int_{q_2^{*a}}^{q_2} \tfrac{-L_1 \sin(\tau)(1-\epsilon)^2}{(1-\epsilon)L_1\cos(\tau)-\epsilon l_2} d\tau\right\} \dot{q}_2^2 \\ &+ \int_{q_2^{*b}}^{q_2} \exp\left\{2\int_{q_2^{*a}}^{s} \tfrac{-L_1 \sin(\tau)(1-\epsilon)^2}{(1-\epsilon)L_1\cos(\tau)-\epsilon l_2} d\tau\right\} \tfrac{2g\sin(\epsilon s)}{(1-\epsilon)L_1\cos(s)-\epsilon l_2} ds \\ &= \left(\tfrac{(1-\epsilon)L_1\cos(q_2)-\epsilon l_2}{(1-\epsilon)L_1\cos(q_2^{*a})-\epsilon l_2}\right)^{2(1-\epsilon)} \dot{q}_2^2 \\ &+ \int_{q_2^{*b}}^{q_2} \left(\tfrac{(1-\epsilon)L_1\cos(s)-\epsilon l_2}{(1-\epsilon)L_1\cos(q_2^{*a})-\epsilon l_2}\right)^{2(1-\epsilon)} \tfrac{2g\sin(\epsilon s)}{(1-\epsilon)L_1\cos(s)-\epsilon l_2} ds, \end{aligned}$$

where the parameters q_2^{*a} and q_2^{*b} can be arbitrarily chosen inside the interval defined by (5.18). The denominators containing q_2^{*a} are scalar factors, so that they can be taken out, and the integral function becomes:

$$\begin{aligned} V_\epsilon(q_2, \dot q_2) = &\left((1-\epsilon)L_1\cos(q_2) - \epsilon l_2\right)^{2(1-\epsilon)} \dot q_2^2 \\ &+2g\int_{q_2^*}^{q_2} \left((1-\epsilon)L_1\cos(s) - \epsilon l_2\right)^{1-2\epsilon} \sin(\epsilon s)\, ds. \end{aligned} \tag{5.19}$$

For any cycle, there exists $\bar V_\epsilon$ such that $V_\epsilon = \bar V_\epsilon$ along the cycle. On the other hand, the span of values for $\bar V_\epsilon$ that we can use is limited to those that are such that $V_\epsilon = \bar V_\epsilon$ corresponds to a cycle.

The expression of the integral is computable, though no obvious analytical expression is available. However, it simplifies when $\epsilon = \frac{1}{2}$, so that, in that case,

$$V_{0.5}(q_2, \dot q_2) = \frac{L_1\cos(q_2) - l_2}{2}\dot q_2^2 - 4g\cos(\frac{q_2}{2}) \tag{5.20}$$

where the q_2^* term has been dropped.

Also, an approximation of (5.19) can be given when ϵ is small by using

$$\sin(\epsilon q_2) \approx \epsilon \sin(q_2)$$

so that the integral becomes

$$\hat V_\epsilon(q_2, \dot q_2) \;=\; \left[(1-\epsilon)L_1\cos(q_2) - \epsilon l_2\right]^{2(1-\epsilon)} (\dot q_2^2 - \frac{\epsilon g}{L_1(1-\epsilon)^2}). \tag{5.21}$$

In the sequel, we will denote by V_ϵ the exact value of the integral function, and by $\hat V_\epsilon$, its approximation.

Differentiating $\hat V_\epsilon$ along the solutions of (5.17) results in

$$\dot{\hat V}_\epsilon = 2\left[\epsilon l_2 - (1-\epsilon)L_1\cos(q_2)\right]^{1-2\epsilon} g\dot q_2(\sin(\epsilon q_2) - \epsilon\sin(q_2))$$

which confirms that $\hat V_\epsilon$ is constant along the cycles as long as the approximation $\epsilon\sin(q_2) \approx \sin(\epsilon q_2)$ is valid.

In Figure 5.8, it appears that the approximation (5.21) is very good when the cycle is small ($\bar V_\epsilon \le -0.05$), which is what we expected: indeed, small cycles imply small amplitudes for the movement of q_2, which is required for the approximation to be valid. It is also to be noted that, when the cycles are large, the approximation gives a good representation of the behavior of the system, though with a slightly larger amplitude.

In the case of ϵ small, we have $\hat V_\epsilon = \left[(1-\epsilon)L_1 - \epsilon l_2\right]^{2(1-\epsilon)} (-\frac{\epsilon g}{L_1(1-\epsilon)^2})$ at the equilibrium $(q_2, \dot q_2) = (0, 0)$ and $\hat V_\epsilon = 0$ on the hypothetical cycle

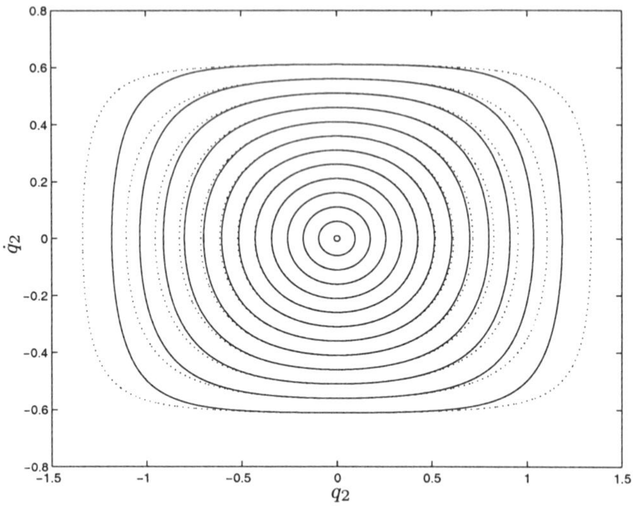

Figure 5.8. Phase plane of the zero dynamics for $\epsilon = 0.02$: comparison of the actual cycles (solid lines) and the approximation (5.21) (dotted line).

that would touch the constraint (5.18). Therefore, $\hat{V}_\epsilon = \bar{V}_\epsilon$ represents a cycle if and only if

$$-\frac{\epsilon g}{L_1(1-\epsilon)^2}\left[(1-\epsilon)L_1 - \epsilon l_2\right]^{2(1-\epsilon)} < \bar{V}_\epsilon < 0.$$

In the case of $\epsilon = 0.5$, Eq. (5.20) indicates that $V_{0.5} = -4g$ at the equilibrium and $V_{0.5} = -4g\sqrt{\frac{l_2+L_1}{2L_1}}$ when an hypothetical cycle touches the constraint. Therefore, $V_{0.5} = \bar{V}_{0.5}$ represents a cycle if and only if

$$-4g < \bar{V}_{0.5} < -4g\sqrt{\frac{l_2+L_1}{2L_1}}.$$

5. Controlled Oscillations

In the previous section we have shown that we could build a linear output such that its regulation would generate neutrally stable oscillations of the Pendubot around the upper vertical. This regulation however would result in an oscillating Pendubot having unknown amplitude, as the phase plane of the zero dynamics is illustrated in Figure 5.7. We now present a control strategy to generate oscillations having pre-specified amplitude and period.

5.1. *Specifications*

As stated in the previous section we have the following specifications for the control problem.

(i) The angle q_1 must oscillate between $\pi - \alpha_s$ and $\pi + \alpha_s$ (with $\alpha_s < \frac{\pi}{2}$).
(ii) The period of oscillation must be equal to T_s.

We will base this analysis on Eq. (5.17). For any ϵ, the angle q_2 can at most oscillate between $-\arccos\left(\frac{\epsilon l_2}{(1-\epsilon)L_1}\right)$ and $+\arccos\left(\frac{\epsilon l_2}{(1-\epsilon)L_1}\right)$, due to the constraint (5.18). This translates into maximal oscillations of q_1 between $\pi - (1-\epsilon)\arccos\left(\frac{\epsilon l_2}{(1-\epsilon)L_1}\right)$ and $\pi + (1-\epsilon)\arccos\left(\frac{\epsilon l_2}{(1-\epsilon)L_1}\right)$. Specification (i) can then only be achieved if ϵ is such that

$$\alpha_s \leq (1-\epsilon)\arccos\left(\frac{\epsilon l_2}{(1-\epsilon)L_1}\right) = \alpha_{max}(\epsilon).$$

It can easily be seen that $\frac{d\alpha_{max}}{d\epsilon} < 0$ for all $\epsilon \in (0,1)$, that $\alpha_{max}(0) = \frac{\pi}{2}$ and that $\alpha_{max}(1) = 0$. Therefore, there exists $\epsilon_{max} > 0$ such that for all $0 \leq \epsilon \leq \epsilon_{max}$, the desired angle of oscillation can be achieved. Therefore, for each $\epsilon < \epsilon_{max}$, there exists $\bar{V}_\epsilon$ such that the oscillation along the level $V_\epsilon(q_2, \dot{q}_2) = \bar{V}_\epsilon$ satisfies specification (i).

We now have to choose, among those $\epsilon < \epsilon_{max}$, the value that will ensure the satisfaction of specification (ii).

Define $T(\epsilon)$ as the period of an oscillation of amplitude $2\alpha_s$ for a given ϵ. This function $T : (0, \epsilon_{max}) \to \mathbb{R}^+$ is continuous inside the interval. We can see that $T(0) = +\infty$; indeed, when $\epsilon = 0$, the oscillation does not really take place: it is replaced by a continuum of equilibria, so that we say that the period of oscillation is infinite. Also, there exists T_{min} such that $T(\epsilon_{max}) = T_{min}$. By continuity of T, for any $T_s > T_{min}$, there exists $\tilde{\epsilon} \in (0, \epsilon_{max})$ such that $T(\tilde{\epsilon}) = T_{min}$. This determines an oscillation that satisfies both (i) and (ii). We will then build a controller that regulates the output $y = q_1 + (1-\tilde{\epsilon})q_2 - \pi$ and ensures convergences of $V_{\tilde{\epsilon}}$ towards $\bar{V}_{\tilde{\epsilon}}$ so that the specifications are satisfied.

5.2. *Control design*

In this section, we build a controller for system (5.12) that forces the output $y = q_1 + (1-\epsilon)q_2 - \pi$ to 0 and convergence of V_ϵ to $\bar{V}_\epsilon$ (we have dropped the $\tilde{}$ sign from $\tilde{\epsilon}$). We first rewrite system (5.12) in new coordinates based on the output: we define $y_1 = y = q_1 + (1-\epsilon)q_2 - \pi$ and $y_2 = \dot{y}_1 = \dot{q}_1 + (1-\epsilon)\dot{q}_2$.

The last two coordinates, based on the coordinates of the zero dynamics, are q_2 and $\dot{q}_2$. The system then becomes:

$$\begin{cases} \frac{dq_2}{dt} = \dot{q}_2 \\ \frac{d\dot{q}_2}{dt} = \frac{(l_2+L_1\cos(q_2))(F(q,\dot{q})+G(q,\dot{q})\tau)+L_1\sin(q_2)(y_2-(1-\epsilon)\dot{q}_2)^2-g\sin(y_1+\epsilon q_2)}{(1-\epsilon)L_1\cos(q_2)-\epsilon l_2} \\ \frac{dy_1}{dt} = y_2 \\ \frac{dy_2}{dt} = F(q,\dot{q}) + G(q,\dot{q})\tau \end{cases}$$

where the expressions of F and G directly come from the model (5.12). It can be seen that $G(q,\dot{q}) \neq 0$ in the region of interest (where (5.18) is satisfied). We can then linearize the y part of the system by feedback by imposing

$$\tau = \frac{1}{G(q,\dot{q})}(v - F(q,\dot{q}))$$

with v the new control variable and the system becomes

$$\begin{cases} \frac{dq_2}{dt} = \dot{q}_2 \\ \frac{d\dot{q}_2}{dt} = \frac{(l_2+L_1\cos(q_2))v+L_1\sin(q_2)(y_2-(1-\epsilon)\dot{q}_2)^2-g\sin(y_1+\epsilon q_2)}{(1-\epsilon)L_1\cos(q_2)-\epsilon l_2} \\ \frac{dy_1}{dt} = y_2 \\ \frac{dy_2}{dt} = v. \end{cases} \tag{5.22}$$

Note that the $\frac{d\dot{q}_2}{dt}$ equation can be rewritten as

$$\begin{aligned} \frac{d\dot{q}_2}{dt} = &\frac{L_1\sin(q_2)(1-\epsilon)^2\dot{q}_2^2-g\sin(\epsilon q_2)}{(1-\epsilon)L_1\cos(q_2)-\epsilon l_2} \\ &+\frac{(l_2+L_1\cos(q_2))v+L_1\sin(q_2)(y_2^2-2(1-\epsilon)\dot{q}_2y_2)-g[\sin(y_1+\epsilon q_2)-\sin(\epsilon q_2)]}{(1-\epsilon)L_1\cos(q_2)-\epsilon l_2} \end{aligned}$$

where the first term contains the zero dynamics, and that the evolution of V_ϵ is as follows:

$$\begin{aligned} \frac{dV_\epsilon}{dt} = &2\dot{q}_2((1-\epsilon)L_1\cos(q_2)-\epsilon l_2)^{1-2\epsilon} \\ &\times(L_1\sin(q_2)(y_2^2-2(1-\epsilon)\dot{q}_2y_2)-g[\sin(y_1+\epsilon q_2)-\sin(\epsilon q_2)]) \\ &+2\dot{q}_2((1-\epsilon)L_1\cos(q_2)-\epsilon l_2)^{1-2\epsilon}(l_2+L_1\cos(q_2))v \end{aligned}$$

and it is easily seen that, when $y_1 = y_2 = 0$, the control law

$$v = -k_V\dot{q}_2(V_\epsilon - \bar{V}_\epsilon)$$

with $k_V > 0$ steers $(q_2, \dot{q}_2)$ to the cycle corresponding to $V_\epsilon = \bar{V}_\epsilon$ (except if $(q_2(0), \dot{q}_2(0)) = (0,0)$).

On the other hand, the control law

$$v = -k_1y_1 - k_2y_2$$

with $k_1, k_2 > 0$ steers (y_1, y_2) to $(0,0)$ and results in neutral stability of the oscillations in the zero dynamics.

In order to achieve both at the same time, we could apply the method that was presented in Shiriaev and Canudas de Wit [16], which requires the analysis of the local controllability of an auxiliary time-varying system around the target orbit, but we prefer to simply sum both control laws and analyze the resulting behavior, *i.e.* we set

$$v = -k_V \dot{q}_2 (V_\epsilon - \bar{V}_\epsilon) - k_1 y_1 - k_2 y_2. \tag{5.23}$$

We would like to check if the cycle corresponding to $V_\epsilon = \bar{V}_\epsilon$ and $y_1 = y_2 = 0$ is exponentially stable in (5.22). Exponential stability of a limit-cycle can be verified [8] by first applying a (diffeomorphic) change of coordinates

$$(q_2, \dot{q}_2, y_1, y_2) \longrightarrow (\theta, \rho)$$

where ρ represents coordinates that are transverse to the considered limit cycle and equal to zero on the limit cycle, and θ represents the evolution of the solution along the limit cycle so that the system is rewritten in the form

$$\begin{cases} \dot{\theta} = 1 + f_1(\theta, \rho) \\ \dot{\rho} = A(\theta)\rho + f_2(\theta, \rho) \end{cases}$$

with $f_1(\theta, 0) = 0$, $f_2(\theta, 0) = 0$ and $\frac{\partial f_2(\theta,0)}{\partial \rho} = 0$. The limit cycle is then an exponentially stable orbit if and only if the transverse linearization

$$\frac{d\rho}{d\theta} = A(\theta)\rho \tag{5.24}$$

is asymptotically stable. Ideally, we should build a Lyapunov function $\rho^T P(\theta)\rho$, with P positive definite, satisfying

$$\frac{dP}{d\theta} = -A(\theta)^T P - PA(\theta) - Q(\theta), \qquad Q(\theta) > 0$$

to show this stability, but we are not able to; however, we can at least show that the matrix A is Hurwitz along the cycle, which is a good indication for stability (note, however, that this is not sufficient for asymptotic stability of the nonautonomous system (5.24), as shown in Khalil [11]).

In our case, the ρ coordinate is already available. It suffices to take $\rho = (V_\epsilon - \bar{V}_\epsilon, y_1, y_2)$. We will not explicitly build the θ coordinates, because it will not change anything for us if we show that $A(\theta)$ or $A(q_2, \dot{q}_2)$ is Hurwitz. Both matrices are identical. Write the ρ dynamics, with $h(q_2, \dot{q}_2) =$

$2((1-\epsilon)L_1\cos(q_2)-\epsilon l_2)^{1-2\epsilon}$ (>0 in the region of interest):

$$\begin{cases} \frac{dW}{dt} = 2\dot{q}_2 h(q_2,\dot{q}_2)\times \\ \qquad \left(L_1\sin(q_2)(y_2^2-2(1-\epsilon)\dot{q}_2 y_2) - g\left[\sin(y_1+\epsilon q_2)-\sin(\epsilon q_2)\right]\right) \\ \qquad +2\dot{q}_2 h(q_2,\dot{q}_2)(l_2+L_1\cos(q_2))(-k_V\dot{q}_2 W - k_1 y_1 - k_2 y_2) \\ \frac{dy_1}{dt} = y_2 \\ \frac{dy_2}{dt} = -k_V\dot{q}_2 W - k_1 y_1 - k_2 y_2 \end{cases}$$

with $W = V_\epsilon - \bar{V}_\epsilon$. We can linearize the ρ dynamics around $(W, y_1, y_2) = (0,0,0)$ and obtain

$$\begin{pmatrix} \dot{W} \\ \dot{y}_1 \\ \dot{y}_2 \end{pmatrix} = \begin{pmatrix} -\dot{q}_2^2 a_{11}(q_2,\dot{q}_2) & -\dot{q}_2 a_{12}(q_2,\dot{q}_2) & -\dot{q}_2 a_{13}(q_2,\dot{q}_2) \\ 0 & 0 & 1 \\ -k_V\dot{q}_2 & -k_1 & -k_2 \end{pmatrix} \begin{pmatrix} W \\ y_1 \\ y_2 \end{pmatrix}$$

where

$$\begin{aligned} a_{11}(q_2,\dot{q}_2) &= k_V(l_2+L_1\cos(q_2))h(q_2,\dot{q}_2) \\ a_{12}(q_2,\dot{q}_2) &= (g\cos(\epsilon q_2)+k_1(l_2+L_1\cos(q_2)))h(q_2,\dot{q}_2) \\ a_{13}(q_2,\dot{q}_2) &= (2(1-\epsilon)L_1\sin(q_2)\dot{q}_2 + k_2(l_2+L_1\cos(q_2)))h(q_2,\dot{q}_2) \end{aligned}$$

and the $\dot{}$ derivative now represents the derivative with respect to the time evolution θ along the target cycle, with q_2 and $\dot{q}_2$ being the value of those states along the target limit-cycle at time θ.

The characteristic polynomial of this linear system for fixed $(q_2, \dot{q}_2)$ on the target orbit is

$$\begin{aligned} p_{CA}(s) &= s^3 + (k_2+\dot{q}_2^2 a_{11})s^2 + (k_1 + k_2\dot{q}_2^2 a_{11} - k_V\dot{q}_2^2 a_{13})s + \\ &\quad k_1\dot{q}_2^2 a_{11} - k_V\dot{q}_2^2 a_{12} \\ &= s^3 + (k_2 + \dot{q}_2^2 k_V(l_2+L_1\cos(q_2))h)s^2 \\ &\quad +(k_1 - 2(1-\epsilon)L_1 k_V\dot{q}_2^3\sin(q_2)h)s - k_V\dot{q}_2^2 g\cos(\epsilon q_2)h. \end{aligned}$$

The first conclusion that can be drawn is that, contrary to what we expected, k_V needs to be negative for the last term to be positive when $\dot{q}_2 \neq 0$ (if $\dot{q}_2 = 0$, the polynomial has one root in $s = 0$). The Routh criterion also indicates that we need to have

$$k_2 > \dot{q}_2^2|k_V|(l_2+L_1\cos(q_2))h$$

for all $(q_2, \dot{q}_2)$ on the target orbit. This can be achieved because, for a given $k_V < 0$ the right-hand side is bounded (it then suffices to take k_2 large enough). The Routh criterion also imposes

$$\begin{aligned} &(k_2+\dot{q}_2^2 k_V(l_2+L_1\cos(q_2))h)(k_1 - 2(1-\epsilon)L_1 k_V\dot{q}_2^3\sin(q_2)h) \\ &\qquad > |k_V|\dot{q}_2^2 g\cos(\epsilon q_2)h \end{aligned}$$

for all $(q_2, \dot{q}_2)$ on the target orbit. This can be achieved by taking k_1 large enough.

To summarize this analysis, the parameters of the control law have to be picked as follows: k_V needs to be taken negative while k_1 and k_2 need to be taken large enough. The efficiency of the control law is confirmed in the following simulations.

5.3. *Simulations*

We will now present simulations of the control law for the Pendubot with the parameters that were given in Canudas de Wit *et al.* [4], that is

$$L_1 = 0.52m \; l_1 = 0.30m \; l_2 = 0.29m \; m_1 = 6kg \; m_2 = 4kg \; g = 9.81\frac{m}{s^2}.$$

We first apply the procedure of Section 5.1 in order to find a cycle that has an angular amplitude of $2\alpha_s = \frac{\pi}{3}$ and a period of 10 seconds. This yields that we should take $\epsilon = 0.0195$ and $\bar{V}_\epsilon = -0.074$ (we obtain that value by computing the approximate value $\hat{V}_\epsilon$ of V_ϵ along the target cycle). The Pendubot then has the desired behavior. If the Pendubot starts at rest in a position close to the vertical $(q_2(0), \dot{q}_2(0), y_1(0), y_2(0)) = (0.01, 0, 0, 0)$, the controlled system (with $k_V = -100$, $k_1 = 10$, $k_2 = 20$ and $\hat{V}_\epsilon$ instead of V_ϵ) behaves as shown in Figures 5.9 and 5.10. The time-evolutions of the states, of V_ϵ and the torque τ are shown in Figure 5.9 while the projection of the solution on the $(q_2, \dot{q}_2)$ phase plane is shown in Figure 5.10. Note that the time evolution of $\hat{V}_\epsilon$ is not monotonous and does not exactly settle at a constant value because we display the approximate value of the integral function ($\hat{V}_\epsilon$) and not its exact value (V_ϵ) and we use this approximate value in the control law. See also that the torque oscillates with time and that (y_1, y_2) stays close to $(0, 0)$ during the whole time-span (of 60 seconds). Figure 5.10 shows that the solution converges directly towards the desired orbit.

6. Conclusion

In this chapter, we have presented two results about the stability of periodic orbits in cascade systems when there is a periodic orbit in the zero dynamics. After giving some results about periodic orbits in two-dimensional systems, we have shown how oscillations can be generated in the Pendubot through the design and regulation of an output. A detailed analysis shows that a prespecified behavior can be achieved by tuning the available parameters (ϵ and $\bar{V}_\epsilon$).

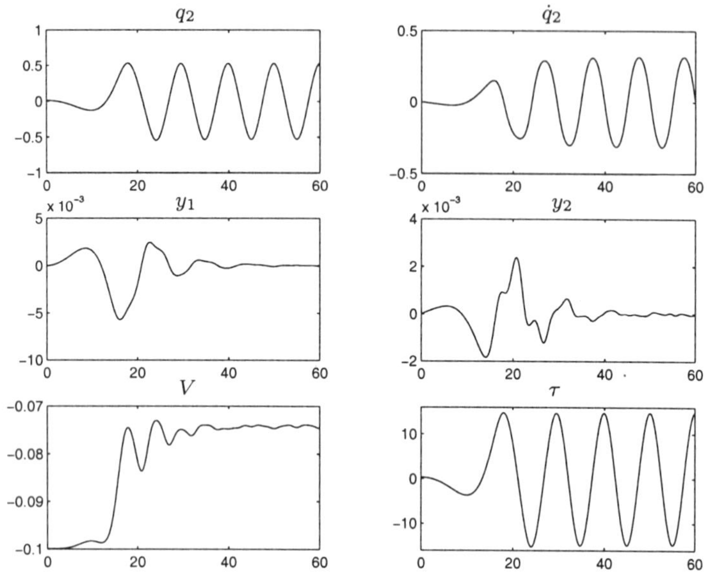

Figure 5.9. Time evolution of the states, the integral function V, and the torque along the solution of the controlled Pendubot with $(q_2(0), \dot{q}_2(0), y_1(0), y_2(0)) = (0.01, 0, 0, 0)$.

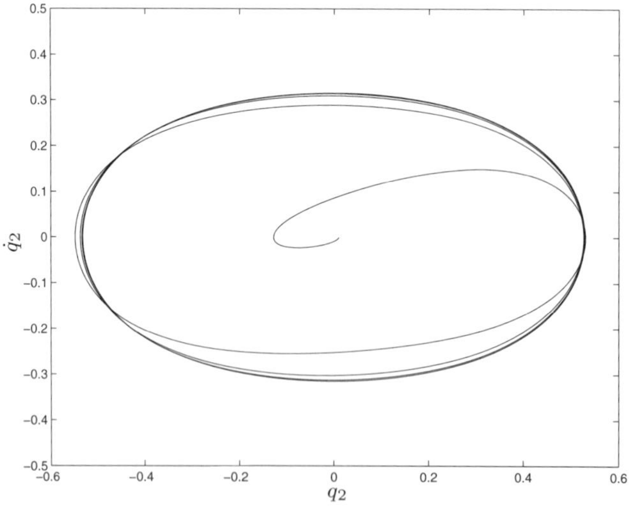

Figure 5.10. Projection of the solution on the $(q_2, \dot{q}_2)$ phase-plane with $(q_2(0), \dot{q}_2(0), y_1(0), y_2(0)) = (0.01, 0, 0, 0)$.

Bibliography

1. J. Aracil, F. Gordillo, and J. Acosta. Stabilization of oscillations in the inverted pendulum. *Proc. 15th IFAC World Congress, Barcelona, Spain*, 2002.
2. G. Besançon. *Contribution à l'étude et à l'observation des systèmes non linéaires avec recours au calcul formel.* Ph.D. Thesis, Grenoble, 1996.
3. C. Canudas-de-Wit and J. Ramirez. Optimal torque control for current-fed induction motors. *IEEE Trans. Automatic Control*, 44(5):1084–1089, 1999.
4. C. Canudas de Wit, B. Espiau, and C. Urrea. Orbital stabilization of underactuated mechanical systems. *Proc. 15th IFAC World Congress, Barcelona, Spain*, 2002.
5. A. De Luca and G. Ulivi. Design of an exact nonlinear controller for induction motors. *IEEE Trans. Automatic Control*, 34(12):1304–1307, 1989.
6. F. Grognard and C. Canudas-de-Wit. Design of orbitally stable zero dynamics for a class of nonlinear systems. *Systems & Control Letters*, 51:89–103, 2004.
7. J. Guckenheimer and P. Holmes. *Nonlinear Oscillations, Dynamical systems, and Bifurcations of Vector Fields.* Springer-Verlag, New York, 1983.
8. J. Hauser and C.C. Chung. Converse Lyapunov functions for exponentially stable periodic orbits. *Systems & Control Letters*, 23:27–34, 1994.
9. A. Isidori. *Nonlinear Control Systems.* Springer-Verlag, Berlin, 1989.
10. M. Janković, R. Sepulchre, and P.V. Kokotović. Constructive Lyapunov stabilization of nonlinear cascade systems. *IEEE Trans. Automatic Control*, 41(12):1723–1735, 1996.
11. H.K. Khalil. *Nonlinear Systems.* Prentice-Hall, 3rd ed., 2002.
12. L. Marconi, A. Isidori, and A. Serrani. Autonomous vertical landing on an oscillating platform: an internal-model based approach. *Automatica*, 38(1):21–32, 2002.
13. F. Mazenc and L. Praly. Adding integrations, saturated controls and global asymptotic stabilization for feedforward systems. *IEEE Trans. Automatic Control*, 41(11):1559–1578, 1996.
14. J. Perram, A. Shiriaev, C. Canudas-de-Wit, and F. Grognard. Explicit formula for a general integral of motion for a class of mechanical systems subject to holonomic constraints. *Proc. IFAC Workshop on Lagrangian and Hamiltonian Systems, Sevilla*, 2003.
15. J. Ramirez. *Contribution à la commande optimale de machines asynchrones.* Ph.D. Thesis, Grenoble, 1998.
16. A. Shiriaev and C. Canudas-de-Wit. Virtual constraints: a tool for orbital stabilization of nonlinear systems. *Proc. 6th IFAC Symp. Nonlinear Control Systems, Stuttgart, Germany*, 2004.
17. R. Sepulchre, M. Janković, and P.V. Kokotović. *Constructive Nonlinear Control.* Springer-Verlag, 1997.
18. R. Sepulchre and G.B. Stan. Feedback mechanisms for global oscillations in Lur'e systems. *Systems & Control Letters*, 54(8):809–818, 2005.
19. E. Westervelt, J.W. Grizzle, and D.E. Koditschek. Hybrid zero-dynamics of planar biped walkers. *IEEE Trans. Automatic Control*, 48(1):42–56, 2003.

CHAPTER 6

Nonlinear Control of a Small Four-Rotor Rotorcraft

P. Castillo[1], R. Lozano[1], P. Garcia[2], P. Albertos[2]

[1]*Heudiasyc, UTC, UMR CNRS 6599, B.P. 20529, 60205 Compiègne, France, E-mail: {castillo,rlozano}@hds.utc.fr*

[2]*Dept. of Systems Engineering and Control, Universidad Politécnica de Valencia, Valencia, Spain, E-mail: {pggil,pedro@isa.upv.es}*

In this chapter we present two control algorithms to stabilize a small helicopter with four rotors. First, we propose a simple nonlinear controller based on Lyapunov analysis for a Planar Vertical Take Off and Landing aircraft (PVTOL). It is proved that the proposed control scheme is globally asymptotically stable. Second, we present a discrete-time controller based on state feedback and the prediction of the state. It is shown that this controller stabilizes possibly unstable continuous-time delay systems. The stability is shown to be robust with respect to uncertainties in the knowledge on the plant parameters, the system delay and the sampling period. Both results have been experimentally tested in a laboratory prototype. Real-time experiments show a satisfactory performance of the proposed controls schemes.

1. Introduction

Flight control problems for small unmanned vehicles (UAVs) have attracted a lot of attention from control researches in the last decade. Classical control strategies for UAVs basically assume a linear model obtained for a particular operating point. The use of modern nonlinear control theory should improve the performance of the controller and allow the tracking of aggressive trajectories.

Generally, the control strategies are based on simplified models which have both a minimum number of states and a minimum number of inputs. These reduced models should retain the main features that must be considered when designing control laws for real aerial vehicles.

Unmanned vehicles are important when it comes to performing a desired task in a dangerous and/or unaccessible environment. Being able to design an unmanned vehicle which is highly maneuverable and extremely unstable is an important contribution to the field of aerial robotics and will lead to numerous applications.

The classical helicopter is one of the most complex flying objects. Such a helicopter is basically composed of a main rotor and a tail rotor. However other types of helicopters exist including the twin rotor or tandem helicopter and the co-axial rotor helicopter. Helicopters, however, are extremely dangerous in practice due to the exposed rotor blades.

The Planar Vertical Take Off and Landing (PVTOL) is a very simple flying machine that evolves in a vertical plane. It has three degrees of freedom (x, y, θ) corresponding to its position and orientation in the plane. It is composed of two independent thrusters that produce a force and a moment on the flying machine, see Figure 6.1. The PVTOL is an underactuated system since it has three degrees of freedom and only two inputs. It poses a very interesting and challenging nonlinear control problem that is a particular case of what is today known as Motion Control.

The study of the PVTOL is clearly motivated by the need to stabilize aircrafts that are able to take-off vertically, such as helicopters and some special airplanes, and a number of results have been reported in the literature. An algorithm to control the PVTOL based on an approximate I-O linearization procedure was proposed in [7]. Their algorithm achieves bounded tracking and asymptotic stability. An extension of this algorithm was presented in [12], where they were able to find a flat output of the system that was used for tracking control of the PVTOL in the presence of unmodeled dynamics.

A nonlinear small gain theorem was proposed in [25] proving the stability of a controller based on nested saturations which can be used to stabilize a PVTOL. The forwarding technique developed in [13] was used in [2] to propose a control algorithm for the PVTOL. This approach leads to a Lyapunov function which ensures asymptotic stability. Other techniques based on linearization were also proposed in [3]. Marconi [11] proposed a control algorithm of the PVTOL for landing on a ship whose deck oscillates. They designed an internal-model-based error feedback dynamic regulator that is robust with respect to uncertainties. Olfati-Saber [20] presented an algorithm to stabilize a PVTOL aircraft with a strong input coupling using a smooth static state feedback.

In this chapter the theoretical nonlinear control of a PVTOL aircraft is presented and its properties are analyzed. Practical implementation of the controller requires us to deal with additional issues: sampling period selection and time delays compensation.

In practical digital implementation of any controller, delays appear due to transport phenomena, computation of the control input, time-consuming information processing in measurement devices, *etc.* The area of control of delayed systems has attracted the attention of many researchers in the past few years [4, 17] because delays may be responsible for instabilities in closed-loop control systems. In order to cope with these delays, a number of algorithms have been reported. A fundamental algorithm based on state-prediction control was proposed in [10], requiring the computation of an integral, used to predict the state. In the ideal case this control scheme leads to a finite pole-placement. However, arbitrary small errors in the computation of the integral term produce instability, as shown in [14,26]. In [5] a continuous-time state-predictive control system is presented and robust stability is proved for uncertainties in the gain and the delay of the plant. Within this framework, [8] presents conditions for a compensator to be realizable with internal stability. Stability of delay systems based on passivity is studied in [18]. Other approaches for the control of systems with delays are available, such as the Smith predictor [15, 16, 21, 24] and its many improved schemes, generically named Process-Model Control schemes [27] or finite spectrum assignment techniques [10]. A close analysis of these methods shows that they all use, in an explicit or implicit manner, prediction of the state in order to generate the control of the system. A common drawback, linked to the internal instability of the prediction, is that they fail to stabilize unstable systems. Additional instability problems may appear in discrete-time pole-placement control algorithms.

The sampling period is constrained by the available on-board hardware, whereas a discrete-time prediction-based state-feedback controller able to cope with delays in unstable systems is discussed afterwards. This predictor has been developed for a linearized model of the aircraft.

A laboratory-scale four-rotor mini-rotorcraft is experimentally controlled. A linear controller takes care of some variables (the yaw and pitch angles, described later) in such a way that the dynamics of the rest (the roll and throttle) can be modeled by the PVTOL equations.

Thus, the proposed controller is experimentally tested in this mini-helicopter having four rotors. The additional issues of the discrete-time implementation of the controller as well as the time delay introduced by the position and orientation sensors are overcome, proving the performances of the new predictor.

2. Nonlinear Control of the PVTOL Aircraft

In this section we present a simple control algorithm for the PVTOL, whose convergence analysis is relatively simple as compared to other controllers proposed in the literature. We present a new approach based on Lyapunov analysis to control the PVTOL which can lead to further developments in nonlinear systems. The controller is first tested in numerical simulations. Then, in Section 4.1 the controller is applied to control the roll angle and the horizontal displacement of a radio-controlled electrical four-rotor mini-rotorcraft. The simplicity of the algorithm is very useful in the practical application and, in particular, it makes the tuning of the controller parameters easy.

2.1. *Dynamic model*

The PVTOL system equations are given by

$$\begin{aligned} \ddot{x} &= -\sin(\phi)u_1 + \varepsilon\cos(\phi)u_2 \\ \ddot{y} &= \cos(\phi)u_1 + \varepsilon\sin(\phi)u_2 - 1 \\ \ddot{\phi} &= u_2 \end{aligned} \tag{6.1}$$

where x is the horizontal displacement, y is the vertical displacement and ϕ is the angle the PVTOL makes with the horizontal line. u_1 is the *collective* input and u_2 is the torque as shown in Figure 6.1. The parameter ε is a small coefficient which characterizes the coupling between the rolling moment and the lateral acceleration of the aircraft. The term -1 is the normalized gravitational acceleration.

Consider the change of coordinates proposed in [19]

$$\begin{aligned} \bar{x} &= x - \varepsilon\sin(\phi) \\ \bar{y} &= y + \varepsilon(\cos(\phi) - 1). \end{aligned} \tag{6.2}$$

The system dynamics, considering these new coordinates, become

$$\begin{aligned} \ddot{\bar{x}} &= -\sin(\phi)\bar{u}_1 \\ \ddot{\bar{y}} &= \cos(\phi)\bar{u}_1 - 1 \\ \ddot{\phi} &= u_2, \end{aligned} \tag{6.3}$$

where $\bar{u}_1 = u_1 - \varepsilon\dot{\phi}^2$.

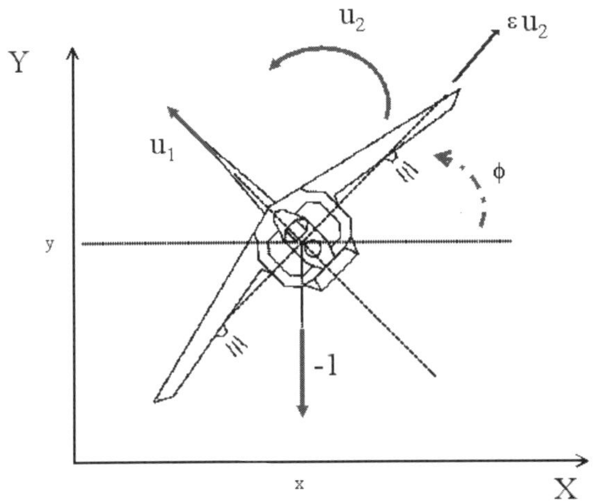

Figure 6.1. The PVTOL aircraft (front view).

2.2. *Control of the vertical displacement*

The vertical displacement $\bar{y}$ will be controlled by forcing the altitude to behave as a linear system. This is done by using the following control strategy

$$\bar{u}_1 = \frac{r_1 + 1}{\cos \sigma_p(\phi)}, \tag{6.4}$$

where $0 < p < \frac{\pi}{2}$ and σ_η, for some $\eta > 0$, is a saturation function

$$\sigma_\eta(s) = \begin{cases} \eta & \text{for } s > \eta \\ s & \text{for } -\eta \le s \le \eta \\ -\eta & \text{for } s < -\eta \end{cases} \tag{6.5}$$

and

$$r_1 = -a_1 \dot{\bar{y}} - a_2(\bar{y} - y_d), \tag{6.6}$$

where y_d is the desired altitude and a_1 and a_2 are positive constants such that the polynomial $s^2 + a_1 s + a_2$ is stable. Assume that after a finite time T_2, $\phi(t)$ belongs to the interval

$$I_{\frac{\pi}{2}} = (-\frac{\pi}{2} + \epsilon, \frac{\pi}{2} - \epsilon) \tag{6.7}$$

for some $\epsilon > 0$ so that $\cos\phi(t) \neq 0$. Introducing (6.4) and (6.6) into (6.3) we obtain for $t > T_2$

$$\begin{aligned} \ddot{\bar{x}} &= -\tan(\phi)(r_1 + 1) \\ \ddot{\bar{y}} &= -a_1\dot{\bar{y}} - a_2(\bar{y} - y_d) \\ \ddot{\phi} &= u_2. \end{aligned} \tag{6.8}$$

Note that in view of the above, $\bar{y} \to y_d$ and $r_1 \to 0$ as $t \to \infty$.

2.3. *Control of the roll angle and the horizontal displacement*

We now design u_2 to control $\dot{\phi}$, ϕ, $\dot{\bar{x}}$ and $\bar{x}$. The control algorithm will be obtained step by step. The final expression for u_2 will be given at the end of this section (see Eq. (6.50)). Roughly speaking, for ϕ close to zero, the $(\bar{x}, \phi)$ subsystem is represented by four integrators in cascade.

We also show that $\phi(t) \in I_{\frac{\pi}{2}}$ (see Eq. (6.7)) after $t = T_2$, independently of the input $\bar{u}_1$ in (6.4).

2.3.1. *Boundedness of* $\dot{\phi}$

In order to establish a bound for $\dot{\phi}$ define u_2 as

$$u_2 = -\sigma_a(\dot{\phi} + \sigma_b(z_1)), \tag{6.9}$$

where $a > 0$ is the desired upper bound for $|u_2|$ and z_1 will be defined later. Let

$$V_1 = \tfrac{1}{2}\dot{\phi}^2, \tag{6.10}$$

then it follows that

$$\dot{V}_1 = -\dot{\phi}\sigma_a(\dot{\phi} + \sigma_b(z_1)). \tag{6.11}$$

Note that if $|\dot{\phi}| > b + \delta$ for some $b > 0$ and some $\delta > 0$ arbitrarily small, then $\dot{V}_1 < 0$. Therefore, after some finite time T_1, we have

$$|\dot{\phi}(t)| \leq b + \delta. \tag{6.12}$$

Assume that b verifies

$$a \geq 2b + \delta. \tag{6.13}$$

Then, from Eqs. (6.8) and (6.9), we obtain for $t \geq T_1$

$$\ddot{\phi} = -\dot{\phi} - \sigma_b(z_1). \tag{6.14}$$

2.3.2. *Boundedness of* ϕ

To establish a bound for ϕ, define z_1 as

$$z_1 = z_2 + \sigma_c(z_3), \tag{6.15}$$

for some z_3 to be defined later, and

$$z_2 = \phi + \dot{\phi}. \tag{6.16}$$

From Eqs. (6.14)–(6.16) we have

$$\dot{z}_2 = -\sigma_b(z_2 + \sigma_c(z_3)). \tag{6.17}$$

Let

$$V_2 = \tfrac{1}{2} z_2^2, \tag{6.18}$$

then

$$\dot{V}_2 = -z_2 \sigma_b(z_2 + \sigma_c(z_3)). \tag{6.19}$$

Note that if $|z_2| > c + \delta$ for some δ arbitrarily small and some $c > 0$, then $\dot{V}_2 < 0$. Therefore, it follows that after some finite time $T_2 \geq T_1$, we have

$$|z_2(t)| \leq c + \delta. \tag{6.20}$$

From Eq. (6.16) we obtain, for $t \geq T_2$,

$$\phi(t) = \phi(T_2)e^{-(t-T_2)} + \textstyle\int_{T_2}^{t} e^{-(t-\tau)} z_2(\tau) d\tau. \tag{6.21}$$

It follows that there exists a finite time T_3 such that, for $t \geq T_3 > T_2$, we have

$$|\phi(t)| \leq \bar{\phi} \triangleq c + 2\delta. \tag{6.22}$$

If

$$c + 2\delta \leq \frac{\pi}{2} - \epsilon, \tag{6.23}$$

then $\phi(t) \in I_{\frac{\pi}{2}}$ (see Eq. (6.7)) for $t \geq T_2$.

Assume that b and c also satisfy

$$b \geq 2c + \delta. \tag{6.24}$$

Then, in view of Eq. (6.20), (6.17) reduces to

$$\dot{z}_2 = -z_2 - \sigma_c(z_3), \tag{6.25}$$

for $t \geq T_3$.

Note now that the following inequality holds for $|\phi| < 1$:

$$|\tan(\phi) - \phi| \leq \phi^2. \tag{6.26}$$

We will use the above inequality in the following development.

2.3.3. *Boundedness of $\dot{\bar{x}}$*

To establish a bound for $\dot{\bar{x}}$, define z_3 as

$$z_3 = z_4 + \sigma_d(z_5), \tag{6.27}$$

where z_4 is defined as

$$z_4 = z_2 + \phi - \dot{\bar{x}}, \tag{6.28}$$

and z_5 will be defined later. From Eqs. (6.8), (6.16) and (6.25) and the above it follows that

$$\dot{z}_4 = (1 + r_1)\tan(\phi) - \phi - \sigma_c(z_4 + \sigma_d(z_5)). \tag{6.29}$$

Define

$$V_3 = \tfrac{1}{2} z_4^2, \tag{6.30}$$

then

$$\dot{V}_3 = z_4 \left[(1 + r_1)\tan(\phi) - \phi - \sigma_c(z_4 + \sigma_d(z_5))\right]. \tag{6.31}$$

Since $r_1 \tan(\phi) \to 0$ (see (6.6) and (6.8)), there exists a finite time $T_5 > T_4$, large enough such that if

$$|z_4| > d + \bar{\phi}^2 + \delta$$

and

$$c \geq \bar{\phi}^2 + \delta, \tag{6.32}$$

for some δ arbitrarily small and $d > 0$, then $\dot{V}_3 < 0$. Therefore, after some finite time $T_6 > T_5$, we have

$$|z_4(t)| \leq d + \delta + \bar{\phi}^2. \tag{6.33}$$

Assume that d and c verify

$$c \geq 2d + \delta + \bar{\phi}^2. \tag{6.34}$$

Thus, after a finite time T_6, Eq. (6.29) reduces to

$$\dot{z}_4 = (1 + r_1)\tan(\phi) - \phi - z_4 - \sigma_d(z_5). \tag{6.35}$$

Finally, by Eqs. (6.16), (6.28) and (6.33) it follows that $\dot{\bar{x}}$ is bounded.

2.3.4. *Boundedness of $\bar{x}$*

To establish a bound for $\bar{x}$, define z_5 as

$$z_5 = z_4 + \phi - 2\dot{\bar{x}} - \bar{x}. \tag{6.36}$$

From Eqs. (6.8), (6.16), (6.28) and (6.35) we get

$$\begin{aligned}\dot{z}_5 &= (1 + r_1)\tan(\phi) - \phi - z_4 - \sigma_d(z_5) + \dot{\phi} + 2\tan(\phi)(r_1 + 1) - \dot{\bar{x}} \\ &= -\sigma_d(z_5) + 3r_1 \tan(\phi) + 3(\tan(\phi) - \phi).\end{aligned} \tag{6.37}$$

Define

$$V_4 = \tfrac{1}{2} z_5^2, \tag{6.38}$$

then

$$\dot{V}_4 = z_5 \left[-\sigma_d(z_5) + 3r_1 \tan(\phi) + 3(\tan(\phi) - \phi)\right]. \tag{6.39}$$

Since $r_1 \tan(\phi) \to 0$, there exists a finite time $T_7 > T_6$, large enough such that if $|z_5| > 3\bar{\phi}^2 + \delta$ for some δ arbitrarily small and

$$d \geq 3\bar{\phi}^2 + \delta, \tag{6.40}$$

then $\dot{V}_4 < 0$. Therefore, after some finite time $T_8 > T_7$, we have

$$|z_5(t)| \leq 3\bar{\phi}^2 + \delta. \tag{6.41}$$

After time T_8, Eq. (6.37) reduces to

$$\dot{z}_5 = -z_5 + 3r_1 \tan(\phi) + 3(\tan(\phi) - \phi). \tag{6.42}$$

Boundedness of $\bar{x}$ follows from Eqs. (6.33), (6.36) and (6.41).

Rewrite all the constraints on the parameters a, b, c, d and $\bar{\phi}$, namely

$$\begin{aligned} &a \geq 2b + \delta \\ &\bar{\phi} \triangleq c + 2\delta \leq 1 \\ &b \geq 2c + \delta \\ &c \geq (c + 2\delta)^2 + 2d + \delta \\ &d \geq 3(c + 2\delta)^2 + \delta. \end{aligned} \tag{6.43}$$

From the above we obtain

$$\begin{aligned} a &\geq 4c + 3\delta \\ b &\geq 2c + \delta \\ c + 2\delta &\leq 1 \\ c &\geq (c + 2\delta)^2 + 2d + \delta \\ &\geq (c + 2\delta)^2 + 2(3(c + 2\delta)^2 + \delta) + \delta \\ &\geq 7(c + 2\delta)^2 + 3\delta \\ d &\geq 3(c + 2\delta)^2 + \delta. \end{aligned} \tag{6.44}$$

2.3.5. *Convergence of ϕ, $\dot{\phi}$, $\bar{x}$ and $\dot{\bar{x}}$ to zero*

Therefore, c and δ should be chosen small enough to satisfy conditions (6.44). The parameters a, b and d can then be computed as a function of c as above.

From Eq. (6.42) it follows that for a time large enough,

$$|z_5(t)| \leq 3\phi^2 + \delta, \tag{6.45}$$

for some δ arbitrarily small. From (6.35) and (6.45) we have that for a time large enough,

$$|z_4(t)| \leq 4\phi^2 + 2\delta, \tag{6.46}$$

for some δ arbitrarily small. From (6.27) and the above we have

$$|z_3(t)| \leq 7\phi^2 + 3\delta. \tag{6.47}$$

Similarly, from (6.25)

$$|z_2(t)| \leq 7\phi^2 + 4\delta, \tag{6.48}$$

and finally for a time large enough and an arbitrarily small δ, from (6.21) and the above we get

$$|\phi| \leq 7\phi^2 + 5\delta. \tag{6.49}$$

Since δ is arbitrarily small, the above inequality implies that either i) $\phi = 0$ or ii) $|\phi| \geq \frac{1}{7}$. If c is chosen small enough such that $\bar{\phi} < \frac{1}{7}$ (see Eq. (6.22)), then the only possible solution is $\phi = 0$. Therefore $\phi \to 0$ as $t \to \infty$. From (6.45)–(6.48) and (6.15) we have that $z_i(t) \to 0$ for $i = 1, 2, ..., 5$. From (6.16) we get $\dot{\phi} \to 0$. From (6.28) and (6.36) it follows respectively that $\dot{\bar{x}} \to 0$ and $\bar{x} \to 0$. The control input u_2 is given by (6.9), (6.15), (6.16), (6.27), (6.28) and (6.36), *i.e.*

$$u_2 = -\sigma_a(\dot{\phi} + \sigma_b(\phi + \dot{\phi} + \sigma_c(2\phi + \dot{\phi} - \dot{\bar{x}} + \sigma_d(3\phi + \dot{\phi} - 3\dot{\bar{x}} - \bar{x})))). \tag{6.50}$$

The amplitudes of the saturation functions should satisfy the constraints in Eqs. (6.44).

2.4. *Simulation results*

In order to validate the results of the proposed control law we have performed some simulations. We started the PVTOL aircraft at the position $(x, y, \phi) = (20, 10, 0.9)$. We have also run simulations with the control including in the system the term ε. For $\varepsilon \leq 0.2$, the results were very similar as for $\varepsilon = 0$. Simulations showed that the performance of the proposed controller is satisfactory.

The simulation results for $\varepsilon = 0.2$ are shown in Figures 6.2 and 6.3.

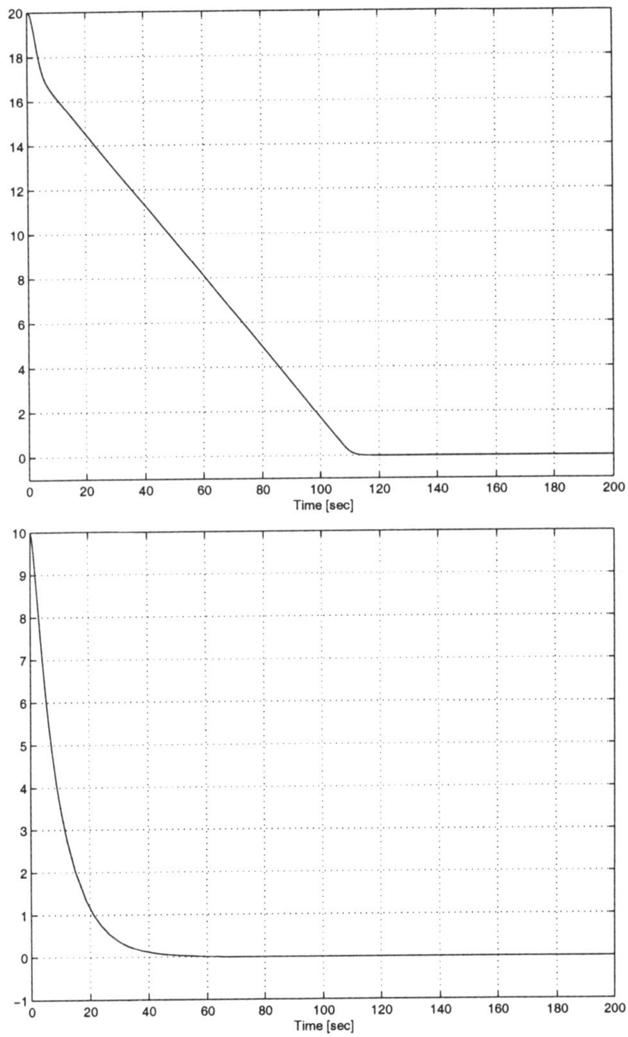

Figure 6.2. Positions x and y of the PVTOL aircraft.

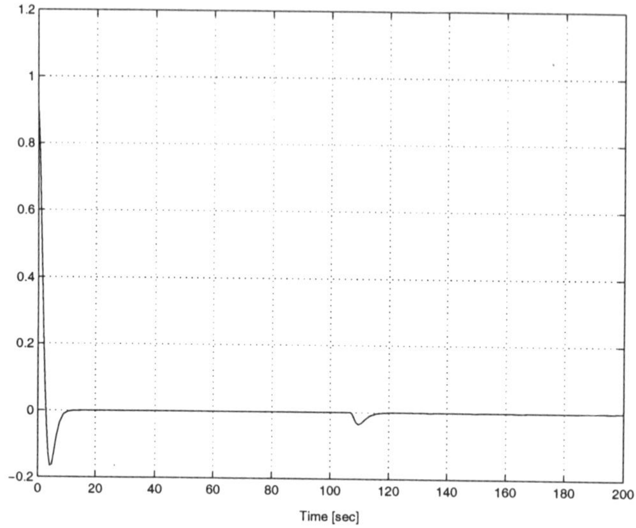

Figure 6.3. Angle ϕ of the PVTOL aircraft.

3. Discrete-Time Controller for Continuous-Time Systems with Delay

In the previous control design an external controller took care of the yaw angle. The decoupled model can be assumed as a linear one. Nevertheless, the discrete time implementation of the control, as well as the use of position sensors introduces unavoidable time delays.

In this section we present the stability analysis of a hybrid control scheme, *i.e.* when the system representation is given in continuous-time while the controller is expressed in discrete-time. The controller is basically a discrete-time state-feedback control in which the actual state is replaced by the prediction of the state. We present a stability proof based on Lyapunov analysis of the hybrid closed-loop system. Convergence of the state to the origin is insured regardless of whether the original system is stable or not. The stability is established in spite of uncertainties in the knowledge of the plant parameters and the delay. Robustness is also proved with respect to small variations of the time elapsed between sampling instants.

The application of the proposed prediction-based controller is presented in the next section.

3.1. *Problem formulation*

Consider the following continuous-time state space representation of a system with input delay

$$\dot{x}(t) = A_c x(t) + B_c u(t - h(t)), \tag{6.51}$$

where the nominal plant parameter matrices are $A_c \in \mathbb{R}^{n \times n}$, $B_c \in \mathbb{R}^{n \times m}$ and $h(t)$ is the time-varying plant delay. Usually, in the discrete-time framework, the sampling time instant t_k is defined as $t_k = kT$ where T is the sampling period and k is an integer. However, since we wish to prove robustness of the control scheme with respect to the time elapsed between sampling time instants, we will not define t_k as a multiple of T. We will rather define t_k as the *k-th* sampling instant and such that

$$t_{k+1} - t_k = T + \varepsilon_k \tag{6.52}$$

where T is the ideal sampling period and ε_k is a small variation of the time elapsed between sampling instants. Furthermore, we will assume that T and h satisfy

$$h(t) = dT + \epsilon(t) \tag{6.53}$$

where d is an integer and $\epsilon(t)$ is a small uncertainty in the knowledge of the delay $h(t)$. Both variations ε_k and $\epsilon(t)$ can be positive or negative but they should be bounded as follows

$$\begin{aligned} |\varepsilon_k| \ &\leq \bar{\varepsilon} \ll T \\ |\epsilon(t)| &\leq \bar{\epsilon} \ll T. \end{aligned}$$

We use the notation $x_k = x(t_k)$. From (6.51) we obtain the following time response equation

$$x_{k+1} = A_1 x_k + \int_{t_k}^{t_{k+1}} e^{A_c(t_{k+1}-\tau)} B_c u(\tau - h(\tau)) d\tau, \tag{6.54}$$

where

$$A_1 = e^{A_c(t_{k+1}-t_k)} = e^{A_c(T+\varepsilon_k)}. \tag{6.55}$$

We define A as

$$A = e^{A_c T} \tag{6.56}$$

and Δ_4 such that

$$A_1 = A + \Delta_4. \tag{6.57}$$

Since we are interested in implementing the control law in a computer, we assume that the input u is constant between sampling instants, *i.e.* $u(t) = u_k$, for all $t \in [t_k, t_{k+1})$.

The following lemma shows how the recursive equation for x_k is modified due to the uncertainties in the plant parameters A_c and B_c, the delay $h(t)$ and the ideal sampling period T.

Lemma 1: *The recursive equation for system (6.51) for time sampling instants defined in (6.52) and the delay in (6.53) is given by*

$$x_{k+1} = Ax_k + Bu_{k-d} + \Delta f_k, \tag{6.58}$$

where $\Delta \in \mathbb{R}^{n\times s}$ *and* $f_k \in \mathbb{R}^s$ *with* $s = 3m + n$ *are defined as*

$$\begin{aligned} \Delta &= [\Delta_1, \Delta_2, \Delta_3, \Delta_4] \\ f_k &= \left[u_{k-d-1}^T, u_{k-d}^T, u_{k-d+1}^T, x_k^T\right]^T, \end{aligned} \tag{6.59}$$

and Δ *is a matrix which is bounded by* $\bar{\varepsilon}$ *and* $\bar{\epsilon}$. *Therefore* Δ *converges to zero as* $\bar{\varepsilon}$ *and* $\bar{\epsilon}$ *converge to zero.*

Proof: See [9]. □

Then (6.58) can be viewed as a general state-space representation for discrete-time systems in which Δ takes into account uncertainties in the matrices A_c and B_c, in the delay h and in the ideal sampling period T. We assume that the nominal plant parameters A_c and B_c and the ideal sampling period T are such that (A, B) is a controllable pair. To prove robustness of the control scheme we will mainly use the property that $\Delta \to 0$ as the uncertainties in A_c, B_c, h (*i.e.* ϵ) and T (*i.e.* ε) go to zero.

3.2. *d-step ahead prediction scheme*

In this subsection we extend the ideas in [6] to compute a d-step ahead prediction of the state in the case of the linear system with uncertainties (6.58). For simplicity of notation we have dropped the subscript k from the uncertainties Δ. From (6.58) the prediction of x_{k+d} is given by

$$\begin{aligned} x_{k+d} = {} & A^d x_k + A^{d-1}Bu_{k-d} + \dots + ABu_{k-2} \\ & + Bu_{k-1} + A^{d-1}\Delta f_k + A^{d-2}\Delta f_{k+1} \\ & + \dots + \Delta f_{k+d-1}, \end{aligned} \tag{6.60}$$

or

$$x_{k+d} = A^d x_k + A^{d-1}Bu_{k-d} + ... + ABu_{k-2} + Bu_{k-1} + \bar{\Delta}\ \bar{f}_{k+d-1}, \tag{6.61}$$

where $\bar{\Delta}$ and $\bar{f}_{k+d-1}$ are given by

$$\bar{\Delta} = \left[A^{d-1}\Delta,\ A^{d-2}\Delta, ..., \Delta\right] \tag{6.62}$$

and

$$\bar{f}_{k+d-1} = \left[f_k^T,\ f_{k+1}^T, ...,\ f_{k+d-1}^T\right]^T. \tag{6.63}$$

Define x_{k+d}^p as the prediction of the state x_{k+d} at time t_k

$$x_{k+d}^p = A^d x_k + A^{d-1}Bu_{k-d} + ... + Bu_{k-1}. \tag{6.64}$$

Note that x_{k+d}^p can be computed with information available at time t_k.

3.3. *Prediction-based state feedback control*

In this subsection we define a prediction-based controller following the ideas of [6]. We prove that our controller is robust with respect to uncertainties that are small enough. Consider the control input

$$u_k = K^T x_{k+d}^p \tag{6.65}$$

or, using (6.64),

$$u_k = K^T(A^d x_k + A^{d-1}Bu_{k-d} + ... + Bu_{k-1}). \tag{6.66}$$

From the above and (6.61) it follows that

$$u_k = K^T(x_{k+d} - \bar{\Delta}\ \bar{f}_{k+d-1}). \tag{6.67}$$

Introducing the above equation into (6.58) we obtain

$$x_{k+1} = (A + BK^T)x_k - BK^T\bar{\Delta}\ \bar{f}_{k-1} + \Delta f_k. \tag{6.68}$$

As will be shown next, for small parameter and delay uncertainties, the stability of the above system will be insured if $A + BK^T$ is stable and if we can show that $\bar{f}_{k-1}$ and f_k are linear combinations of the elements of the closed-loop system state

$$z_k = \left[x_k^T, ..., x_{k-d}^T, u_{k-d-1}^T, ..., u_{k-2d-1}^T\right]^T, \tag{6.69}$$

where $z_k \in \mathbb{R}^l$ with $l = (d+1)(n+m)$. Recall from (6.58) and (6.59) that

$$\Delta f_k = \Delta_1 u_{k-d-1} + \Delta_2 u_{k-d} + \Delta_3 u_{k-d+1} + \Delta_4 x_k. \tag{6.70}$$

In the above equation $u_{k-d-1}$and x_k are clearly elements of z_k in (6.69). Using (6.66), u_{k-d} can be expressed in terms of $x_{k-d}, u_{k-2d}, ...,$ and u_{k-d-1} which are elements of z_k . Similarly, u_{k-d+1} can be expressed in terms of $x_{k-d+1}, u_{k-2d+1}, ...,$ and u_{k-d}. As before, u_{k-d} can be expressed in terms of elements of z_k. Therefore f_k in (6.68) can be expressed as a function of the elements of z_k. Note also that we can prove similarly that f_{k-1} is a function of z_k.

From (6.62) and (6.63) we have

$$\bar{\Delta}\ \bar{f}_{k-1} = A^{d-1}\Delta f_{k-d} + A^{d-2}\Delta f_{k-d+1} + ... + \Delta f_{k-1}. \tag{6.71}$$

In view of (6.59), $f_{k-d}, f_{k-d+1}, .., $ and f_{k-2} in the above equation are functions of z_k in (6.69). As explained before f_{k-1} is also a function of z_k and we conclude that $\bar{f}_{k-1}$ in (6.71) is a function of z_k. Therefore the term $-BK^T\bar{\Delta}\ \bar{f}_{k-1} + \Delta f_k$ in (6.68) can be expressed as

$$-BK^T\bar{\Delta}\ \bar{f}_{k-1} + \Delta f_k = \Delta' z_k \tag{6.72}$$

where Δ' is a matrix whose elements vanish as Δ goes to zero. From (6.67) we get

$$u_{k-d} = K^T(x_k - \bar{\Delta}\ \bar{f}_{k-1}), \tag{6.73}$$

or

$$u_{k-d} = K^T x_k + \Delta'' z_k, \tag{6.74}$$

where $\Delta'' = K^T\bar{\Delta}$ is a matrix whose elements vanish as Δ goes to zero. From (6.68), (6.72) and (6.74), the closed-loop system can be written as

$$\begin{bmatrix} x_{k+1} \\ x_k \\ \vdots \\ x_{k-d+1} \\ u_{k-d} \\ u_{k-d-1} \\ \vdots \\ u_{k-2d} \end{bmatrix} = \begin{bmatrix} (A+BK^T) & 0 & 0 & \cdots & \cdots & \cdots & \cdots & 0 \\ 1 & 0 & 0 & \cdots & \cdots & \cdots & \cdots & 0 \\ \vdots & \ddots & \ddots & \cdots & \cdots & \cdots & \cdots & \vdots \\ 0 & 0 & 1 & \ddots & \cdots & \cdots & \cdots & 0 \\ K^T & 0 & \ddots & \ddots & \ddots & \cdots & \cdots & 0 \\ \vdots & \cdots & \cdots & \ddots & 1 & \ddots & \cdots & \vdots \\ \vdots & \cdots & \cdots & \cdots & \ddots & \ddots & \ddots & \vdots \\ 0 & 0 & \cdots & \cdots & \cdots & 0 & 1 & 0 \end{bmatrix} \begin{bmatrix} x_k \\ x_{k-1} \\ \vdots \\ x_{k-d} \\ u_{k-d-1} \\ u_{k-d-2} \\ \vdots \\ u_{k-2d-1} \end{bmatrix} + \begin{bmatrix} \Delta' \\ 0 \\ \vdots \\ 0 \\ \Delta'' \\ 0 \\ \vdots \\ 0 \end{bmatrix} z_k. \tag{6.75}$$

With obvious notation we rewrite the above system as

$$z_{k+1} = \bar{A} z_k + \bar{B} z_k, \tag{6.76}$$

where $\bar{B} \to 0$ as $\Delta \to 0$ and $\bar{A} \in \mathbb{R}^{l\times l}$, $\bar{B} \in \mathbb{R}^{l\times l}$ with $l = (d+1)(n+m)$. Note that from (6.53) it follows that $d \to \infty$ as $T \to 0$. This means that as $T \to 0$ the closed-loop system in (6.75) becomes infinite-dimensional. In the following section we present a stability analysis of the closed-loop system (6.75) when $T \neq 0$, *i.e.* when the dimension of z_k in (6.75) is finite.

3.4. *Stability of the closed-loop system*

We now prove the stability of the closed-loop system in (6.75) or (6.76), and the robustness with respect to small uncertainties in A_c, B_c, h and T in the system (6.51). It can be seen from (6.75) and (6.76) that the eigenvalues of $\bar{A}$ are given by the set of the n eigenvalues of $(A + BK^T)$ and $(l - n)$ eigenvalues at the origin. If K is chosen such that $(A + BK^T)$ is a Schur matrix, then $\bar{A}$ is also a Schur matrix, *i.e.* $\bar{A}$ has all its eigenvalues strictly inside the unity circle. It then follows that for every $Q > 0$ there exists $P > 0$ such that the following Lyapunov equation holds

$$\bar{A}^T P \bar{A} - P = -Q. \tag{6.77}$$

Define the candidate Lyapunov function V_k as

$$V_k = z_k^T P z_k. \tag{6.78}$$

From (6.76), (6.77) and (6.78) we have

$$\begin{aligned} V_{k+1} &= z_{k+1}^T P z_{k+1} \\ &= (\bar{A} z_k + \bar{B} z_k)^T P (\bar{A} z_k + \bar{B} z_k) \\ &= z_k^T \bar{A}^T P \bar{A} z_k + 2 z_k^T \bar{B}^T P \bar{A} z_k + z_k^T \bar{B}^T P \bar{B} z_k \\ &= V_k - z_k^T Q z_k + z_k^T (2\bar{B}^T P \bar{A} + \bar{B}^T P \bar{B}) z_k. \end{aligned} \tag{6.79}$$

If the uncertainties are small enough, *i.e.* are such that

$$-Q + \left\| 2\bar{B}^T P \bar{A} + \bar{B}^T P \bar{B} \right\| < -\eta Q \tag{6.80}$$

for some $\eta > 0$, then

$$V_{k+1} - V_k < -\eta z_k^T Q z_k. \tag{6.81}$$

It then follows that $z_k \to 0$, exponentially, as $k \to \infty$. Given that x and u converge to zero at the sampling instants (see (6.69)), it follows that $u(t)$ converges to zero as $t \to \infty$. From (6.51) it follows that $x(t)$ converges to zero as $t \to \infty$.

4. Experimental Results

In this section we present two real-time experiments on a four-rotor mini-rotorcraft. In the first experiment, the nonlinear controller developed in Section 2 is applied to control the roll angle and the throttle input. In the second experiment the linearized yaw control is presented. Delays are introduced to the system due to the position/orientation measuring system and also due to the computation of the control input. In both experiments we first describe the architecture of the platform, the characteristics of this rotorcraft and the hardware used.

Figure 6.4. Photo of the rotorcraft (front view).

4.1. *Experimental platform for the roll control*

The four-rotor mini-rotorcraft (or quad-rotor) used is shown in Figure 6.4. In this type of rotorcraft the front and the rear motors rotate counter clockwise, while the other two rotate clockwise. Pitch movement is obtained by increasing the speed of the rear motor while reducing the speed of the front motor. The roll movement is obtained similarly using the lateral motors. The yaw movement is obtained by increasing the speed of the front and rear motors while decreasing the speed of the lateral motors.

Note that when the yaw and pitch (or roll) angles are set to zero, the quad-rotor reduces to a PVTOL (see Figure 6.5).

In this experiment the pitch and yaw angles are assumed to be independently controlled, *e.g.* by an experienced pilot. The remaining controls, *i.e.* the collective input (or throttle input) and the roll control, are con-

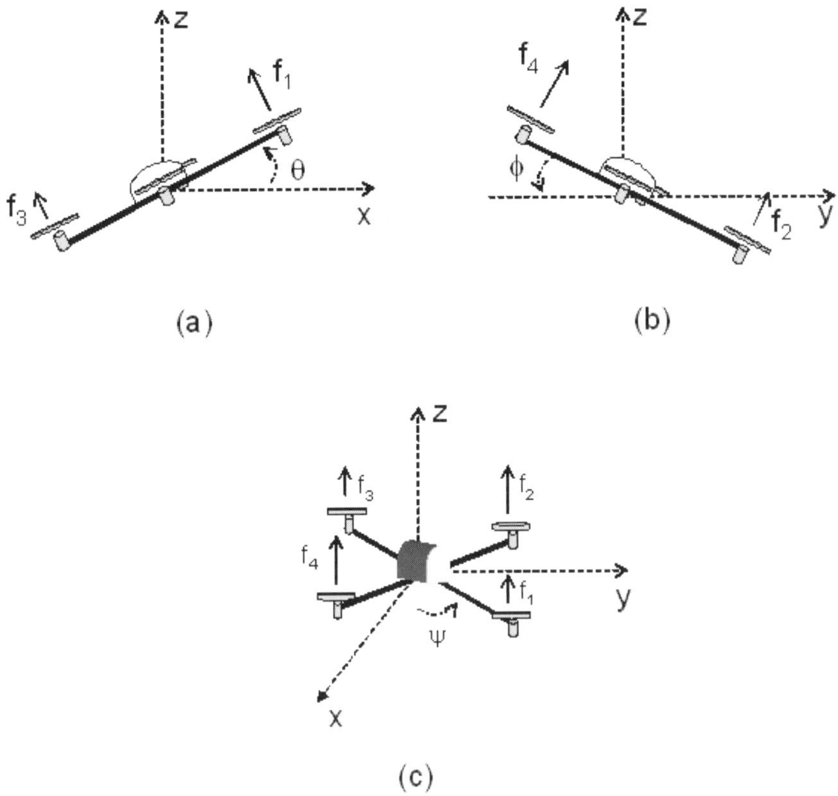

Figure 6.5. Configurations of the quad-rotor. (a) Pitch, (b) roll and (c) yaw angles.

trolled using the control strategy presented in Section 2. In the four-rotor helicopter, the throttle input is the sum of the thrust of each motor.

The two control signals are transmitted by a Futaba Skysport 4 radio. The control signals are referred as the throttle control input, $\bar{u}_1$, and the roll control input, u_2. These control signals are constrained to satisfy

$$\begin{aligned} 0.66\ [\mathrm{V}] &< \bar{u}_1 < 4.70\ [\mathrm{V}] \\ 1.23\ [\mathrm{V}] &< u_2 < 4.16\ [\mathrm{V}]. \end{aligned} \tag{6.82}$$

The radio and the PC (INTEL Pentium III) are connected using data acquisition cards (ADVANTECH PCL-818HG and PCL-726). The connection in the radio is directly made to the joystick potentiometers for the collective and roll controls.

The rotorcraft evolves freely in three-dimensional space without any flying stand. To measure the position (x, y, z) and orientation (ψ, θ, ϕ) of the rotorcraft we use the 3D tracker system (POLHEMUS) [22]. The Polhemus is connected via RS232 to the PC. This type of sensor is very sensitive to electromagnetic noise and we had to install it as far as possible from the electric motors and their drivers.

4.2. *Experiment and controller parameters tuning*

The controller parameters are selected using the following procedure. The parameters of the roll control input (u_2) are assigned while the throttle is in manual mode. The parameters of the roll control are adjusted in the following sequence. We first select the gain concerning the roll angular velocity $\dot{\phi}$. Due to the on-board gyros, this gain is relatively small. We next select the controller gain concerning the roll displacement ϕ. We wish the roll error to converge to zero fast, but without undesirable oscillations. The roll control input should also satisfy the constraints (6.82).

The controller gain concerning $\dot{x}$ and the amplitude of the saturation function are selected in such a way that the mini-aircraft reduces its speed in the x-axis fast enough. To complete the tuning of the roll control parameters we choose the gains concerning the x-displacement to obtain a satisfactory performance.

Finally we tuned the parameters of the throttle control to obtain a desired altitude. One of the controller parameters is used to compensate the gravity force which is estimated off-line using experimental data.

The computation of the control input requires the knowledge of the various angular and linear velocities. The sensor that is at our disposal only measures position and orientation. We have thus computed estimates of the angular and linear velocities by using the following approximation $\dot{q}_t \approx \frac{q_t - q_{t-T}}{T}$ where q is a given variable and T is the sampling period. In our experiment $T = 71\ ms$ due to limitations imposed by the measuring device. In order to obtain a good estimate of the angular and linear velocities and avoid abrupt changes in these signals we have introduced numerical filters.

Notice that since this mini-rotorcraft has soft blades, the tuning of the parameters can be done while holding the rotorcraft in the hand and wearing eye-protection glasses. This can certainly not be done with larger flying machines and therefore more simulations have to be performed before actually applying the controller to the real system.

The choice of the values for a, b, c, d were carried out satisfying the inequalities (6.44). However these parameters have been tuned experimentally in the sequence as they appear in the control input u_2 (see Table 6.1).

We wish to use our control law with a quad-rotor rotorcraft, this helicopter evolves in three-dimensional space and its movements are defined by the variables $(x, y, z, \psi, \theta, \phi)$. We are going to assimilate the altitude of the quad-rotor rotorcraft to the altitude of the PVTOL. It means that we will see y of the PVTOL as z of the quad-rotor rotorcraft.

The control objective is to make the rotorcraft hover at an altitude of 30 cm, *i.e.* we wish to reach the position $(x, z) = (0, 30)$ in centimeters while $\phi = 0°$. We also make the aircraft follow a simple horizontal trajectory. The gain values used for the control law are as in Table 6.1.

Table 6.1. Gain values used in the control law.

Phase	Control parameter	Value
1.- Altitude control	a_1	0.001
	a_2	0.002
2.- Roll control	a	0.523
	b	0.261
	c	0.13
	d	0.06
$1-2$	T	71 ms

Figure 6.6 shows the performance of the controller when applied to the helicopter. Take-off and landing were performed autonomously. Hovering at 30 cm, as well as the tracking of an horizontal trajectory were performed satisfactorily.

4.3. *Comments*

We have been able to test the control algorithm for stabilizing the PVTOL in a real-time application. We applied it to control the altitude, the roll angle and the horizontal displacement of a radio-controlled electrical four-rotor mini-helicopter. The simplicity of the algorithm helps in the implementation of the control algorithm. The results showed that the algorithm performs well. We were able to perform autonomously the tasks of take-off, hover and landing.

We aim at using visual servoing control for the mini-helicopter in fu-

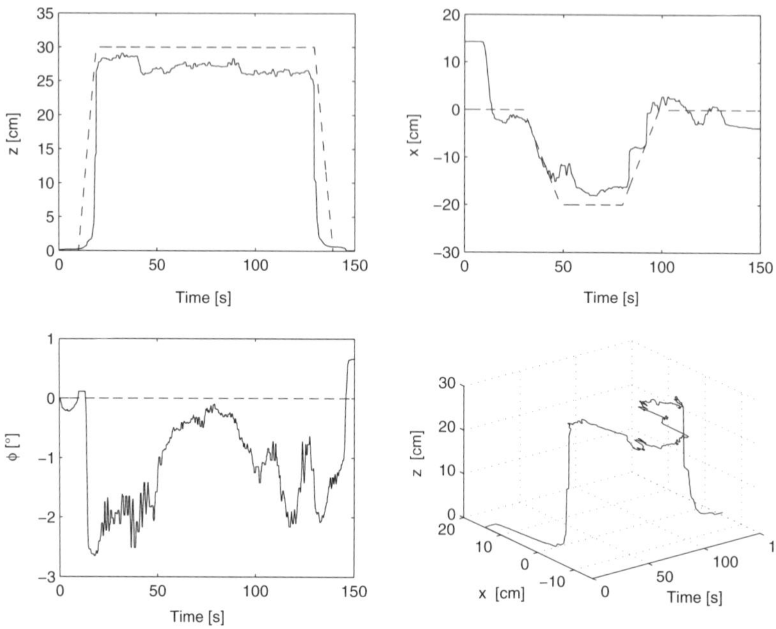

Figure 6.6. Positions x and z and orientation ϕ of the quad-rotor.

ture work. Image processing will introduce a considerable delay and the prediction-based control algorithm presented in Section 3 could be used to avoid instabilities in the position and orientation control of a flying vehicle.

In order to test this predictor, a new experiment has been carried out, as described next.

4.4. *Experimental control based on state prediction*

In this subsection we show that the a linear state-predictor based controller has a satisfactory behavior when applied to control the yaw displacement of a mini-helicopter. We use a mini-helicopter which has four rotors as shown in Section 4.1 (see Figure 6.7).

The control of the rotors is performed by sending the actions to the four motors through a digital/analog converter. Additionally, the system will receive commands from a small keyboard and will send periodically the system status to a host to monitor the system's variables and status.

The experimental validation of the proposed algorithm has been carried out on a novel real-time system, MaRTE OS, which allows the implementation of minimum real-time systems according to the standard POSIX.13 of the IEEE [23].

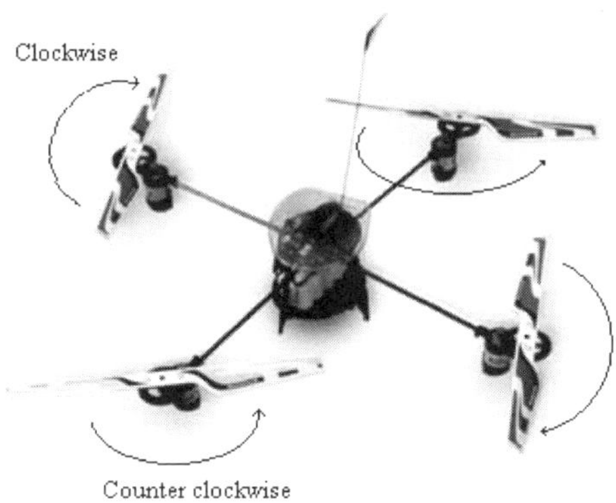

Figure 6.7. The four-rotor helicopter.

4.4.1. *Real-Time implementation*

We present, in this section, the characteristics and implementation of the real-time control system environment that we have used. We use an embedded system based on the MaRTE OS environment.

MaRTE OS [1] is a real-time kernel for embedded applications that follows the Minimal Real-Time POSIX.13 subset [23], providing both the C and Ada language POSIX interfaces. It allows cross-development of Ada and C real-time applications. Mixed Ada-C applications can also be developed, with a globally consistent scheduling of Ada tasks and C threads. MaRTE OS works in a cross development environment. The host computer is a Linux PC with the *gnat* and *gcc* compilers. The target platform is any bare machine based on any 386 PC or higher, with a floppy disk (or equivalent) for booting the application, but not requiring a hard disk. The kernel has a low-level abstract interface for accessing the hardware. This interface encapsulates operations for interrupt management, clock and timer management, and thread control.

The main applications of this kernel are industrial embedded systems developed in Ada. The hardware access facilities allow the implementation of specific device drivers in Ada style.

The final embedded system can be connected to other computers using RS232 or Ethernet drivers. Using these facilities, data can be sent to other applications to be monitored and analyzed. Also, commands from other applications can be received, using the same drivers, to modify the system behavior.

The development and the execution environments are shown in Figure 6.8. Figure 6.9 shows the interaction between the system and the external devices.

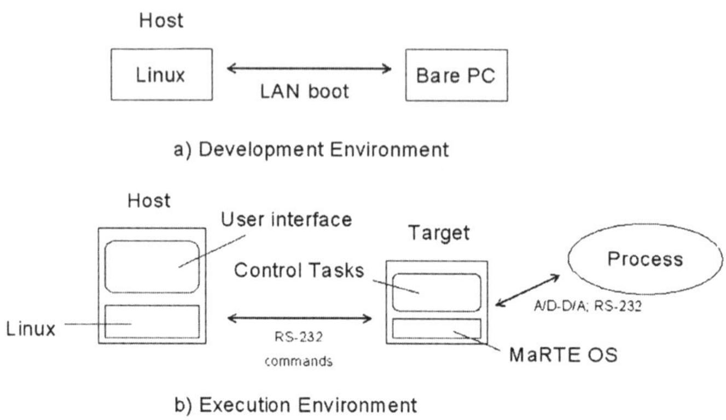

Figure 6.8. Environment of MaRTE OS.

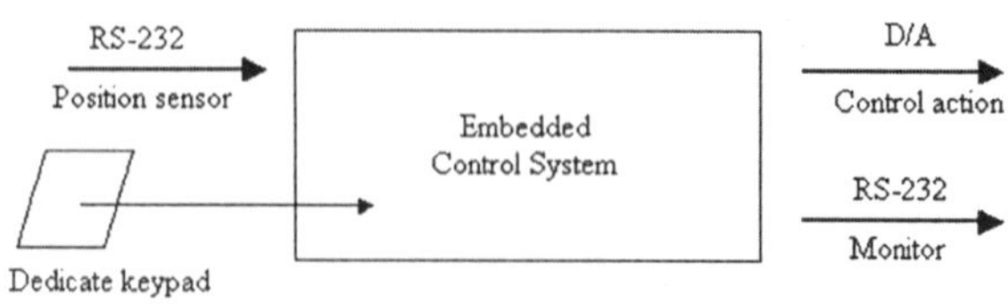

Figure 6.9. Interaction between the system and the external devices.

To design the real time control five main tasks have been defined.

- Control_Task: this periodic task gets information of the helicopter position and calculate the actions to be sent to the motors. This task has a period of 80 ms. The control actions are sent to a shared protected object which stores the system information. The actions are not sent directly to the motors.
- Send_Actions: this is a periodic task which is in charge of extracting the information from the control status and send the motor actions using the digital/analog converter. This task can introduce forced delays in the actions to be sent to the motors in order to test different control algorithms. The forced delays are introduced by getting actions calculated in previous periods when the delay is greater than the control period. If the delay is less than the period then an internal delay is executed.
- Monitor: This is a periodic task for control status monitoring. The task gets information from the shared object control status and send it to a RS232 line to be used by the host to visualize the control variables.
- User_Commands_Task: this task reads user commands from the keyboard and execute them. User commands can change the monitoring period, change control parameters or start and stop the control.
- Control_Status: this is a shared protected object where the tasks get or put information about the process.

Several drivers have been implemented to handle the RS-232 serial line, keyboard, and the digital/analog converters. Figure 6.10 shows the global application architecture and the modules involved in the final design.

In conclusion, there are three periodic tasks Control_Task, Send_Actions and Monitor and a sporadic task: User_Commands (see Table 6.2).

Table 6.2. Periodic tasks for the real-time application.

Task	Period	Priority	Offset
Control_Task	80ms	1	0
Send_Actions	80ms	2	10
Monitor	user defined	3	0

The control system has been implemented in Ada and developed in a Linux based host producing code for the target using the MaRTE OS. The

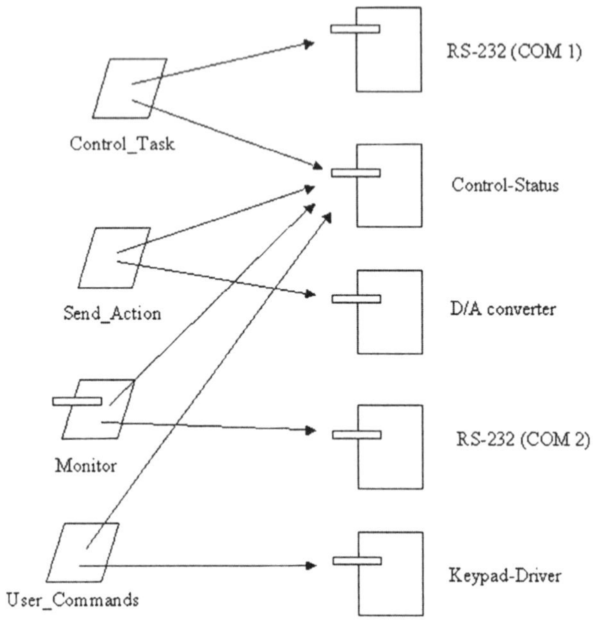

Figure 6.10. Application architecture.

image obtained is less than 300Kb and runs in a bare 386 machine.

4.4.2. *Experimental results*

The transfer function from the yaw-control input to the yaw-displacement has been identified by introducing a pulse input while the mini-helicopter was hovering. The obtained pulse response is shown in Figure 6.11.

We know that the mini-helicopter has an built-in gyro that introduces an angular velocity feedback. The transfer function without the gyro is basically a double integrator. However, the transfer function of the system including the angular velocity feedback, has a pole at the origin and a negative real pole.

We assumed that the system was represented by a second-order system with two parameters. Trying different values for the parameters we observed that the following model has a behavior that is close to the behavior of the

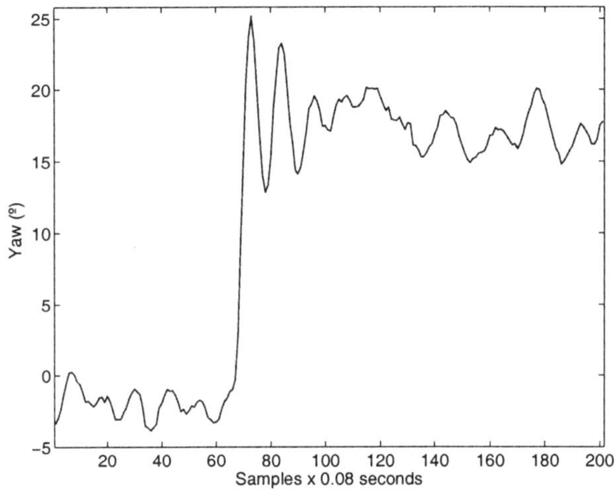

Figure 6.11. Pulse response of the system without measurement delay.

real system:

$$G(s) = \frac{200}{s(s+4)}. \tag{6.83}$$

The simple controller

$$u_k = 0.08(y^* - y_k), \tag{6.84}$$

where y^* is a reference signal, can be used to stabilize the model (6.83).

However, when there is a delay of 3 sampling periods (0.24 seconds) in the measurement of the yaw angular position, the controller becomes

$$u_k = 0.08(y^* - y_{k-3}) \tag{6.85}$$

and the closed loop system behavior is unstable as can be seen in Figure 6.12.

The discrete-time state-space representation for the model in (6.83), for $T = 0.08$ sec, is given by:

$$\begin{bmatrix} x_{k+1}^1 \\ x_{k+1}^2 \end{bmatrix} = \begin{bmatrix} 0.7261 & 0 \\ 0.2739 & 1 \end{bmatrix} \begin{bmatrix} x_k^1 \\ x_k^2 \end{bmatrix} + \begin{bmatrix} 0.5477 \\ 0.0923 \end{bmatrix} u_k \tag{6.86}$$

$$y_k = \begin{bmatrix} 0 & 6.25 \end{bmatrix} \begin{bmatrix} x_k^1 \\ x_k^2 \end{bmatrix}. \tag{6.87}$$

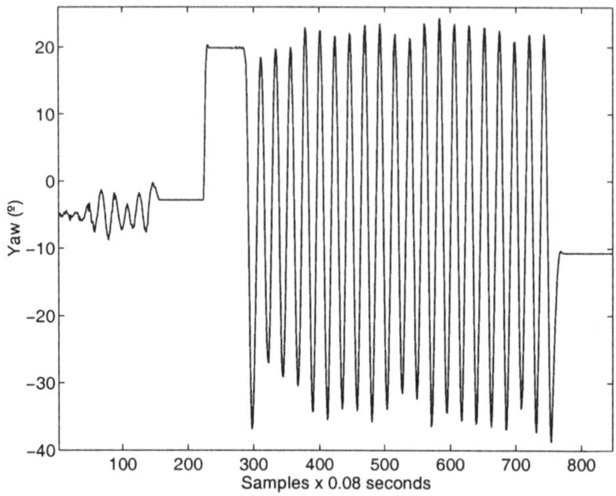

Figure 6.12. Output of the delayed system when using the controller 6.85 without prediction.

Since the state x_k is not measurable, we use the following observer

$$\begin{bmatrix} \hat{x}^1_{k+1} \\ \hat{x}^2_{k+1} \end{bmatrix} = \begin{bmatrix} 0.7261 & -2.7940 \\ 0.2739 & -1.0261 \end{bmatrix} \begin{bmatrix} \hat{x}^1_k \\ \hat{x}^2_k \end{bmatrix} + \begin{bmatrix} 0.4470 \\ 0.3242 \end{bmatrix} y_k + \begin{bmatrix} 0.5477 \\ 0.0923 \end{bmatrix} u_{k-3}. \tag{6.88}$$

The prediction of the state 3-steps ahead can be done using the following equation (see also Eq. (6.64))

$$x_p(k) = \begin{bmatrix} 0.3829 & 0 & 0.2888 & 0.3977 & 0.5477 \\ 0.6171 & 1 & 0.3512 & 0.2423 & 0.0923 \end{bmatrix} \begin{bmatrix} \hat{x}^1_k \\ \hat{x}^2_k \\ u_{k-3} \\ u_{k-2} \\ u_{k-1} \end{bmatrix}. \tag{6.89}$$

The control law in (6.84) (see also (6.65)) becomes:

$$u_k = 0.08(y^* - \begin{bmatrix} 0 & 6.25 \end{bmatrix} x_p(k)). \tag{6.90}$$

The yaw angular displacement of the mini-helicopter when using the above control law is shown in Figure 6.13. We have chosen y^* as a square wave function. As it can be seen, the system is stable.

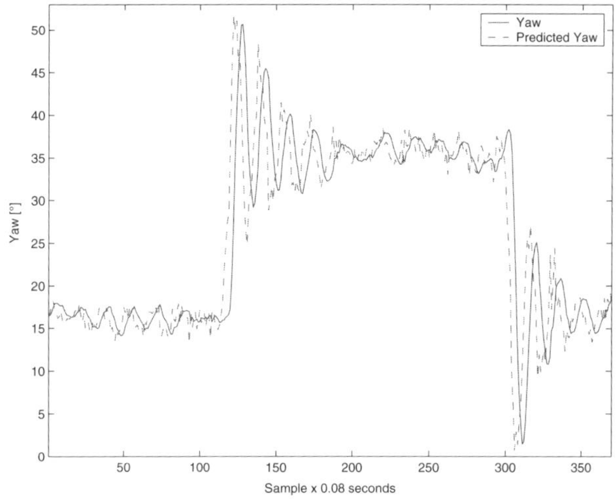

Figure 6.13. Closed-loop behavior using the prediction-based controller.

5. Conclusion

In this chapter we have investigated the control of mini-helicopters as an application to the control of continuous-time systems with delay. Two different problems have been addressed. First, a new control for the nonlinear model of the PVTOL has been presented and its properties tested by simulation and experiments. The design is based on Lyapunov analysis opening a path to further developments in nonlinear systems.

Second, we have proposed a discrete-time controller based on state feedback using the prediction of the state. A convergence analysis has been presented. This shows that the state converges to the origin in spite of uncertainties in the knowledge of the plant parameters, the system delay and even variations of the sampling period.

Both results have been experimentally tested in a laboratory prototype. The proposed control scheme has been implemented to control the yaw displacement of a real four-rotor mini-helicopter. Real-time experiments have shown a satisfactory performance of the proposed control scheme.

Bibliography

1. M. Aldea and M. González. MaRTE OS: An ada kernel for real-time embedded applications. *Proc. Int. Conf. Reliable Software Technologies, Ada-Europe*, 2001. Lecture Notes in Computer Science, LNCS 2043, Leuven, Belgium.

2. I. Fantoni and R. Lozano. Control of nonlinear mechanical systems. *European Journal of Control*, 7:328–348, 2001.
3. I. Fantoni and R. Lozano. *Non-linear control for underactuated mechanical systems*. Springer-Verlag, London, 2001.
4. C. Foias, H. Özbay, and A. Tannenbaum. *Robust Control of Infinite Dimensional Systems: Frequency Domain Methods*. Springer-Verlag, London, 1996.
5. E. Furutani and M. Araki. Robust stability of state-predictive and Smith control systems for plants with a pure delay. *Int. J. Robust and Nonlinear Control*, 8(18):907–919, 1998.
6. G.C. Goodwin and K. Sang Sin. *Adaptive Filtering: Prediction and Control*. Prentice-Hall, New Jersey, 1984.
7. J. Hauser, S. Sastry, and G. Meyer. Nonlinear control design for slightly nonminimum phase systems: application to V/STOL aircraft. *Automatica*, 28(4):665–679, 1992.
8. J.J. Loiseau, K. Mori, V. Van-Assche, and J.F. Lafay. Feedback realization of compensators for a class of time-delay systems. *Proc. 38th IEEE Conf. Decision and Control, Phoenix, Arizona*, 1999.
9. R. Lozano, P. Castillo, P. García, and A. Dzul. Robust prediction-based control for unstable delay systems: Application to the yaw control of a mini-helicopter. *Automatica*, 40(4):603–612, 2004.
10. A.Z. Manitius and A.W. Olbrot. Finite spectrum assignment problem for systems with delays. *IEEE Trans. Automatic Control*, 24(4):541–553, 1979.
11. L. Marconi, A. Isidori, and A. Serrani. Autonomous vertical landing on an oscillating platform: an internal-model based approach. *Automatica*, 38:21–32, 2002.
12. P. Martin, S. Devasia, and B. Paden. A different look at output tracking: control of a VTOL aircraft. *Automatica*, 32(1):101–107, 1996.
13. F. Mazenc and L. Praly. Adding integrations, satured controls, and stabilization for feedforward systems. *IEEE Trans. Automatic Control*, 41(11):1559–1578, 1996.
14. S. Mondié, M. Dambrine, and O. Santos. Approximation of control laws with distributed delays: a necessary condition for stability. *IFAC Conf. Systems, Structure and Control, Prague, Czech Rep.*, 2001.
15. S. Mondié, P. García, and R. Lozano. Resetting Smith predictor for the control of unstable systems with delay. *Proc. 15th IFAC World Congress, Barcelona, Spain*, 2002.
16. S. Mondié, R. Lozano, and J. Collado. Resetting process-model control for unstable systems with delay. *Proc. 40th IEEE Conf. Decision and Control, Orlando, Florida*, 2001.
17. S. Niculescu. *Delay effects on stability: a robust control approach*. Springer-Verlag, Heidelberg, 2001.
18. S. Niculescu and R. Lozano. On the passivity of linear delay systems. *IEEE Trans. Automatic Control*, 46(3):460–464, 2001.
19. R. Olfati-Saber. Global configuration stabilization for the VTOL aircraft with strong input coupling. *Proc. 39th IEEE Conf. Decision and Control, Sydney, Australia*, pages 3588–3589, 2000.

20. R. Olfati-Saber. Global configuration stabilization for the VTOL aircraft with strong input coupling. *IEEE Trans. Automatic Control*, 47(11):1949–1952, 2002.
21. Z.J. Palmor. Time delay compensation-Smith predictor and its modifications. In *The Control Handbook, W.S. Levine (Ed.)*, CRSC Press, pages 224–237, 1996.
22. Fastrack 3Space Polhemus. *User's Manual.* Colchester, Vermont, USA, 2001.
23. POSIX. IEEE standard for information technology - standardized application environment profile - POSIX Realtime Application Support (AEP). *IEEE Std 1003.13-1998*, 1999.
24. O.J.M. Smith. Closer control of loops with dead time. *Chem. Eng. Prog.*, 53:217–219, 1959.
25. A.R. Teel. A nonlinear small gain theorem for the analysis of control systems with saturation. *IEEE Trans. Automatic Control*, 41(9):1256–1270, 1996.
26. V. Van-Assche. *Etude et mise en oeuvre de commandes distribuées.* Ph.D. Thesis, École Central de Nantes, Université de Nantes, France, 2002.
27. K. Watanabe and M. Ito. A process model control for linear systems with delay. *IEEE Trans. Automatic Control*, 26(6):1261–1268, 1981.

CHAPTER 7

Global Attitude Control of Spacecraft Using Magnetic Actuators

A. Astolfi[1] and M. Lovera[2]

[1]*Department of Electrical and Electronic Engineering, Imperial College London, SW7 2AZ, United Kingdom, E-mail: a.astolfi@imperial.ac.uk*

[2]*Dipartimento di Elettronica e Informazione, Politecnico di Milano, Piazza Leonardo da Vinci 32, 20133 Milano, Italy, E-mail: lovera@elet.polimi.it*

The problem of global stabilization of the attitude dynamics for a magnetically actuated spacecraft is considered, and the cases of inertial pointing and Earth pointing are discussed. For the inertial pointing case, an almost global solution to the problem is obtained by means of static attitude and rate feedback. A local solution based on dynamic attitude feedback is also presented. For the Earth pointing case, almost global stabilization is achieved by means of an adaptive PD-like state feedback control law. Simulation results are presented to illustrate the performance of the proposed approaches.

1. Introduction

The attitude regulation problem for a rigid spacecraft can be posed in a number of different ways depending on the assumptions on the available actuators. If a rigid spacecraft, *i.e.* a spacecraft modeled by the Euler's equations and by a suitable parameterization of the attitude, is equipped with three independent actuators, a complete solution to the set point and tracking control problems is available. These problems have been mainly faced by means of PD-like control laws [10, 25], *i.e.* control laws which make use of the angular velocity and of the attitude. In other contributions the same problems have been solved using dynamic output feedback control laws [1, 4, 6, 13]. Similarly, if the spacecraft is underactuated [5], *i.e.* if only two independent actuators are available, the problem of attitude regulation is not solvable by means of continuous (static or dynamic) time-invariant control laws, whereas a time-varying control law, achieving local asymptotic (non-exponential) stability, has been proposed [19].

On the other hand, when the case of a spacecraft equipped with magnetic torquers is considered, a number of different issues arise. The operation of magnetic actuators is based on the interaction between a set of three orthogonal current-driven coils with the geomagnetic field; this has a number of implications which make the magnetic spacecraft control problem significantly different from the conventional attitude regulation one. First of all, magnetic actuators cannot provide three independent control torques at each time instant. In addition, the behavior of these actuators is intrinsically time-varying, as the control mechanism hinges on the variations of the Earth magnetic field along the spacecraft orbit. Nevertheless, attitude stabilization is possible because *on average* the system possesses strong controllability properties for a wide range of orbit inclinations.

The problems of analysis and design of magnetic control laws in the linear case, *i.e.* control laws for *nominal* operation of a satellite near its equilibrium attitude, have received significant attention in recent years. In particular, nominal and robust stability and performance have been studied, using either tools from periodic control theory exploiting the (quasi) periodic behavior of the system near an equilibrium [17, 18, 20, 21, 28] or other techniques aiming at developing suitable time varying controllers [8, 23].

On the other hand, global formulations of the magnetic attitude control problem have not been investigated to with the same attention and a number of problems remain open. If inertial pointing is considered, the global stabilization problems by means of full (or partial) state feedback is still theoretically unsolved. Note, in passing, that from a practical point of view these problems have an engineering solution, as demonstrated by the increasing number of applications of this approach to attitude control.

Similarly, the attitude regulation problem for Earth pointing spacecraft has been addressed exploiting periodicity assumptions on the system, hence resorting to standard passivity arguments to prove local asymptotic stabilizability of stable open loop equilibria [3, 9, 27]. Similar arguments have been used to analyze a state feedback control law for the particular case of an inertially spherical spacecraft [24] .

In the light of the above discussion, the aims of this chapter can be summarized as follows [16]:

- To obtain stability conditions for state feedback control laws achieving inertial pointing for magnetically actuated spacecraft. This result can be achieved by means of arguments similar to those in [25], provided that time-varying feedback laws are used and that the control gains satisfy certain scaling properties. In particular, with respect to previous work dealing only with the case of a magnetically actuated, isoinertial spacecraft [15], this chapter deals with a generic magnetically actuated satellite. For this problem, an almost global[a] stabilization result is given for the case of full state feedback.

- To extend the applicability of the partial (*i.e.* attitude only) state feedback results [1, 4, 13] from the case of a spacecraft with three independent controls to the case of a magnetically controlled spacecraft. For the case of partial state feedback, however, almost global stabilization can be guaranteed only in the case of isoinertial spacecraft.
- To show how similar stability conditions can be derived for control laws achieving Earth pointing for magnetically actuated spacecraft, taking also into account the effect of gravity gradient torques. For this problem, an almost global stabilization result is given for the case of full state feedback, resorting to an adaptive control approach.

Finally, note that the results presented herein do not rely on the (frequently adopted) periodicity assumption for the geomagnetic field along the considered orbit, which is correct only to first approximation [26].

The chapter is organized as follows. In Section 2 the considered model of a magnetically actuated spacecraft is presented. In Section 3 some results on the state and output feedback stabilization of inertially pointing magnetically actuated spacecraft are presented. The case of the stabilization of the relative Earth pointing equilibrium is discussed in Section 4. Finally, Section 5 and Section 6 present some simulation results and concluding remarks.

2. Mathematical Model of a Magnetically Actuated Satellite

The model of a rigid spacecraft with magnetic actuation can be described in various reference frames [26]. For the purpose of the present analysis, the following reference systems are adopted.

- Earth Centered Inertial reference axes (ECI). The origin of these axes

[a]Given a system $\dot{x} = f(x)$ we say that an equilibrium x_0 is almost globally asymptotically stable if it is locally asymptotically stable, all the trajectories of the system are bounded and the set of initial conditions giving rise to trajectories which do not converge to x_0 has zero Lebesgue measure.

is in the Earth's center. The x-axis is parallel to the line of nodes, that is the intersection between the Earth's equatorial plane and the plane of the ecliptic, and is positive in the Vernal equinox direction (Aries point). The z-axis is defined as being parallel to the Earth's geographic north-south axis and pointing north. The y-axis completes the right-handed orthogonal triad.

- Satellite body axes. The origin of these axes is in the satellite center of mass; the axes are assumed to coincide with the body's principal inertia axes.

The attitude dynamics can be expressed by the well known Euler's equations [26]

$$I\dot{\omega} = S(\omega)I\omega + T_{coils} + T_{dist} \tag{7.1}$$

where $\omega \in \mathbb{R}^3$ is the vector of spacecraft angular rates, expressed in body frame, $I \in \mathbb{R}^{3\times 3}$ is the inertia matrix, $S(\omega)$ is given by

$$S(\omega) = \begin{bmatrix} 0 & \omega_z & -\omega_y \\ -\omega_z & 0 & \omega_x \\ \omega_y & -\omega_x & 0 \end{bmatrix}, \tag{7.2}$$

$T_{coils} \in \mathbb{R}^3$ is the vector of external torques induced by the magnetic coils and $T_{dist} \in \mathbb{R}^3$ is the vector of external disturbance torques.

In turn, the attitude kinematics can be described by means of a number of possible parameterizations (see, *e.g.* [26]). The most common parameterization is given by the four Euler parameters (or quaternions), which lead to the following representation for the attitude kinematics

$$\dot{\boldsymbol{q}} = W(\omega)\boldsymbol{q} \tag{7.3}$$

where $\boldsymbol{q} = \begin{bmatrix} q_1 \; q_2 \; q_3 \; q_4 \end{bmatrix}^T = \begin{bmatrix} \mathbf{q}^T \; q_4 \end{bmatrix}^T$ is the vector of unit norm ($\boldsymbol{q}^T\boldsymbol{q} = 1$) Euler parameters and

$$W(\omega) = \frac{1}{2}\begin{bmatrix} 0 & \omega_z & -\omega_y & \omega_x \\ -\omega_z & 0 & \omega_x & \omega_y \\ \omega_y & -\omega_x & 0 & \omega_z \\ -\omega_x & -\omega_y & -\omega_z & 0 \end{bmatrix}. \tag{7.4}$$

It is useful to point out that Eq. (7.3) can be equivalently written as

$$\dot{\boldsymbol{q}} = \tilde{W}(\boldsymbol{q})\omega \tag{7.5}$$

where

$$\tilde{W}(\boldsymbol{q}) = \frac{1}{2}\begin{bmatrix} q_4 & -q_3 & q_2 \\ q_3 & q_4 & -q_1 \\ -q_2 & q_1 & q_4 \\ -q_1 & -q_2 & -q_3 \end{bmatrix}. \tag{7.6}$$

Note that the attitude of inertially pointing spacecraft is usually referred to the ECI reference frame.

The magnetic attitude control torques are generated by a set of three magnetic coils, aligned with the spacecraft principal inertia axes, which generate torques according to the law

$$T_{coils} = m_{coils} \times \tilde{b}(t), \tag{7.7}$$

where $\times$ denotes the vector cross product, $m_{coils} \in \mathbb{R}^3$ is the vector of magnetic dipoles for the three coils (which represent the actual control variables for the coils), $\tilde{b}(t) \in \mathbb{R}^3$ is the vector formed with the components of the Earth's magnetic field in the body frame of reference. Note that the vector $\tilde{b}(t)$ can be expressed in terms of the attitude matrix $A(\boldsymbol{q})$ (see [26] for details) and of the magnetic field vector expressed in the ECI coordinates, namely $\tilde{b}_0(t)$, as

$$\tilde{b}(t) = A(\boldsymbol{q})\tilde{b}_0(t), \tag{7.8}$$

and that the orthogonality of $A(\boldsymbol{q})$ implies $\|\tilde{b}(t)\| = \|\tilde{b}_0(t)\|$. The dynamics of the magnetic coils reduce to a very short electrical transient and can be neglected. The cross product in Eq. (7.7) can be expressed more simply as a matrix-vector product as

$$T_{coils} = S(\tilde{b}(t))m_{coils}. \tag{7.9}$$

Note that since $S(\tilde{b}(t))$ is structurally singular, as mentioned in the introduction, magnetic actuators do not provide full controllability of the system at each time instant. In particular, it is easy to see that $\text{rank}[S(\tilde{b}(t))] = 2$ (since $\|\tilde{b}_0(t)\| \neq 0$ along all orbits of practical interest for magnetic control) and that the kernel of $S(\tilde{b}(t))$ is given by the vector $\tilde{b}(t)$ itself, *i.e.* at each time instant it is *not* possible to apply a control torque along the direction of $\tilde{b}(t)$.

If a preliminary feedback of the form

$$m_{coils} = \frac{1}{\|\tilde{b}_0(t)\|^2} S^T(\tilde{b}(t))u \tag{7.10}$$

is applied to the system, where $u \in \mathbb{R}^3$ is a new control vector, the overall dynamics can be written as

$$\begin{aligned} \dot{\boldsymbol{q}} &= \tilde{W}(\boldsymbol{q})\omega \\ I\dot{\omega} &= S(\omega)I\omega + \Gamma(t)u \end{aligned} \tag{7.11}$$

where $\Gamma(t) = S(b(t))S^T(b(t)) \geq 0$ and $b(t) = \frac{1}{\|\tilde{b}_0(t)\|}\tilde{b}(t) = \frac{1}{\|\tilde{b}(t)\|}\tilde{b}(t)$. Similarly, let $\Gamma_0(t) = S(b_0(t))S^T(b_0(t)) \geq 0$ and $b_0(t) = \frac{1}{\|\tilde{b}_0(t)\|}\tilde{b}_0(t)$. Note, also, that $\Gamma(t)$ can be written as $\Gamma(t) = \mathcal{I}_3 - b(t)b(t)^T$, where $\mathcal{I}_3$ is the 3×3 identity matrix and $\Gamma(t) \geq 0$. We now prove a preliminary result which will be exploited in the next section.

Lemma 1: *Consider the system (7.11) and assume that the considered orbit for the spacecraft satisfies the condition*

$$\bar{\Gamma}_0 = \lim_{T\to\infty} \frac{1}{T} \int_0^T S(b_0(t))S^T(b_0(t))dt > 0.$$

Then, there exists $\omega_M > 0$ such that if $\|\omega\| < \omega_M$ for all $t > \bar{t}$, for some $0 < \bar{t} < \infty$, then

$$\bar{\Gamma} = \lim_{T\to\infty} \frac{1}{T} \int_0^T S(b(t))S^T(b(t))dt > 0 \tag{7.12}$$

along the trajectories of the system (7.11).

Proof: Consider first the particular case $\omega = 0$, which implies that $\boldsymbol{q} = \bar{\boldsymbol{q}} = \text{const}$. If $\bar{\Gamma}$ is singular there exists a nonzero vector $\bar{v}$ such that

$$\bar{v}^T \bar{\Gamma} \bar{v} = 0 \tag{7.13}$$

and $v_0 = A(\bar{\boldsymbol{q}})^T \bar{v}$. However, Eqs. (7.13) and (7.8) imply that

$$v_0^T \bar{\Gamma}_0 v_0 = 0, \tag{7.14}$$

which contradicts the assumption.

Finally, continuity arguments suffice to guarantee that (7.12) holds provided that ω is sufficiently small for all $t > \bar{t}$. □

Lemma 1 lends itself to a very simple physical interpretation. Condition $\det(\bar{\Gamma}) = 0$ defines the set of all trajectories along which average controllability is lost. Clearly this represents a non-generic condition, as it implies that the combination of the natural, on-orbit variability of $b_0(t)$ with the attitude motion of the satellite gives rise to a *constant* magnetic field vector ($b(t) = \bar{b}$) in the body reference frame. Such a condition, however, can only arise whenever the angular rate of the spacecraft is sufficiently large, hence average controllability in the sense of (7.12) is guaranteed for sufficiently small ω.

3. Magnetic Attitude Control for Inertially Pointing Satellites

3.1. *State feedback stabilization*

In this section a general stabilization result for a spacecraft with magnetic actuators is given in the case of full state feedback (attitude and rate). Without loss of generality in the following we assume that the equilibrium to be stabilized is given by $(\bar{\boldsymbol{q}}, 0)$, where $\bar{\boldsymbol{q}} = \begin{bmatrix} 0 & 0 & 0 & 1 \end{bmatrix}^T$ and we denote by $\mathrm{CN}(A)$ the condition number of the matrix A.

Proposition 1: *Consider the magnetically actuated spacecraft described by (7.11) and the control law*

$$u = -(\varepsilon^2 k_p \mathbf{q} + \varepsilon k_v I \omega). \tag{7.15}$$

Suppose that $0 < \bar{\Gamma}_0 < \mathcal{I}_3$. *Then there exist* $\varepsilon^\star > 0$, $k_p > 0$ *and* $k_v > 0$ *with*

$$k_v^2 > k_p \frac{\sigma_{min}^2(\bar{\Gamma})}{\sigma_{min}(I)} \sqrt{\mathrm{CN}(\bar{\Gamma})} \tag{7.16}$$

such that for any $0 < \varepsilon < \varepsilon^\star$ *the control law (7.15) ensures that* $(\bar{\boldsymbol{q}}, 0)$ *is a locally exponentially stable equilibrium for the closed-loop system (7.11)–(7.15). Moreover, all trajectories of (7.11)–(7.15) are such that* $\mathbf{q} \to 0$ *and* $\omega \to 0$.

Proof: To begin with we prove that for all $k_p > 0$ and $k_v > 0$ there exists $\varepsilon > 0$ such that for the closed-loop system (7.11)–(7.15) $\bar{\Gamma} > 0$. To this end, consider the ω-subsystem only and the function

$$V_1 = \frac{\lambda}{2} \omega^T I^2 \omega - \frac{1}{2} \omega^T I A(\bar{q}) M(t) A(\bar{q})^T \omega, \tag{7.17}$$

where $\lambda > 0$,

$$M(t) = \int_0^t (b_0(\tau) b_0(\tau)^T - N) d\tau \tag{7.18}$$

and $N \geq 0$ is a constant matrix. The assumption $\bar{\Gamma}_0 < \mathcal{I}_3$ implies that it is possible to select N such that $-\sigma \mathcal{I}_3 \leq M(t) \leq \sigma \mathcal{I}_3$ for some positive σ.

Note that V_1 is positive definite for sufficiently large λ. The time derivative of V_1 is given by

$$\dot{V}_1 = -\omega^T I A(\boldsymbol{q}) Q A(\boldsymbol{q})^T I \omega - \varepsilon^2 k_p \omega^T I (\lambda \mathcal{I}_3 - A(\boldsymbol{q}) M(t) A(\boldsymbol{q})^T) \Gamma(t) \mathbf{q}, \tag{7.19}$$

where

$$Q = (\varepsilon k_v \lambda \Gamma_0(t) - \frac{\varepsilon k_v}{2} M(t) \Gamma_0(t) - \frac{\varepsilon k_v}{2} \Gamma_0(t) M(t) + b_0 b_0^T - N). \tag{7.20}$$

Introduce now the time varying vectors $b_1(t)$ and $b_2(t)$ such that $b_i^T b_j = \delta_{ij}$, where δ_{ij} is the Kronecker delta and $i, j = 0, 1, 2$, and let

$$\tilde{Q} = \begin{bmatrix} b_0^T \\ b_1^T \\ b_2^T \end{bmatrix} Q \begin{bmatrix} b_0 & b_1 & b_2 \end{bmatrix}. \tag{7.21}$$

Then, it is easy to show that for any $\varepsilon > 0$ there exists a $\lambda > 0$ and sufficiently large such that $\tilde{Q}$ (and, therefore, Q) is positive definite. As a result

$$\begin{aligned} \dot{V}_1 \le & -\psi(\varepsilon, \lambda) \omega^T \omega + \frac{\varepsilon^2 k_p \lambda}{2} \omega^T I^2 \omega + \frac{\varepsilon^2 k_p \lambda}{2} \mathbf{q}^T \Gamma^2(t) \mathbf{q} \\ & + \frac{\varepsilon^2 k_p}{2} \omega^T I^2 \omega + \frac{\varepsilon^2 k_p}{2} \mathbf{q}^T \Gamma(t) A(\boldsymbol{q}) M^2(t) A(\boldsymbol{q})^T \Gamma(t) \mathbf{q}, \end{aligned} \tag{7.22}$$

where $\psi(\varepsilon, \lambda) > \varepsilon$ for all $\varepsilon > 0$ and $\lambda > 0$ and sufficiently large, from which one has

$$\dot{V}_1 \le -(\varepsilon - \varepsilon^2 k_p \lambda \sigma_{max}^2(I) - \varepsilon^2 k_p \sigma_{max}^2(I)) \omega^T \omega + \frac{\varepsilon^2 k_p \lambda}{2} + \frac{\varepsilon^2 k_p \sigma^2}{2}. \tag{7.23}$$

This, in turn, implies that for any $\omega_M > 0$ there exists $\varepsilon > 0$ such that $\|\omega\| < \omega_M$ for sufficiently large t, and therefore, by Lemma 1, $\bar{\Gamma} > 0$.

Introduce now the coordinates transformation

$$\boldsymbol{z}_1 = \boldsymbol{q} \qquad z_2 = \frac{\omega}{\varepsilon} \tag{7.24}$$

(so that $\mathbf{z}_1 = \mathbf{q}$ and $z_{14} = q_4$) in which the system (7.11) is described by the equations

$$\begin{aligned} \dot{\boldsymbol{z}}_1 &= \varepsilon \tilde{W}(\boldsymbol{z}_1) z_2 \\ I \dot{z}_2 &= \varepsilon S(z_2) I z_2 + \varepsilon \Gamma(t)(-k_p \mathbf{z}_1 - k_v I z_2). \end{aligned} \tag{7.25}$$

System (7.25) satisfies all the hypotheses[b] for the applicability of the generalized averaging theory [12, Theorem 10.5], which yields the averaged

[b] In particular, it is easy to verify that the Jacobian of the difference between the right hand sides of Eqs. (7.25) and (7.26) has zero average.

system

$$\begin{aligned} \dot{\boldsymbol{z}}_1 &= \varepsilon \tilde{W}(\boldsymbol{z}_1) z_2 \\ I\dot{z}_2 &= \varepsilon S(z_2) I z_2 + \varepsilon \bar{\Gamma}(-k_p \mathbf{z}_1 - k_v I z_2). \end{aligned} \tag{7.26}$$

As a result, there exists $\varepsilon^\star > 0$ such that for any $0 < \varepsilon < \varepsilon^\star$ the trajectories of system (7.26) are close to the trajectories of system (7.25). Consider now the positive definite function

$$V_2(z_2) = \frac{1}{2} z_2^T I^2 z_2, \tag{7.27}$$

and its time derivative

$$\dot{V}_2 = \varepsilon z_2^T \bar{\Gamma}(-k_p \mathbf{z}_1 - k_v I z_2) \tag{7.28}$$

and note that, for any $\alpha > 0$,

$$\dot{V}_2 \leq -\varepsilon(k_v - \frac{k_p}{2\alpha}) z_2^T I \bar{\Gamma} I z_2 + \varepsilon k_p \frac{\alpha}{2} \mathbf{z}_1^T \bar{\Gamma} \mathbf{z}_1. \tag{7.29}$$

The positive definiteness of $\bar{\Gamma}$ and the boundedness of $\boldsymbol{z}_1$ imply that, for a proper selection of k_p, k_v and α

$$\dot{V}_2 \leq -r V_2 + d \tag{7.30}$$

for some constants $r > 0$ and $d > 0$. In particular, for[c] $\alpha = \alpha^\star = \frac{k_p}{k_v}$, Eq. (7.30) implies that along the trajectories of the closed-loop system one has

$$\|z_2(t)\|^2 \leq \frac{k_p^2}{k_v^2} \frac{\mathrm{CN}(\bar{\Gamma})}{\sigma_{min}^2(I)} = K^2, \tag{7.31}$$

for all $t \geq t^\star$ and for some $0 \leq t^\star \leq \infty$. As a result, for any $K > 0$ the set

$$Z_K = \{(\boldsymbol{z}_1, z_2) : \|z_2\| < K\} \tag{7.32}$$

is attractive and positively invariant. Observe that K can be made arbitrarily small by a suitable choice of k_p and k_v.

We now prove that all trajectories of system (7.26) starting in the set (7.32) are such that $\mathbf{z}_1 \to 0$ and $z_2 \to 0$. To this end, consider the Lyapunov function

$$V_3(\boldsymbol{z}_1, z_2) = \frac{1}{2} k_p (\mathbf{z}_1^T \mathbf{z}_1 + (z_{14} - 1)^2) + \frac{1}{2} z_2^T \bar{\Gamma}^{-1} z_2, \tag{7.33}$$

and its time derivative

$$\dot{V}_3 = \varepsilon z_2^T \bar{\Gamma}^{-1} S(z_2) I z_2 - \varepsilon k_v z_2^T I z_2. \tag{7.34}$$

[c]It is easy to see that this selection of α is *optimal*.

Note that

$$\dot{V}_3 \leq \varepsilon K \sigma_{min}^2(\bar{\Gamma}) z_2^T I z_2 - \varepsilon k_v z_2^T I z_2, \tag{7.35}$$

which is negative if condition (7.16) holds. As a result, $z_2 \to 0$ and, applying LaSalle's invariance principle, $\mathbf{z}_1 \to 0$.

Finally, consider the linear approximation of system (7.26) around the equilibrium $(\bar{\boldsymbol{q}}, 0)$, which is given by

$$\begin{aligned} \dot{\mathbf{z}}_1 &= \tfrac{1}{2}\varepsilon z_2 \\ I\dot{z}_2 &= -\varepsilon\bar{\Gamma}(k_p \mathbf{z}_1 + k_v I z_2). \end{aligned} \tag{7.36}$$

It is easy to verify that

$$V_L(\boldsymbol{z}_1, z_2) = 2k_p \mathbf{z}_1^T \mathbf{z}_1 + z_2^T \bar{\Gamma}^{-1} z_2 \tag{7.37}$$

is a Lyapunov function for the linear system (7.36), so the convergence of the trajectories of the closed-loop system is locally exponential. □

Remark 1: Proposition 1 clarifies the main difference between magnetic attitude control and the fully actuated case. It has been shown [25] that whenever three independent torques are available, the state feedback problem can be solved via a PD control law and that almost global stability of the closed-loop system can be guaranteed for *any* choice of $k_p > 0$ and $k_v > 0$. This is not the case for magnetic attitude control, as the proportional and derivative actions must meet the scaling condition defined by ε in order to guarantee closed loop stability. In this respect, this result provides a very useful guideline for the design of magnetic controllers in practical cases, as it combines the simplicity of a state feedback control law [25] with an explicit stability condition. On the other hand, the choice of a suitable value for ε cannot be carried out on the basis of Proposition 1 only, but is likely to require some iterations of the tuning process.

When considering the problem of nonlinear attitude control, actuators saturations usually play an important role. In this respect it is worth pointing out that the above state feedback magnetic control law can be readily modified to deal with saturation of the magnetic coils, as expressed in the following statement.

Corollary 1: *Consider the system (7.11) and the state feedback control law*[d]

$$u = -\varepsilon^2 k_p \mathbf{q} - \varepsilon\beta \mathrm{sat}(k_v \frac{I\omega}{\beta}), \tag{7.38}$$

[d] By sat(·) we indicate a continuous saturating function limited between −1 and 1.

with $\beta > 0$ *and suppose that* $0 < \bar{\Gamma}_0 < \mathcal{I}_3$. *Then for any* $\rho > 0$ *there exist* $\varepsilon^\star > 0$, $k_p > 0$ *and* $k_v > 0$, *satisfying condition (7.16), such that for any* $0 < \varepsilon < \varepsilon^\star$ *the control law (7.38) ensures that* $(\bar{\mathbf{q}}, 0)$ *is a locally exponentially stable equilibrium for the closed-loop system (7.11)–(7.38), all trajectories of (7.11)–(7.38) are such that* $\mathbf{q} \to 0$, $\omega \to 0$ *and*

$$|u_i| \leq \rho. \tag{7.39}$$

Proof: The proof of the first two statements is similar to the proof of Proposition 1. To prove the bound (7.39), note that

$$|u_i| \leq \varepsilon^2 k_p + \varepsilon\beta, \tag{7.40}$$

and this can be made arbitrarily small by a proper selection of the design parameters. □

Remark 2: The bound (7.39) on the signals u implies the bound

$$|m_{coils_i}| \leq \|b_o\| \, \rho \tag{7.41}$$

on the actual control inputs m_{coils}.

Remark 3: The parameter β in the control law (7.38) is used only to assign the amplitude of the saturation function.

Remark 4: An interesting particular case is the one of a spacecraft that has an inertia matrix which is proportional to the identity matrix $\mathcal{I}_3$, *i.e.*

$$I = \kappa \mathcal{I}_3 \tag{7.42}$$

for some $\kappa > 0$, so that $S(\omega)I\omega = 0$ for all ω. In this case one can achieve convergence of the trajectories of the closed-loop system (7.11)–(7.15) for any positive k_p and k_v and $0 < \varepsilon < \varepsilon^\star$, as the derivative of the Lyapunov function V_3 reduces to

$$\dot{V}_3 = -\varepsilon k_v z_2^T I z_2. \tag{7.43}$$

The same considerations apply to the closed-loop system (7.11)–(7.38), *i.e.* in the presence of saturations.

3.2. *Stabilization without rate feedback*

The ability of ensuring attitude regulation without rate feedback is of great importance from a practical point of view. In this section, an approach similar to the one for the case of a fully actuated spacecraft [1] is followed for the case of magnetic attitude control, and an *almost global* result is given in the case of an isoinertial spacecraft. In addition, a local stability result is derived for a generic satellite.

Proposition 2: *Consider the system (7.11) with I such that (7.42) holds, and the control law*

$$\begin{aligned} \dot{\delta} &= \alpha(\boldsymbol{q} - \varepsilon\lambda\delta) \\ u &= -\varepsilon^2(k_p\mathbf{q} + k_v\alpha\lambda\tilde{W}^T(\boldsymbol{q})(\boldsymbol{q} - \varepsilon\lambda\delta)). \end{aligned} \tag{7.44}$$

Suppose that $0 < \bar{\Gamma}_0 < \mathcal{I}_3$. Then there exists $\varepsilon^\star > 0$, $k_p > 0$, $k_v > 0$, $\alpha > 0$ and $\lambda > 0$ such that for any $0 < \varepsilon < \varepsilon^\star$ the control law renders the equilibrium $(\bar{\boldsymbol{q}}, 0, \frac{1}{\varepsilon\lambda}\bar{\boldsymbol{q}})$ of the closed-loop system (7.11)–(7.44) locally exponentially stable. Moreover, the equilibrium is almost globally asymptotically stable.

Proof: As in the case of the proof of Proposition 1, introduce the coordinates transformation

$$\boldsymbol{z}_1 = \boldsymbol{q} \qquad z_2 = \frac{\omega}{\varepsilon} \qquad z_3 = \varepsilon\delta \tag{7.45}$$

in which the system (7.11)–(7.44) is described by the equations

$$\begin{aligned} \dot{\boldsymbol{z}}_1 &= \varepsilon\tilde{W}(\boldsymbol{z}_1)z_2 \\ \dot{z}_2 &= -\tfrac{\varepsilon}{\kappa}\Gamma(t)(k_p\mathbf{z}_1 + k_v\alpha\lambda\tilde{W}^T(\boldsymbol{z}_1)(\boldsymbol{z}_1 - \lambda z_3)) \\ \dot{z}_3 &= \varepsilon\alpha(\boldsymbol{z}_1 - \lambda z_3). \end{aligned} \tag{7.46}$$

System (7.46) satisfies all the hypotheses for the applicability of generalized averaging theory ([12, Theorem 10.5]), which leads to the averaged system

$$\begin{aligned} \dot{\boldsymbol{z}}_1 &= \varepsilon\tilde{W}(\boldsymbol{z}_1)z_2 \\ \dot{z}_2 &= -\tfrac{\varepsilon}{\kappa}\bar{\Gamma}(k_p\mathbf{z}_1 + k_v\alpha\lambda\tilde{W}^T(\boldsymbol{z}_1)(\boldsymbol{z}_1 - \lambda z_3)) \\ \dot{z}_3 &= \varepsilon\alpha(\boldsymbol{z}_1 - \lambda z_3). \end{aligned} \tag{7.47}$$

Consider now the Lyapunov function

$$\begin{aligned} V_4(\boldsymbol{z}_1, z_2, z_3) = {} & \frac{1}{2}k_p(\mathbf{z}_1^T\mathbf{z}_1 + (z_{14} - 1)^2) + \frac{1}{2}z_2^T\bar{\Gamma}^{-1}z_2 \\ & + \frac{1}{2}k_v(\boldsymbol{z}_1 - \varepsilon\lambda z_3)^T(\boldsymbol{z}_1 - \varepsilon\lambda z_3), \end{aligned} \tag{7.48}$$

yielding

$$\dot{V}_4 = -k_v\lambda(\boldsymbol{z}_1 - \varepsilon\lambda z_3)^T(\boldsymbol{z}_1 - \varepsilon\lambda z_3). \tag{7.49}$$

As a result, $\boldsymbol{z}_1 - \varepsilon\lambda z_3 \to 0$, hence $z_2 \to 0$ and $\mathbf{z}_1 \to 0$, provided that

$$\bar{\Gamma} = \lim_{T\to\infty}\frac{1}{T}\int_0^T S(b(t))S^T(b(t))dt > 0.$$

Note now that, by Lemma 1, any trajectory starting sufficiently close to the equilibrium $(\bar{q}, 0, \frac{1}{\varepsilon\lambda}\bar{q})$ is such that $\bar{\Gamma} > 0$, and this together with the above Lyapunov arguments proves local exponential stability of the equilibrium.

To complete the proof, we need to show that the set of initial conditions yielding *bad* trajectories, *i.e.* trajectories that do not converge to the equilibrium $(\bar{\boldsymbol{q}}, 0, \frac{1}{\varepsilon\lambda}\bar{\boldsymbol{q}})$, has zero (Lebesgue) measure. These *bad* trajectories are those converging to the equilibrium $(-\bar{\boldsymbol{q}}, 0, -\frac{1}{\varepsilon\lambda}\bar{\boldsymbol{q}})$ and those for which $\bar{\Gamma}$ is singular. Note that the equilibrium $(-\bar{\boldsymbol{q}}, 0, -\frac{1}{\varepsilon\lambda}\bar{\boldsymbol{q}})$ is unstable, hence the set of initial conditions yielding trajectories converging to $(-\bar{\boldsymbol{q}}, 0, -\frac{1}{\varepsilon\lambda}\bar{\boldsymbol{q}})$ is composed only by the associated stable manifold, which has zero measure. Finally, the trajectories such that $\bar{\Gamma}$ is singular are non-generic, hence the equilibrium is almost globally asymptotically stable. □

Remark 5: The signal u generated by the output feedback control law (7.44) is bounded, provided that $\delta(0)$ is properly selected. Indeed, if

$$\varepsilon\lambda\delta_i(0) \in [-1, 1] \tag{7.50}$$

then

$$|\varepsilon\lambda\delta_i(t)| \in [-1, 1] \tag{7.51}$$

and therefore

$$|u_i(t)| \leq \varepsilon^2(k_p + 2k_v). \tag{7.52}$$

In particular, condition (7.50) holds if $\delta(0) = \boldsymbol{q}(0)$ and $\varepsilon\lambda \leq 1$, or if $\delta(0) = 0$.

Proposition 2 holds only in the case of an isoinertial spacecraft, *i.e.* if I is such that (7.42) holds. In the general case, it is possible to prove the following weaker result.

Proposition 3: *Consider the system (7.11) and the control law*

$$\begin{aligned} \dot{\delta} &= \alpha(\boldsymbol{q} - \varepsilon\lambda\delta) \\ u &= -\varepsilon^2(k_p I^{-1}\mathbf{q} + k_v\alpha\lambda\tilde{W}^T(\boldsymbol{q})(\boldsymbol{q} - \varepsilon\lambda\delta)), \end{aligned} \tag{7.53}$$

and suppose that $0 < \bar{\Gamma}_0 < \mathcal{I}_3$. *Then there exists* $\varepsilon^\star > 0$ *such that for any* $0 < \varepsilon < \varepsilon^\star$ *the control law renders the equilibrium* $(\bar{\boldsymbol{q}}, 0, \frac{1}{\varepsilon\lambda}\bar{\boldsymbol{q}})$ *of the closed-loop system (7.11)–(7.53) locally exponentially stable.*

Proof: The claim can be proved by introducing the coordinates transformation (7.45) and considering the Lyapunov function

$$V_5(\boldsymbol{z}_1, z_2, z_3) = \frac{1}{2}k_p\mathbf{z}_1^T\mathbf{z}_1 + \frac{1}{2}z_2^T I\bar{\Gamma}^{-1}I z_2 + \frac{1}{2}k_v\alpha\lambda(\mathbf{z}_1 - \varepsilon\lambda z_{3r})^T I(\mathbf{z}_1 - \varepsilon\lambda z_{3r})$$

for the linear approximation of system (7.11)–(7.53) around the equilibrium $(\bar{q}, 0, \frac{1}{\varepsilon\lambda}\bar{q})$. □

4. Magnetic Attitude Control for Earth Pointing Satellites

4.1. *Mathematical model*

With respect to the case of inertial pointing, some modifications to the mathematical model must be made. First of all, the orbital reference frame is defined, as follows. The origin of these axes is in the Earth's center. The x-axis is parallel to the line of nodes, that is the intersection between the Earth's equatorial plane and the plane of the ecliptic, and is positive in the Vernal equinox direction (Aries point). The z-axis is defined as being parallel to the Earth's geographic north-south axis and pointing north. The y-axis completes the right-handed orthogonal triad..

Only the case of a spacecraft in a circular orbit is considered; the (constant) orbital angular rate will be denoted by ω_0. In the following the unit vectors corresponding to the orbital axes will be denoted with e_x, e_y and e_z respectively, with the superscript ^{o} (^{b}) when considering the components of the unit vectors along the orbital (body) axes.

We devote specific attention to gravity gradient torques, as they play a major role in defining the equilibria of relative motion for Earth pointing spacecraft. For a satellite in circular orbit, the gravity gradient torque can be written as

$$T_{gg} = 3\omega_0^2 S(Ie_x^b)e_x^b \tag{7.54}$$

where vector e_x^b defines the local Nadir direction in the body frame.

The focus will be on the *relative* kinematics rather than on the inertial kinematics. In other words, we will be concerned with representations of the attitude of the spacecraft with respect to the (rotating) orbital axes. Therefore, the attitude kinematics will be described in terms of the following representation

$$\dot{q} = W(\omega_r)q \tag{7.55}$$

where ω_r is the satellite angular rate relative to the orbital axes, in body frame, *i.e.*

$$\omega_r = \omega - \omega_t, \tag{7.56}$$

and $\omega_t = -\omega_0 e_z^b$. Letting now $A(q)$ the attitude matrix relating the orbital and the body frames, one has that

$$e_x^b = A(q)e_x^o = A(q)\begin{bmatrix}1\\0\\0\end{bmatrix} \tag{7.57}$$

and similarly for e_y^b,e_z^b. Finally, note that $A(q) = \mathcal{I}_3$ for $q = \pm\bar{q} = \pm\left[0\,0\,0\,1\right]^T$.

The overall dynamics for the attitude of an Earth pointing satellite can be written as

$$\begin{aligned}\dot{\boldsymbol{q}} &= W(\omega_r)\boldsymbol{q}\\ I\dot{\omega} &= S(\omega)I\omega + 3\omega_0^2 S(e_x^b)Ie_x^b + \Gamma(t)u.\end{aligned} \tag{7.58}$$

4.2. *State feedback control*

In this section an almost globally convergent adaptive PD-like control law for Earth pointing magnetic attitude regulation is proposed. In particular, Lemma 1 shows that for sufficiently small angular rates the system (7.58) has "average" controllability properties as expressed by the full rank of the matrix $\bar{\Gamma}$. This fact plays a major role in the derivation of the following, preliminary result.

Proposition 4: *Consider the system (7.58) and the control law*

$$u = -\varepsilon k_v \omega_r. \tag{7.59}$$

Suppose that $0 < \bar{\Gamma}_0 < \mathcal{I}_3$. *Then, for all* $\varepsilon > 0$ *and* $k_v > 0$ *there exists* $\bar{t} > 0$ *such that for all* $t > \bar{t}$

$$\bar{\Gamma}(t) = \frac{1}{t}\int_0^t \Gamma(\tau)d\tau > 0. \tag{7.60}$$

Proof: Consider the function [11, 25]

$$\begin{aligned}V_1 = &\frac{\lambda}{2}\left[\omega_r^T I\omega_r + 3\omega_0^2(e_x^T I e_x - I_x) + \omega_0^2(I_z - e_z^T I e_z)\right]\\ &-\frac{1}{2}\omega_r^T I A(\boldsymbol{q})M_0(t)A(\boldsymbol{q})^T I\omega_r,\end{aligned} \tag{7.61}$$

where $\lambda > 0$,

$$M_0(t) = \int_0^t (b_0(\tau)b_0(\tau)^T - N_0)d\tau, \tag{7.62}$$

and $N_0 \geq 0$ is a constant matrix. The assumption $\bar{\Gamma}_0 < \mathcal{I}_3$ implies that it is possible to select N_0 such that $-\sigma\mathcal{I}_3 \leq M_0(t) \leq \sigma\mathcal{I}_3$ for some positive σ.

Note that V_1 is positive definite for sufficiently large λ. The time derivative of V_1 is given by

$$\dot{V}_1 = -\omega_r^T A(\boldsymbol{q})QA(\boldsymbol{q})^T\omega_r - \omega_r^T IM(t)(S(I\omega_t) + S(\omega_t)I + IS(\omega_t))\omega_r \\ -\omega_r^T IM(t)(S(I\omega_t)I\omega_t + T_{gg}) \tag{7.63}$$

where

$$Q = (\varepsilon k_v \lambda \Gamma_0(t) + A(\boldsymbol{q})^T I(-\frac{\varepsilon k_v}{2}M(t)\Gamma(t) - \frac{\varepsilon k_v}{2}\Gamma(t)M(t) + bb^T - N)IA(\boldsymbol{q})).$$

Introduce the time varying vectors $b_1(t)$ and $b_2(t)$ such that $b_i^T b_j = \delta_{ij}$, where δ_{ij} is the Kronecker delta and $i, j = 0, 1, 2$, and let

$$\tilde{Q} = \begin{bmatrix} b_0^T \\ b_1^T \\ b_2^T \end{bmatrix} Q \begin{bmatrix} b_0 \ b_1 \ b_2 \end{bmatrix}. \tag{7.64}$$

Then, it can be shown that there exists a $\lambda > 0$ and sufficiently large such that $\tilde{Q}$ (and, therefore, Q) is positive definite. This, in turn, implies that for any $\omega_M > 0$ there exists $\lambda > 0$ such that $\|\omega_r\| < \omega_M$ for sufficiently large t, and therefore, by Lemma 1, $\bar{\Gamma} > 0$. □

The main result concerning Earth pointing attitude regulation is given in the following proposition.

Proposition 5: *Consider the system (7.11) and the control law*

$$u = \begin{cases} -\varepsilon k_v \omega_r & t \leq \bar{t} \\ -\hat{\Gamma}_{av}^{-1}(\varepsilon^2 k_p q_r + \varepsilon k_v \omega_r) & t > \bar{t} \end{cases} \tag{7.65}$$

where

$$\dot{\hat{\Gamma}}_{av} = \frac{1}{t}\Gamma - \frac{1}{t}\hat{\Gamma}_{av}, \quad t > 0 \tag{7.66}$$

and

$$\hat{\Gamma}_{av}(0) = \Gamma(0). \tag{7.67}$$

Then there exist $\varepsilon^\star > 0$, $k_p > 0$, $k_v > 0$ such that for any $0 < \varepsilon < \varepsilon^\star$ the control law renders the equilibrium $(q, \omega_r) = (\bar{q}, 0)$ of the closed loop system (7.11)–(7.65) locally exponentially stable. Moreover, all trajectories of the closed-loop system (7.11)–(7.65) converge to the points $(q, \omega_r) = (\pm\bar{q}, 0)$.

Proof: Proposition 4 ensures that the application of the control law (7.65) for $t \leq \bar{t}$ leads to $\bar{\Gamma}(t) > 0$ for all $t > \bar{t}$. Note that the solution of Eq. (7.66) is given by

$$\hat{\Gamma}_{av}(t) = \frac{1}{t}\int_0^t \Gamma(\tau)d\tau, \tag{7.68}$$

so Proposition 4 also implies that $\lim_{t\to\infty}\hat{\Gamma}_{av}(t) = \Gamma_{av}$.

As in Proposition 1, introduce now the coordinates transformation (7.24) in which the closed-loop system (7.58)–(7.65) for $t > \bar{t}$ is described by the equations

$$\begin{aligned} \dot{\boldsymbol{z}}_1 &= \varepsilon\tilde{W}(\boldsymbol{z}_1)z_{2r} \\ I\dot{z}_2 &= \varepsilon S(z_2)Iz_2 + \varepsilon 3z_0^2 S(Ie_x^b)e_x^b + \varepsilon\Gamma(t)\hat{\Gamma}_{av}^{-1}(t)(-k_p\boldsymbol{z}_{1r} - k_v z_{2r}). \end{aligned} \tag{7.69}$$

System (7.69) satisfies all the hypotheses for the applicability of the generalized averaging theory [12, Theorem 10.5], which yields the averaged system

$$\begin{aligned} \dot{\boldsymbol{z}}_1 &= \varepsilon\tilde{W}(\boldsymbol{z}_1)z_{2r} \\ I\dot{z}_2 &= \varepsilon S(z_2)Iz_2 + 3z_0^2 S(Ie_x^b)e_x^b + \varepsilon\bar{K}(-k_p\boldsymbol{z}_{1r} - k_v z_2), \end{aligned} \tag{7.70}$$

where

$$\bar{K} = \lim_{T\to\infty}\frac{1}{T}\int_{\bar{t}}^T \Gamma(t)\hat{\Gamma}_{av}^{-1}(t)dt. \tag{7.71}$$

We now prove that $\bar{K} = \mathcal{I}_3$. To this end, note that from Eq. (7.66) one has

$$\hat{\Gamma}_{av}(t) = \Gamma_{av} + \Delta(t), \tag{7.72}$$

with

$$\lim_{t\to\infty}\|\Delta(t)\| = 0 \tag{7.73}$$

and

$$\hat{\Gamma}_{av}^{-1}(t) = \Gamma_{av}^{-1} - E(t), \tag{7.74}$$

with

$$\lim_{t\to\infty}\|E(t)\| = 0, \tag{7.75}$$

so that $\bar{K}$ can be written as

$$\bar{K} = \mathcal{I}_3 + \frac{1}{T}\int_{\bar{t}}^T \Gamma(t)E(t)dt \tag{7.76}$$

and the boundedness of $E(t)$ ensures that

$$\lim_{T\to\infty}\frac{1}{T}\int_{\bar{t}}^{T}\Gamma(t)E(t)dt = 0.$$

Finally, consider the function

$$V_3 = \frac{1}{2}[z_{2r}^T I z_{2r} + 3z_0^2(e_x^T I e_x - I_x) + z_0^2(I_z - e_z^T I e_z) + 2k_p(1 - z_{14})] \quad (7.77)$$

and note that for sufficiently large k_p function V_3 is positive definite. Its time derivative along the trajectories of the closed-loop system (7.58)–(7.65) is given by [25]

$$\dot{V}_3 = z_{2r}^T u - k_p z_{2r}^T z_{1r} = -k_v z_{2r}^T z_{2r}. \quad (7.78)$$

As $\dot{V}_3 \leq 0$, one has that $z_{2r} \to 0$ and therefore for sufficiently large k_p also $z_{1r} \to 0$. □

5. Simulation Results

In order to assess the performance of the magnetic attitude control laws discussed in this chapter, a number of simulated case studies has been considered. The simulations presented herein have been carried out using the tools presented in [2, 14], on the basis of the models for the space environment described in the classical references [22, 26].

5.1. *Inertial pointing*

5.1.1. *State feedback control*

The considered spacecraft has an inertia matrix given by $I = \text{diag}[27, 17, 25]$ $\text{kg}\,\text{m}^2$, and operates in a near polar (87° inclination) orbit with an altitude of 450 km and a corresponding orbit period of about 5600 s. It is worth, first of all, to check that the assumption $0 < \bar{\Gamma}_0 < \mathcal{I}_3$, which plays a major role in the formulation of the magnetic attitude control problem, is satisfied in practice. In order to illustrate this, in Figure 7.1 a time history of the eigenvalues of $\frac{1}{T}\int_0^T \Gamma_0(t)dt$ computed for the considered orbit is presented. As can be seen from the figure, $\frac{1}{T}\int_0^T \Gamma_0(t)dt$ converges to a $\bar{\Gamma}_0$ which satisfies the assumption.

For the considered spacecraft, a simulation related to the acquisition of the target attitude $\bar{q}$ from an initial condition characterized by a high initial angular rate has been carried out. In order to take into account the effect of disturbance torques on the behavior of the controlled spacecraft,

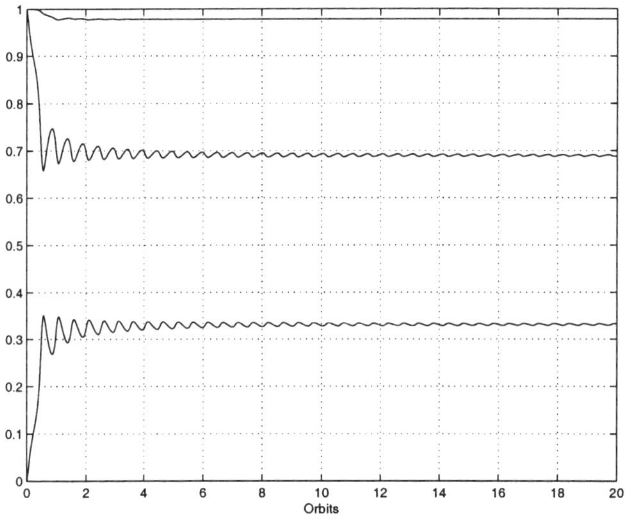

Figure 7.1. Eigenvalues of $\frac{1}{T}\int_0^T \Gamma_0(t)dt$ for the considered orbit.

a residual magnetic dipole $m_0 = \begin{bmatrix}0.5 \; 0.5 \; 0.5\end{bmatrix}^T$ (chosen according to guidelines available in the literature [7]) has been considered, together with the effect of gravity gradient torques. The results of a simulation of the attitude acquisition of the desired attitude $\bar{q}$ from an initial condition characterized by very high angular rate are displayed in Figure 7.2, from which the good performance of the unsaturated, state feedback control law can be seen. The controller parameters are given by $k_p = k_v = 50$ and $\varepsilon = 0.001$. Note, in particular, that, as expected, the disturbance torques affect only the steady state behavior of the system. In particular, the steady state offset in the desired attitude can be eliminated locally via a suitable nominal (linear) control law with integral action [17, 20, 21].

In order to illustrate the performance of the state feedback controller in the presence of saturation on the control action, the simulation related to attitude acquisition has been repeated, using the control law given in Eq. (7.38) with $\beta = 0.15$. From the results, shown in Figure 7.3, it appears that the saturated control law can still guarantee the convergence of the closed-loop system to the desired equilibrium. Clearly, the transient behavior is much slower than in the case of Figure 7.2, but the amplitude of the control inputs is significantly smaller.

5.1.2. *Output feedback control*

The output feedback attitude control has been applied to a spacecraft with an inertia matrix given by $I = \text{diag}[10, 10, 10]$ kg m^2, operating along the same orbit as in the case discussed in Section 5.1.1. The results of the simulations which have been carried out for this case are illustrated in Figure 7.4, taking into account the effect of a magnetic disturbance torque as induced by a residual dipole of strength $m_0 = \left[0.1\ 0.1\ 0.1\right]^T$. The controller parameters are given by $k_p = 50$, $k_v = 100$, $\lambda = 5$ and $\varepsilon = 0.001$. In particular, concerning the attitude acquisition, notice that the control action is actually bounded, as described in Remark 5. This leads to an acquisition transient which is clearly separated in a detumbling phase (saturated derivative action) and a re-orientation phase (linear operation of the controller). Similarly, it is possible to demonstrate the ability of the proposed output feedback control law to deal (at least locally) with the case of a non-isoinertial spacecraft, however the simulation results have been omitted for brevity.

5.2. *Earth pointing*

The considered spacecraft has an inertia matrix given by $I = \text{diag}[5, 60, 70]$ kg m^2, and operates in the same orbit as in the previous case.

For the considered spacecraft two simulations have been carried out: the first one is related to the acquisition of the target attitude $\bar{q}$ from an initial condition characterized by a high initial angular rate; the second one illustrates the behavior of the proposed control strategy when recovering the desired target attitude from an initial condition corresponding to the initial attitude $\left[0\ 0\ 1\ 0\right]^T$ and zero relative angular rate. In both cases, according to Proposition 2, the satellite is initially subject to a purely derivative control law.

The results of the attitude acquisition simulation are displayed in Figure 7.5, from which the good performance of the control law, with parameters $\varepsilon = 0.001$, $k_p = k_v = 20$, can be seen.

The initial condition of the second simulation corresponds to an "upside down" initial attitude, *i.e.* the spacecraft is initially in one of the undesired stable open loop equilibria of relative motion (see [11]) and the controller has to recover the target attitude. As can be seen from Figure 7.6, the proposed adaptive control law, with parameters $\varepsilon = 0.001$, $k_p = 100$, $k_v = 20$, can bring the satellite to the desired attitude. In particular, note that the transient of the attitude quaternion (Figure 7.6) shows that the transition from the initial to the final orientation of the satellite is carried out via an almost pure rotation around the z body axis, *i.e.* the (initially correct) orientation of the x and y-axis is only minimally perturbed. Finally, in Figure 7.7 the behavior of the elements of the (symmetric) matrix $\hat{\Gamma}_{av}$ is shown. As can be seen from the figure, the elements of the estimated average gain converge to constant values for $t \to \infty$.

6. Conclusions

The problem of global stabilization of the attitude dynamics for a magnetically actuated spacecraft is considered, and the cases of inertial pointing and Earth pointing are discussed. For the inertial pointing case, an almost global solution to the problem is obtained by means of static attitude and rate feedback. A local solution based on dynamic attitude feedback is also presented. For the Earth pointing case, almost global stabilization is achieved by means of an adaptive PD-like state feedback control law. Simulation results are presented to illustrate the performance of the proposed approaches.

Acknowledgments

This work was also partly supported by the ASI project "Global attitude determination and control using magnetic sensors and actuators".

Bibliography

1. M.R. Akella. Rigid body attitude tracking without angular velocity feedback. *Systems & Control Letters*, 42:321–326, 2001.
2. G. Annoni, E. De Marchi, F. Diani, M. Lovera, and G.D. Morea. Standardising tools for attitude control system design: the MITA platform experience. In *Data Systems in Aerospace (DASIA), Lisbon, Portugal*, 1999.
3. C. Arduini and P. Baiocco. Active magnetic damping attitude control for gravity gradient stabilised spacecraft. *J. Guidance, Control and Dynamics*, 20(1):117–122, 1997.
4. S. Battilotti. Global output regulation and disturbance attenuation with global stability via measurement feedback for a class of nonlinear systems. *IEEE Trans. Automatic Control*, 41(3):315–327, 1996.
5. C.I. Byrnes and A. Isidori. On the attitude stabilization of rigid spacecraft. *Automatica*, 27(1):87–95, 1991.
6. F. Caccavale and L. Villani. Output feedback control for attitude tracking. *Systems & Control Letters*, 39:91–98, 1999.
7. V. Chobotov. *Spacecraft Attitude Dynamics and Control.* Orbit Books, 1991.

8. F. Curti and F. Diani. Study on active magnetic attitude control for the Italian spacecraft bus MITA. *Proc. 14th Int. Conf. Spaceflight Dynamics, Iguassu Falls, Brasil*, 1999.
9. C.J. Damaren. Comments on "Fully magnetic attitude control for spacecraft subject to gravity gradient". *Automatica*, 38(12):2189, 2002.
10. O.E. Fjellstad and T.I. Fossen. Comments on "The attitude control problem". *IEEE Trans. Automatic Control*, 39(3):699–700, 1994.
11. P. Hughes. *Spacecraft attitude dynamics*. John Wiley and Sons, 1986.
12. H.K. Khalil. *Nonlinear Systems*. Prentice-Hall, 3rd ed., 2001.
13. F. Lizarralde and J.T. Wen. Attitude control without angular velocity measurement: a passivity approach. *IEEE Trans. Automatic Control*, 41(3):468–472, 1996.
14. M. Lovera. Modelling and simulation of spacecraft attitude dynamics. *Proc. 4th Int. Symp. Mathematical Modelling, Vienna, Austria*, 2003.
15. M. Lovera and A. Astolfi. Global attitude regulation using magnetic control. *Proc. 40th IEEE Conf. Decision and Control, Orlando, Florida*, 2001.
16. M. Lovera and A. Astolfi. Spacecraft attitude control using magnetic actuators. *Automatica*, 40(8):1405–1414, 2004.
17. M. Lovera, E. De Marchi, and S. Bittanti. Periodic attitude control techniques for small satellites with magnetic actuators. *IEEE Trans. Control Systems Technology*, 10(1):90–95, 2002.
18. M. Lovera and A. Varga. Optimal discrete-time magnetic attitude control of satellites. *Proc. 16th IFAC World Congress, Prague, Czech Rep.*, 2005.
19. P. Morin, C. Samson, J.B. Pomet, and Z.P. Jiang. Time-varying feedback stabilization of the attitude of a rigid spacecraft with two controls. *Systems & Control Letters*, 25:375–385, 1995.
20. M. Pittelkau. Optimal periodic control for spacecraft pointing and attitude determination. *J. Guidance, Control and Dynamics*, 16(6):1078–1084, 1993.
21. M. Psiaki. Magnetic torquer attitude control via asymptotic periodic linear quadratic regulation. *J. Guidance, Control and Dynamics*, 24(2):386–394, 2001.
22. M. Sidi. *Spacecraft dynamics and control*. Cambridge University Press, 1997.
23. W.H. Steyn. Comparison of low Earth orbit satellite attitude controllers submitted to controllability constraints. *J. Guidance, Control and Dynamics*, 17(4):795–804, 1994.
24. P. Wang and Y. Shtessel. Satellite attitude control using only magnetic torquers. *Proc. AIAA Guidance, Navigation, and Control Conf. and Exhibit, Boston, Massachusetts*, 1998.
25. J.T.Y. Wen and K. Kreutz-Delgado. The attitude control problem. *IEEE Trans. Automatic Control*, 36(10):1148–1162, 1991.
26. J. Wertz. *Spacecraft attitude determination and control*. D. Reidel Publishing Company, 1978.
27. R. Wisniewski and M. Blanke. Fully magnetic attitude control for spacecraft subject to gravity gradient. *Automatica*, 35(7):1201–1214, 1999.
28. R. Wisniewski and L.M. Markley. Optimal magnetic attitude control. *Proc. 14th IFAC World Congress, Beijing, China*, 1999.

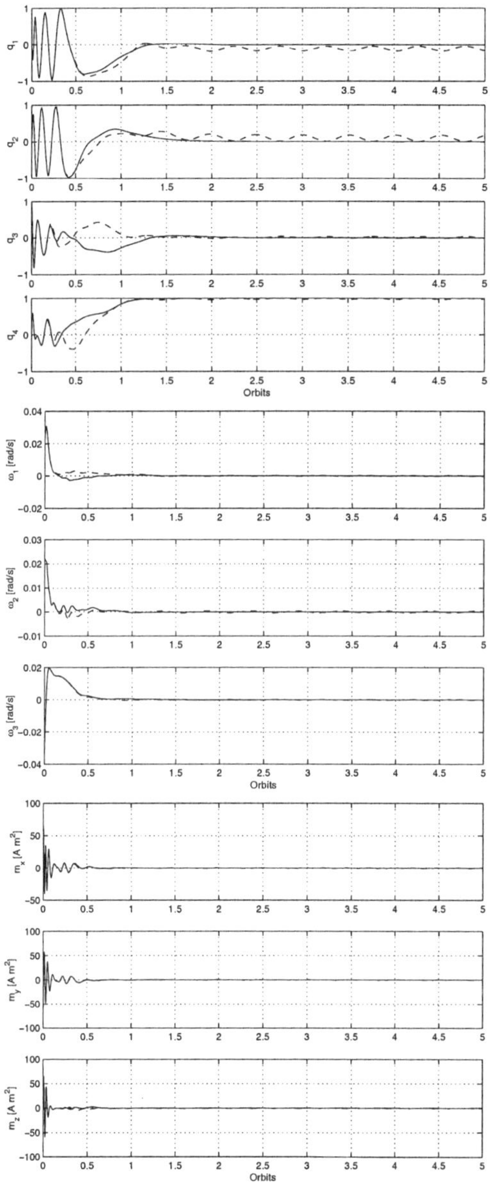

Figure 7.2. Quaternion, angular rates and control dipole moments for the inertial attitude acquisition: state feedback controller – simulations without (solid lines) and with (dashed lines) disturbance torques.

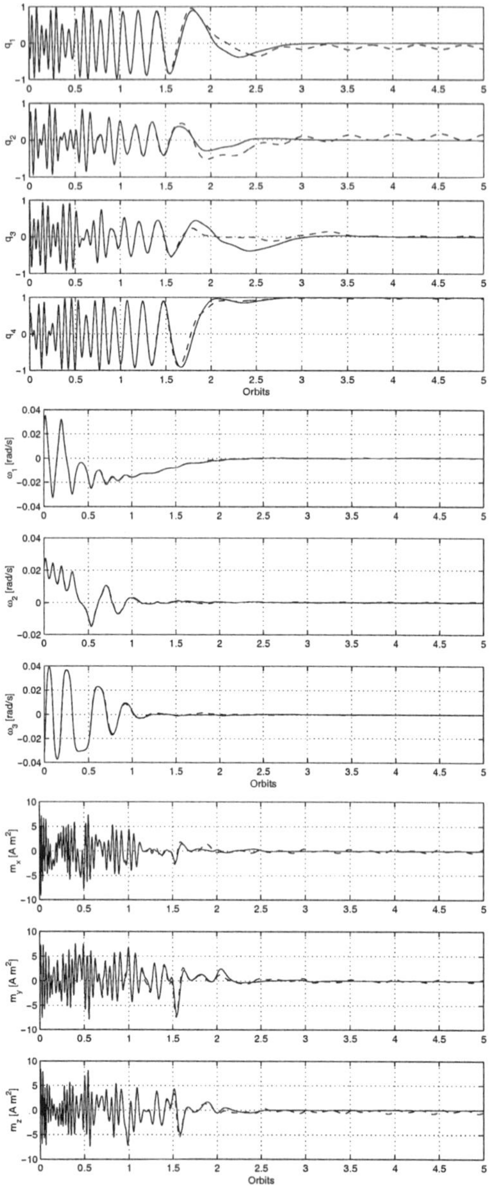

Figure 7.3. Quaternion, angular rates and control dipole moments for the inertial attitude acquisition: state feedback controller with saturation – simulations without (solid lines) and with (dashed lines) disturbance torques.

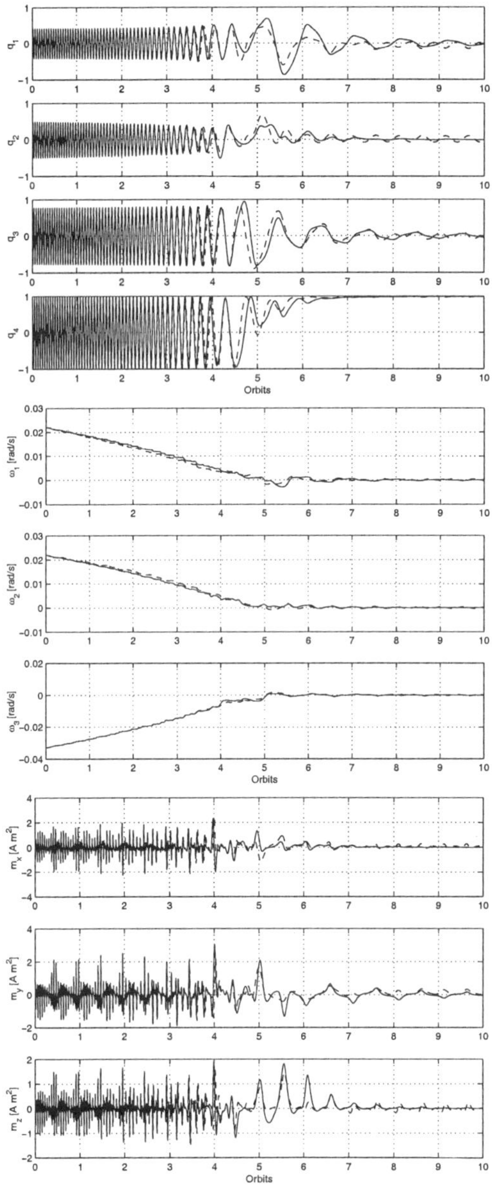

Figure 7.4. Quaternion, angular rates and control dipole moments for the inertial attitude acquisition: output feedback controller, I = diag[10, 10, 10] kg m^2 – simulations without (solid lines) and with (dashed lines) disturbance torques.

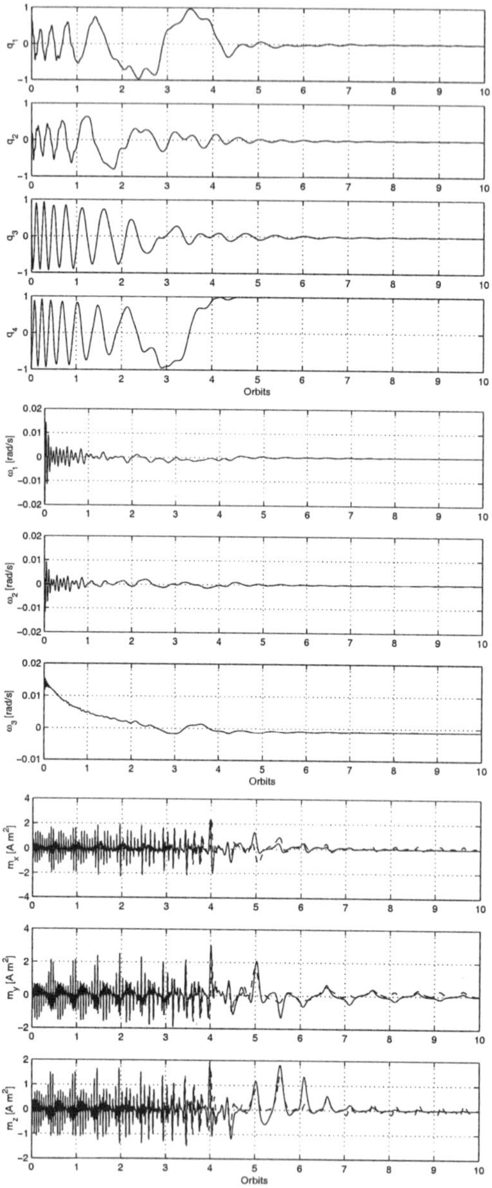

Figure 7.5. Quaternion and angular rates for the Earth pointing attitude acquisition.

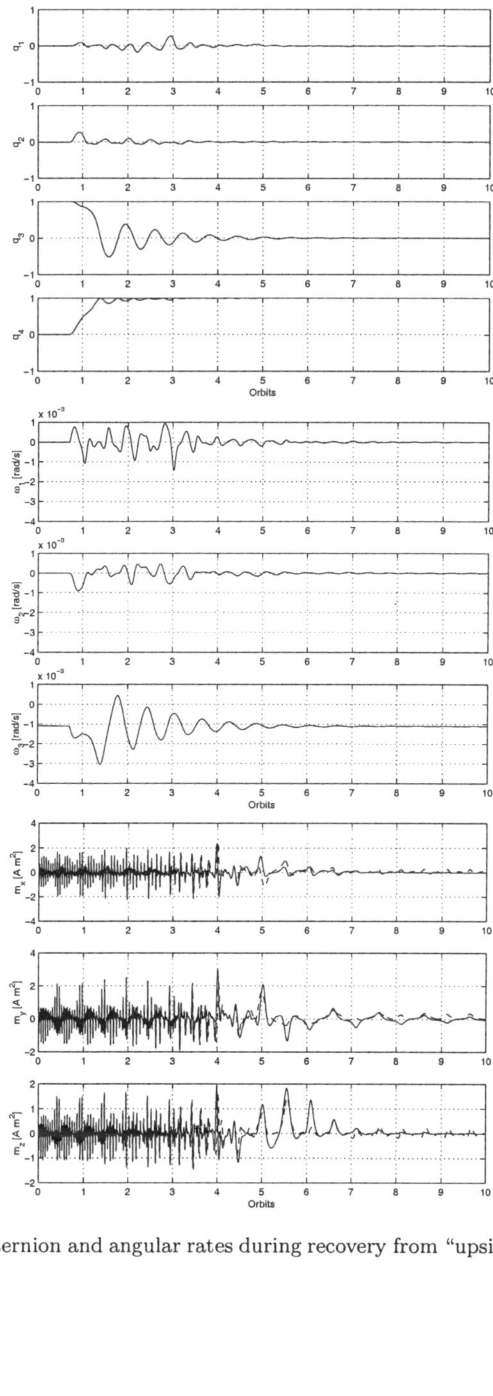

Figure 7.6. Quaternion and angular rates during recovery from "upside down" attitude.

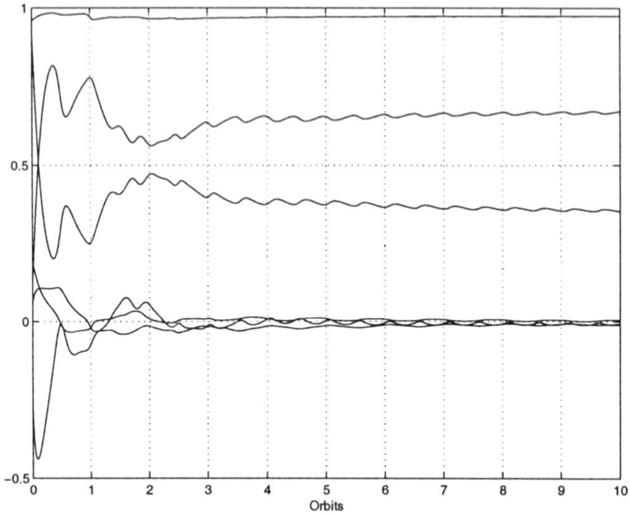

Figure 7.7. Time evolution of elements of the estimated average gain during recovery from "upside down" attitude.

CHAPTER 8

Control of Fed-Batch Bioreactors. Part I

J. Picó[1], E. Picó-Marco[1], J. L. Navarro[1] and H. DeBattista[2]

[1]*DISA, Technical University of Valencia, Camino de Vera, s/n 46022 Valencia, Valencia, Spain, E-mail: jpico@ai2.upv.es, {enpimar,joseluis}@isa.upv.es*

[2]*Faculty of Engineering, National University of La Plata, La Plata, Argentina, E-mail: deba@ing.unlp.edu.ar*

This chapter addresses the control of the microbial specific growth rate in fed-batch bioreactors. Practical constraints concerning the available signals and information on the process are dealt with. Then, for a large class of bioreactions, namely pure cultures with one limiting substrate and oxygen in excess, the control problem is regarded as a problem of coordinating control involving the invariance and attractivity of nontrivial sets in state space. Both monotonous and non-monotonous kinetic functions are dealt with. The dissolved oxygen signal can give valuable information about the cell culture, when the oxygen supply is in balance with the consumption. This is utilized for a probing feeding strategy, treated in Chapter 9.

1. Introduction

Fed-batch processes are extensively used in the expanding biotechnological industry. The requirements to optimize the production and improve the product quality obtained from the bioreaction processes are encouraging the development of robust and reliable controllers. For this reason, fed-batch process control is receiving great attention by the research community. From the control viewpoint, fed-batch fermentation processes are a challenging problem. The control designer must deal with strong modeling approximations, parameter uncertainties, external disturbances, nonlinear and possibly nonminimum-phase dynamics, lack of accurate on-line measurements of important variables involved in the process, *etc.*

A fed-batch bioreactor can be defined as a tank with no outgoing flow, where several microbial growth and enzyme-catalyzed reactions occur simultaneously in a liquid medium. The growth of biomass (bacteria, yeasts, *etc.*) proceeds by consumption of nutrients or substrates (carbon, nitrogen, oxygen, *etc.*) provided the environmental conditions (pH, temperature, illumination, *etc.*) are favorable. Simultaneously, some reactants are transformed into products or metabolites through the enzyme-catalyzed reactions mentioned above.

In the literature, many different models for biotechnological processes can be found. They vary not only in the functions used to model the reaction kinetics but also in the structure and kind of the model equations. For control purposes, a set of standard simplifications are commonly performed. From a biological standpoint, an important goal is to force and keep microorganisms into a given physiological state in which production of a certain species is optimal [4, 5]. This specification usually translates into the following control objective: the regulation of the biomass growth rate.

The survey papers [6–9], describe the history and state of the art in the field of fed-batch processes control. Recent advances in the design and implementation of closed-loop controllers for fed-batch bioreactors can be found in [10, 11, 31, 33, 35]. Although much progress has been made, the design and implementation of closed-loop controllers for fed-batch bioreactors still presents some shortcomings to overcome. In fact, the sensitiveness to the high level of noise corrupting the estimation and the strong dependence on process parameters hamper the implementation of many of these controllers in real processes.

This chapter is organized as follows. Section 2 deals with the search for a limited set of model structures representing most cases of industrial interest.

Section 3 introduces the main control specification to be considered in the chapter, namely regulating the microorganisms specific growth rate. A goal usually associated by the biologists with the maintenance of a definite physiological state.

Section 4 translates the control specifications into a problem of coordinating control where coordination conditions, given in the form of a relation of output variables, define a smooth goal submanifold in the state space. The solution of the control problem is then connected with the invariance and attractivity of the resulting goal manifold. Moreover, in the case of fed-batch bioreactors, partial equilibria and unbounded reference signals must be dealt with.

In Section 5, with the aim of controlling the trajectory towards the goal manifold, and to robustify the controller against parameter uncertainties, two approaches are analyzed. Firstly, feedback of the specific growth rate error. Secondly a globally stabilizing adaptive algorithm is developed based on variable structure control theory and the associated sliding regimes.

Finally, Section 6 presents some simulation and experimental results, and Section 7 summarizes the chapter with some conclusions.

2. Models

2.1. *Introduction*

In this section we recall that there are two standard models or structures that represent most *pure cultures* of industrial interest. It must be taken into account that most often the microorganisms to be genetically modified are chosen for their simple and/or well-known behavioral patterns, so there is a limited number of *typical* microorganisms. Hence, a reduced set of models may be used and their particularities exploited for control purposes.

As a general statement a bioreactor can be defined as *a tank in which several microbial growth and enzyme-catalyzed reactions occur simultaneously in a liquid medium.* A complete model of a bioreactor may have to address mass transfer, growth and biochemistry, physical chemical equilibria and various combinations of each of these. It becomes hard to write simple equations, but it is possible to reduce a system to its main components and formulate mass balances and rate equations that integrate overall behavior. Depending on the simplifications considered there are different kinds of models [3].

- If the cell is regarded as a *black box* and only the main extracellular species consumed or excreted in the medium are considered, without delving into the intracellular mechanisms, then the model is said to be *non-structured.*
- If the model is built supposing an average cell or individual then it is said to be *non-segregated.*
- Only the limiting substrate takes part in the model.
- Only one product is generally considered in the model: either the metabolite of interest or, if it exists, the inhibitor.

Another simplification often considered is *homogeinity.* That is, the conditions and concentrations in the tank are supposed to be homogeneous, which is a good approximation for lab-scale and pre-industrial fermentors.

From these hypotheses we proceed to develop models complete enough to account for the process behavior but not so complex that they become extremely difficult to handle.

Growth of microorganisms (bacteria, yeasts, *etc.*) proceeds by consumption – under favorable environmental conditions (temperature, pH, *etc.*) – of a combination of carbon sources, nitrogen sources, vitamins and other nutrient elements, so-called *substrates*, which altogether form the culture medium. These substrates serve different physiological purposes. Typically, all but one are found in excess both in the medium and the inflow. The nutrient in short supply relative to the others will be exhausted first and will thus *limit cellular growth*, in the sense that the biomass specific growth rate is controlled by the extracellular concentration of that substrate[a]. Therefore, only the *limiting substrate* will take part in the equations, the other ones being disregarded.

The case where several limiting substrates exist is an open research topic [15, 16], and will not be considered in the sequel. In the case of aerobic reactions – those where oxygen must be supplied – the dissolved oxygen concentration may in practice act as a second limiting substrate. Thus, shortage in the oxygen supply will affect microbial growth. This poses constraints on the process that are dealt with in Chapter 9.

The mass of living microorganisms or cells is called the *biomass*, although most often this term refers to their concentration. Usually, populations formed only by one species or strain are dealt with. These are the so-called *pure cultures*. In some not so common instances, there may be more than one species. These are the so-called *mixed cultures*.

Associated with cell growth, but often proceeding at a different rate, are the enzyme catalyzed reactions in which some reactants are transformed into *products* (sometimes called metabolites) through the catalytic action of intracellular or extracellular enzymes. These metabolites include the enzymes, proteins, antibiotics, ... we are interested in. It must be noted that there may be *inhibitory* products, *i.e.* those affecting growth directly.

2.2. *Standard models*

The *standard models* presented in this section are unstructured non-segregated models that represent pure cultures with one limiting substrate. In this models gas exchange is not considered and oxygen is assumed to be

[a]In microbiology the term *limitation of growth* is also used in a stoichiometric sense [14], *i.e.* it indicates that a certain amount of biomass is synthesized from a particular nutrient. This is reflected in the growth yield constants for the different elements.

in excess, unless it is the limiting substrate. It is also standard practice to consider only one product. Either the metabolite of interest or, if it exists, an *inhibitor*, a product that somehow affects microbial growth. According to the way in which product is formed the *standard models* can be classified as follows [1, 2].

- Growth-linked: product is formed in parallel with microbial growth. Two possibilities arise:
  - product is not inhibitor;
  - product inhibits growth.
- Non-growth-linked: product formation takes place either in the final phase of growth or in the secondary way, which is not directly connected with growth.

These two models are commonly taken in the literature as the standard ones for representing fermentation processes [23].

2.2.1. *Model 1a*

In this case the biomass is an autocatalyst, *i.e.* a catalyst of its own production. The more biomass there is, the more biomass (and product) can be produced, as product does not inhibit growth. The following state space model can be derived:

$$\Sigma_{1a} = \begin{cases} \dot{x} = \mu x - Dx \\ \dot{s} = -y_s \mu x + D(s_i - s) \\ \dot{p} = y_p \mu x - Dp \\ \dot{v} = F \end{cases} \tag{8.1}$$

where x, s and p are the biomass, substrate and product concentrations, respectively; v is the volume, $D = F/v$ is the dilution rate, F is the influent flow; s_i is the influent substrate concentration; y_s and y_p are yield coefficients; μ is the specific growth rate, a function of the species in the bioreactor. In this case μ does not depend on the product p. Therefore, the equation corresponding to the product can be disregarded in (8.1). Processes for the production of single-cell protein, alcohol and gluconic acid belong to this category.

2.2.2. *Model 1b*

The model for this case follows the same Eqs. (8.1) but, in this case, the equation for the product p cannot be disregarded, as $\mu = \mu(s,p)$.

2.2.3. *Model 2*

In this case the following state space model can be derived:

$$\Sigma_{2a} = \begin{cases} \dot{x} = \mu x - Dx \\ \dot{s} = -y_{sx}\mu x - y_{sp}\pi x + D(s_i - s) \\ \dot{p} = \pi x - Dp \\ \dot{v} = F \end{cases} \tag{8.2}$$

where π is the specific production rate. Many antibiotics (streptomycin, penicillin), lactic acid, citric acid, itaconic acid, glucoamylase and some amino acids are produced by this type of fermentation.

A more general model may be [17]

$$\Sigma_{2b} = \begin{cases} \dot{x} = \mu x - Dx \\ \dot{s} = -\sigma x + D(s_i - s) \\ \dot{p} = \pi x - kp - Dp \\ \dot{v} = F \end{cases} \tag{8.3}$$

where k is the hydrolysis (or degradation) constant for product; μ, σ and π are the specific rates of growth, substrate consumption and product formation, respectively. In addition, there may be terms for biomass decay and maintenance substrate consumption in the form $k_{d/m}x$, but usually these are not taken into account. The specific rates may depend on substrate, cell, and product concentrations, or they may be related to each other.

See [17] for an interesting classification of fermentations according to the form of μ and π.

Model (8.3) represents various fed-batch fermentations such as

- microbial cell productions involving bacteria (No metabolite production, both μ and σ only depend on s.);
- lysine production, $k = 0$, $\mu(s)$, $\pi(\mu)$ and $\sigma(\mu)$;
- alcohol production, $k = 0$, $\mu(s,p)$, $\pi(s,p)$ and $\sigma(s,p)$;
- antibiotic production, $k = 0$, $\mu(x,s)$, $\pi(s)$ and $\sigma(s)$.

Note that in this list μ and the other specific rates depend always on s and may be on x, but not on the product except for the case of alcohol production. Consequently, it may happen that production of, say, an antibiotic which in principle follows model (8.3), may actually be modeled for control purposes by a simpler set of equations discarding the equation for product.

In case of aerobic fermentations, *i.e.* those in which microorganisms need oxygen to develop properly, the dissolved oxygen (DO) dynamics in the bioreactor is described as follows:

$$\dot{O} = OTR - OUR - DC \tag{8.4}$$

where O is the DO concentration in the reactor, OTR is the oxygen transfer rate and OUR is the oxygen uptake rate. Expressions for these terms can be found in [1, 31] and in Chapter 9. This additional equation must be considered if oxygen is not supplied in excess, as the dissolved oxygen concentration O will appear in the expression of the specific growth rate μ. For a deeper view on the effects and control treatment of oxygen limitation see Chapter 9.

2.3. *Kinetic functions*

As it has been mentioned previously, the kinetic functions μ, σ and π depend on several factors such as the concentrations of substrate and product, but also the pH, temperature, *etc.*Usually, they are expressed as a product of several terms and each one depends on one of the factors previously cited. Thus

$$\mu = \mu_s(s)\mu_p(p)\mu_{pH}(pH)\mu_T(T)\ldots \tag{8.5}$$

Temperature, pH and other environmental variables are usually kept constant. As for the other factors, the relevant characteristics of the kinetic functions from the point of view of control purposes are their boundedness and their monotonicity or non-monotonicity. Moreover, the differences among the different kinetic functions are less relevant if one keeps in mind measuring errors and remaining modeling errors [16]. Therefore, the simplest forms, such as Monod and Haldane, explained below are mostly chosen. In more specific cases kinetic functions can be approximated by products/sums of relatively simple rational functions. For a comprehensive list see [1] and also [16].

Monod's kinetic functions (see Figure 8.1) are monotonous. They take the form:

$$\mu(s) = \frac{\mu_m s}{k_s + s} \tag{8.6}$$

where μ_m is the maximum growth rate, and k_s a transport constant.

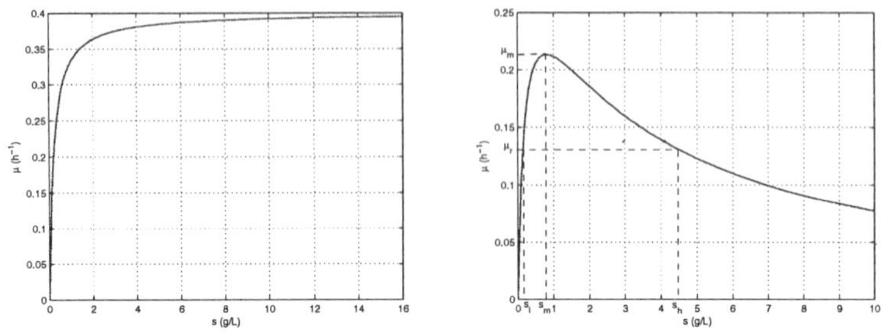

Figure 8.1. Monod (left) and Haldane (right) kinetic functions.

On the contrary Haldane's ones – typically used to model inhibition of growth by substrate – are non-monotonous (see Figure 8.1), and take the form

$$\mu(s) = \frac{\mu_o s}{k_s + s + \frac{s^2}{k_i}}. \tag{8.7}$$

This Haldane's function presents a maximum μ_m at $s_m = \sqrt{k_s k_i}$. In addition, a desired growth rate $\mu_r < \mu_m$ can be obtained at different substrate concentration set-points s_l and s_h satisfying $s_l < s_m < s_h$. Depending on the objective of the process, control could be aimed at regulating $\mu = \mu_r$ at s_l or s_h. Hereafter, the desired substrate set-point will be referred to as s_r and the other one as $\bar{s_r}$.

Finally, the term $\mu(p)$ referring to product is usually monotonously decreasing in order to represent an inhibition.

2.4. *Sources of uncertainty*

These systems show both important parametric and unstructured uncertainties. Unstructured uncertainties are mainly due to the following causes.

- The use of unstructured and non-segregated models, along with assumptions such as homogeneity. Thus, ignoring part of the system dynamics, which is *lumped* into some key factors.
- Further model simplifications. Thus, for instance, terms used to take into account the natural death of microorganisms or *maintenance terms*, representing the amount of substrate used for the biomass survival, are often disregarded.
- The common assumption of excess for some compounds required for growth, such as oxygen.

On the other hand parametric uncertainties may be due to the following causes.

- Identifiability problems.
- The fact that no two populations are equal, because of environmental effects, the preparation of the inoculum.....
- Aging of cells, which is reflected in slight variations of certain parameters during an experiment. For example, the yield coefficient.
- In general, any change in the environment or in the broth can potentially affect the system. Microorganisms are living things that continuously adapt themselves to changing conditions.

In addition to all these factors, the system is also affected by perturbations such as those acting on volume due to evaporation. In some cases the substrate concentration in the inflow may suffer variations. Uncertainty in the actuators (*e.g.* in the peristaltic pumps) may be important at very low flows.

Concerning the measurement of species, in general it is difficult to measure the substrate/s on line. Often the available measures are off-line, and their value may be of the order of magnitude of the measurement noise. On the other hand, there are a few sensors for the biomass [25], even though in general it is not possible to differentiate between different populations within the same reactor. Other state variables such as products are seldom available on-line. Use of observers is not trivial either, and estimations of the specific growth rate μ tend to be too noisy. A comprehensive survey of available measurements for bioreactors can be found in [9].

2.5. *Production modes*

On the basis of liquid medium one-stage bioreactors the following modes are found.

(1) *Batch.* There is no material exchange with the environment except for gases (oxygen, carbon dioxide,..), *i.e.* $D = 0$. All substrates are in excess within the reactor from the beginning of fermentation.
(2) *Continuous bioreactors.* The reactor is continuously fed with a substrate influent. There is also an outflow whose rate is equal to the inflow rate, hence the volume is constant for a fixed dilution. Typically the nutrient is fed at a constant rate, *i.e.* the dilution is $D = F/v = \text{const}$, which implies, in steady state, a constant cell division rate and thus a constant physiological state.

Continuous reactors have two important disadvantages. First, the low efficiency with respect to substrate usage. Second, a higher risk of contamination. Hence, continuous reactors are less used in industry, with the exception of waste treatment processes. Researchers often use them to determine certain physiological parameters of microorganisms.

(3) *Fed-batch*. Usually the production is carried out in fed-batch mode since substrate usage is optimized, and risk of contamination and strain variations are diminished. The basic scheme is shown in Figure 8.2.

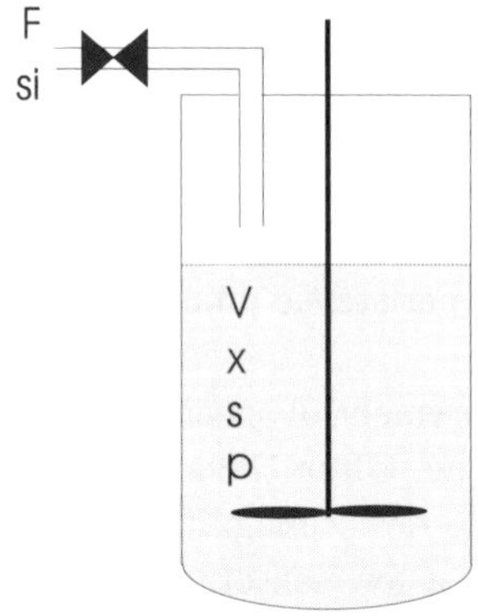

Figure 8.2. Diagram of a fed-batch operated bioreactor.

In the sequel, only fed-batch operated bioreactors will be considered.

3. Control Specifications

The control specifications are either directly or indirectly related to the optimization of production in some way. Thus, this optimization may typically imply the maximization of some cost index, or the regulation of some function of the species in the bioreactor so that the microorganism behaves optimally in some physiological sense.

When the goal is to maximize the amount of either biomass or product, optimal control provides a solution consisting of a feeding profile with nonsingular and singular control phases [11–13]. Actually, achieving the optimal feeding profile reduces to solving two subproblems.

(1) Find the time instant at which the feeding is changed from batch to fed-batch mode.

(2) Keep the specific growth rate at the specified value during the singular phase.

The desired specific growth rate during the singular phase might correspond to the maximum of some kinetic function. In such a case, when both uncertainties and non-monotonous kinetics are involved, extremum seeking strategies are used [34]. If limitations are present – *e.g.* due the production of additional toxic or inhibitory metabolites – the optimal specific growth rate may not correspond to a maximum of the kinetic rates. Thus, for instance, the optimal production of biomass with *Saccharomyces cerevisiae* is attained for a feeding profile that keeps the production of ethanol at a minimum [32] and is well below the maximum attainable specific growth rate.

Besides limitation due to the production of surplus metabolites, oxygen limitation may affect aerobic bioreactions. For small scale vessels oxygen can be easily supplied in excess by increasing the aeration rate and the stirrer velocity[b]. Yet, for large bioreactors there is a limit in the oxygen concentration that can be kept. To cope with shortages in the oxygen supply different strategies have been proposed [31,35], basically trying to attain a specific growth rate as close to the desired one as possible, given the current oxygen availability. See Chapter 9 for further details.

Thus, the following specifications will be considered in the next sections:

(1) Pure cultures with only one limiting substrate and assuming oxygen is in excess will be considered.
(2) Only biomass and volume are measured on-line. No estimation of the substrate will be considered.
(3) The main control specification will be to keep a constant growth rate.
(4) The flow F will be used as control signal.

4. Invariant Control

4.1. *Partial state feedback control and goal manifold*

This section addresses the computation of a partial state feedback that, assuming ideal conditions and perfect model, keeps the specific growth rate μ constant provided the initial conditions are adequate. The problem of keeping a desired specific growth rate is translated into that of staying on

[b]A limit in the stirrer velocity is imposed by the structural integrity of the cells under shear stress.

a particular goal manifold in the state space. Next it is shown that the partial state feedback defines an invariant control with respect to this goal manifold. Finally, it is shown that for systems with model Σ_{1a} (see (8.1)) and Monod-like kinetics, the invariant control renders the goal manifold globally asymptotically attractive, locally for Haldane-like ones. The results can be extended to models Σ_{1b}, Σ_{2a} (see (8.2)) and Σ_{2b} (see (8.3)) by adding a new control signal consisting of a separation of the product in the broth.

An open-loop *exponential feeding* for fed-batch bioreactors has been suggested in many occasions in the past, *e.g.* [6, 10, 18–21]. It can be deduced heuristically by regarding the average individual in a population as a processing unit that needs a determined quantity of energy so as to maintain a given level of activity. Then it follows that the quantity of substrate supplied, and consequently the feeding flux, must be proportional to the total population. The use of a closed-loop version, measuring biomass and volume on-line

$$F = \lambda x v \qquad \text{for some} \quad \lambda = \text{const} \tag{8.8}$$

is suggested for the first time in [22]. See [23] for a detailed history.

A simple deduction of (8.8) could be as follows. Consider model (8.1) with kinetic function μ depending only on the substrate concentration. In order to keep a constant $\mu(s)$, the substrate concentration should be kept constant at a value $s = s_r$ for which $\mu(s_r) = \mu_r$. Taking the flow rate $F = \lambda x v$, the equation for substrate becomes

$$\dot{s} = (-y\mu(s) + \lambda(s_i - s))x.$$

Provided $s = s_r$ from the beginning, it is possible to keep $\dot{s} = 0$ using

$$\lambda = \frac{y\mu_r}{s_i - s_r} = \text{const}, \tag{8.9}$$

independently of the initial conditions for x and v.

Under the previous conditions, the trajectories followed by x and v would be defined by an "exosystem"

$$\Sigma_r = \begin{cases} \dot{x} = \mu_r x - \lambda x^2 \\ \dot{v} = \lambda x v. \end{cases} \tag{8.10}$$

Conversely, if biomass x and volume v follow a trajectory defined by (8.9)–(8.10) then necessarily $\dot{s} = 0$ and $s = s_r$. This is related to the fact that system (8.1) is flat with flat outputs x and v.

The trajectories of (8.10) along with $s = s_r$ define a manifold in the state space on which $\mu(s) = \mu_r$. In order to get explicit expressions for the manifold to be tracked, it must be noticed that the first equation in (8.10) is a logistic one, with solution:

$$x(t) = \frac{\frac{\mu_r}{\lambda}}{1 + (\frac{\mu_r}{\lambda x_0} - 1)e^{-\mu_r t}}. \tag{8.11}$$

The volume trajectory is easily obtained after realizing that absolute mass $\bar{x} = xv$ follows an exponential trajectory and $\dot{v} = \lambda \bar{x}$. Hence

$$v(t) = v_0 + \frac{\lambda x_0 v_0}{\mu_r}(e^{\mu_r t} - 1). \tag{8.12}$$

Solving for t the goal manifold is defined as:

$$\mathcal{Z}^\star = \begin{cases} x - \frac{\mu_r}{\lambda} - (x_{0r}v_{0r} - \frac{\mu_r}{\lambda}v_{0r})\frac{1}{v} = 0. \\ s - s_r = 0. \end{cases} \tag{8.13}$$

This manifold, depicted in Figure 8.3, will be very important in the developments of Section 5.

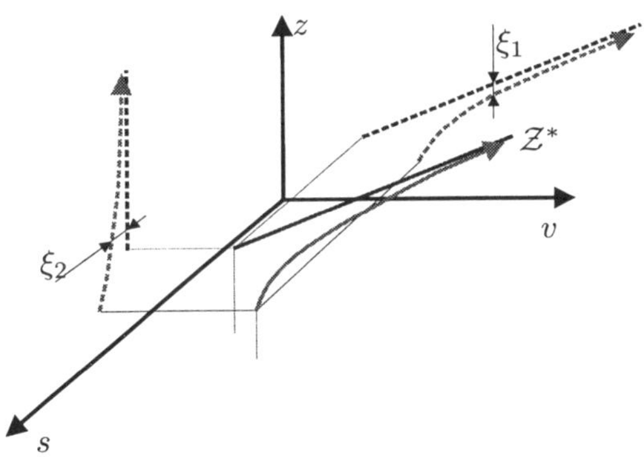

Figure 8.3. Goal manifold $\mathcal{Z}^\star$ showing the coordinates ξ_1 and ξ_2 transversal to it and $z = xv$ along it.

4.2. *Invariance*

For systems with structure Σ_{1a} the partial state feedback (8.8) defines an invariant control with respect to the manifold $\mathcal{Z}^\star$ defined by (8.13). This can be easily checked by casting Σ_{1a} into the form $\dot{x} = f(x) + g(x)u$ and solving the algebraic equation

$$\frac{\partial \varphi}{\partial \mathbf{x}} f(\mathbf{x}) + \frac{\partial \varphi}{\partial \mathbf{x}} g(\mathbf{x})u(\mathbf{x}) = 0 \qquad \mathbf{x} \in \mathcal{Z}^\star, \tag{8.14}$$

where φ is a vector containing the expressions in the equations defining $\mathcal{Z}^\star$. Two equations are obtained, the first one is fulfilled for every λ. The second one gives

$$\lambda = \frac{y\mu_r}{s_i - s_r}, \tag{8.15}$$

as it was expected [26].

This result can be extended to processes with structure Σ_{1b}, Σ_{2a} and Σ_{2b}. To that end, let us introduce a new control action α, representing a product separation from the broth:

$$\Sigma_{2af} = \begin{cases} \dot{x} = \mu(s,p)x - \dfrac{F}{v}x \\ \dot{s} = -y_{sx}\mu(s,p)x - y_{sp}\pi(s,p)x + \dfrac{F}{v}(s_i - s) \\ \dot{p} = \pi(s,p)x - \dfrac{\alpha}{v}p - \dfrac{F}{v}p \\ \dot{v} = F. \end{cases} \tag{8.16}$$

In this case, as it can be easily checked, the invariant control is:

$$\begin{aligned} F &= \lambda x v \\ \alpha &= \alpha' x v, \end{aligned} \tag{8.17}$$

where λ and α' are appropriate constants and the invariant manifold is defined by (8.13) and

$$p - p_r = 0 \qquad p_r = \text{const.} \tag{8.18}$$

This specification is justified by the need of keeping the microorganism in a given physiological state, in which production of a specific metabolite is optimum. Product p may be an inhibitor and/or the metabolite of interest. The specification of $p = \text{const}$, in turn, forces the introduction of the new control action α [23].

4.3. *Stability*

Fed-batch bioreactors correspond to the case of stability of *partial* equilibrium positions. Indeed, from the previous sections it is clear that the volume will tend to infinity, the biomass will track a trajectory, and only the substrate and possibly the product concentrations are to be regulated to a given value. This kind of setting is considered within the framework of partial stability analysis.

Partial stability is defined as the stability of dynamic systems with respect *not* to all but just to a given part of the state variables [27–29].

Assume a growth-linked type fed-batch bioreactor without inhibitor product, *i.e.* model (8.1). In order to analyze partial stability, the following result is used [27].

Theorem 1: Consider the nonlinear autonomous dynamical system:

$$\begin{aligned} \dot{x}_1 &= f_1(x_1, x_2), \qquad x_1(0) = x_{10}, \qquad t \in I_{x_0} \\ \dot{x}_2 &= f_2(x_1, x_2), \qquad x_2(0) = x_{20}, \end{aligned} \tag{8.19}$$

where $x_1 \in D \subseteq R^{n_1}$, D is an open set with $0 \in D$, $x_2 \in R^{n_2}$ and $f_1 : D \times R^{n_2} \longrightarrow R^{n_1}$ is such that $\forall x_2 \in R^{n_2}$

$$f_1(0, x_2) = 0,$$

and $f_1(., x_2)$ is locally Lipschitz in x_1. Besides, $f_2 : D \times R^{n_2} \longrightarrow R^{n_2}$ is such that for every $x_1 \in D$, and $f_2(x_1, .)$ is locally Lipschitz in x_2.

If there exists a continuously differentiable, positive definite function $V : D \mapsto R$ such that

$$\frac{\partial V(x_1)}{\partial x_1} f_1(x_1, x_2) \le 0, \quad \forall (x_1, x_2) \in D \times R^{n_2},$$

then system (8.19) is Lyapunov stable with respect to x_1, uniformly in x_2. If in addition there exists a class $\mathcal{K}$ function $\gamma(\cdot)$ such that

$$\frac{\partial V(x_1)}{\partial x_1} f_1(x_1, x_2) \le -\gamma(\| x_1 \|), \quad \forall (x_1, x_2) \in D \times R^{n_2},$$

then system (8.19) is asymptotically stable with respect to x_1, uniformly in x_2.

Now consider the assignments:

$$f_1 \left(\tilde{s}, (x, v)\right) : \dot{\tilde{s}} = (-y\mu(\tilde{s} + s_r) + \lambda(s_i - s_r - \tilde{s}))x \quad \text{with} \quad \lambda = \frac{y\mu_r}{s_i - s_r}$$

$$f_2 \left(\tilde{s}, (x, v)\right) : \begin{cases} \dot{x} = \mu(s)x - \lambda x^2 \\ \dot{v} = \lambda x v \end{cases}$$

with $\tilde{s} \triangleq s - s_r$ and $f_2\left(0, (x, v)\right) \equiv \Sigma_r$ (see Eq. (8.10)). Whenever $\tilde{s} = 0$, we have $f_1 = 0$ for all $x_2 = (x, v)$. Take the candidate partial Lyapunov function:

$$V(x_1) = \frac{1}{2}(s - s_r)^2.$$

Its derivative, taking $F = \lambda x v$ and assuming perfect knowledge of the model parameters, is:

$$\dot{V} = yx\tilde{s}(-\mu(\tilde{s} + s_r) + \mu_r \frac{s_i - \tilde{s} - s_r}{s_i - s_r}) \tag{8.20}$$

with $y = \text{const} > 0$ and $x > 0$. Clearly, whenever $\tilde{s} > 0$ the curve defined by $\mu(s)$ must be over the straight line defined by $\mu_r(s_i - s)/(s_i - s_r)$ and *vice versa*. This always happens whenever the kinetic function is monotonous or Monod-like – see Figure 8.4 – whereas in the Haldane-like case the parameter s_i may have to be properly chosen. By the Mean Value Theorem,

$$\mu(s) = \mu(s_r) + \frac{\partial \mu}{\partial s}|_{\bar{s} \in [s_r, s]} \tilde{s}.$$

Therefore,

$$\dot{V} = -yx(\frac{\partial \mu}{\partial s}|_{\bar{s}} + \frac{\mu_r}{s_i - s_r})\tilde{s}^2. \tag{8.21}$$

In the Monod-like case the derivative of μ will always be positive and hence the system will be partially asymptotically stable, as

$$\dot{V} \leq -yx\frac{\mu_r}{s_i - s_r}\tilde{s}^2 = -\lambda x \tilde{s}^2. \tag{8.22}$$

In the Haldane-like case the derivative may be, for some values of the substrate concentration s, negative and greater than the term $\mu_r/(s_i - s_r)$ hence the system may be only locally partially asymptotically stable. Reduction of λ and/or appropriate selection of s_i are then required [23].

5. Dealing with Uncertainties

5.1. *Introduction*

As described in Section 2.4, uncertainties, process variations and lack of measurements have traditionally been the main obstacles for the application of closed-loop control strategies to bioreactors. Thus, application of the partial state feedback control $F = \lambda_n x v$, with

$$\lambda_n = \frac{\mu y_s + m}{s_i - s_r^\star} \tag{8.23}$$

requires knowledge of:

- the structural model coefficients yield y_s and maintenance term m;
- the value of substrate concentration $s_r^\star$ such that $\mu(s_r^\star) = \mu_r$. This actually amounts to having a perfect model for the kinetics (8.7).

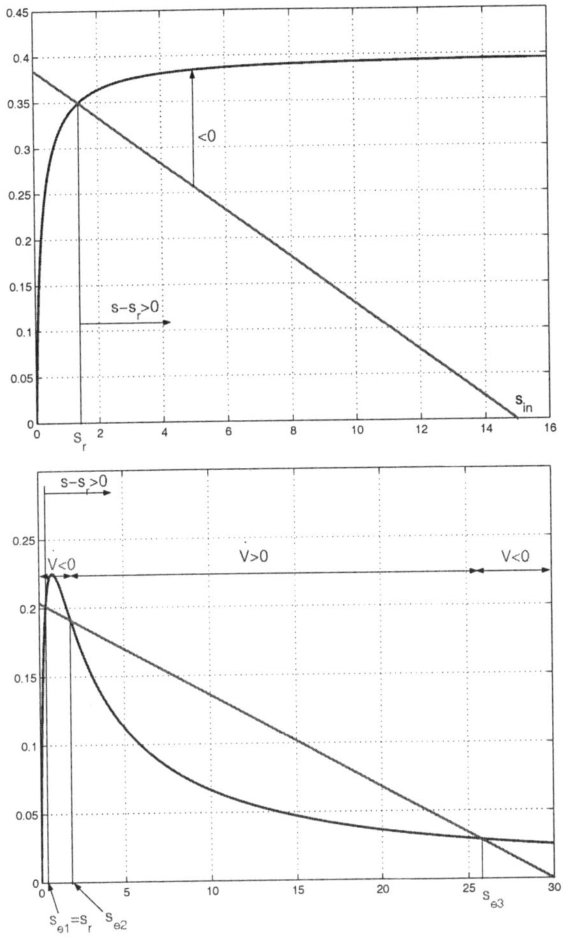

Figure 8.4. Monod (top), Haldane (bottom) and lines $\mu_r(s_i - s)/(s_i - s_r)$

In practice, however, some estimates $\bar{y}_s, \bar{m}$ and s_r will be available, so that $\bar{y}_s = y_s + \delta y_s, \bar{m} = m + \delta m$ and $s_r = s_r^\star + \delta s_r$. Therefore the feedback gain applied will be

$$\begin{aligned}\lambda_0 &= \frac{\mu \bar{y}_s + \bar{m}}{s_i - s_r} = \lambda_n \frac{s_i - s_r^\star}{s_i - s_r} + \frac{\mu(s)\delta y_s + \delta m}{s_i - s_r^\star} \frac{s_i - s_r^\star}{s_i - s_r} \\ &= \lambda_n + \frac{\delta\lambda_0}{s_i - s_r^\star},\end{aligned} \tag{8.24}$$

where $\delta\lambda_0 \triangleq \lambda_0 \delta s_r + \mu(s)\delta y_s + \delta m$.

To cope with uncertainties this section analyzes two alternative approaches, both based on the appropriate on-line modification of the gain λ in the partial state feedback defined above. The modification being a function of some error signal. A diagram of the resulting control loop is depicted in Figure 8.5.

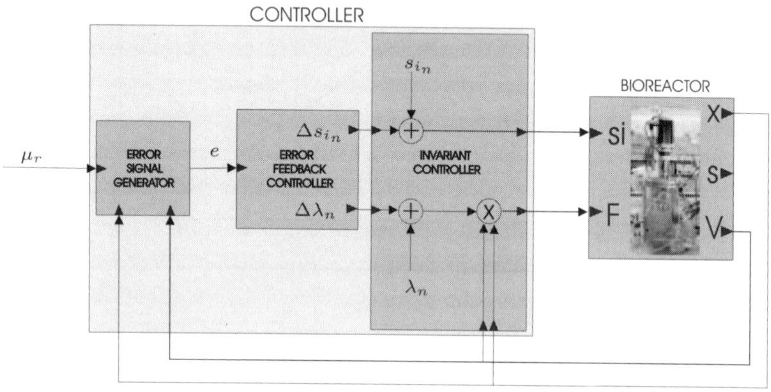

Figure 8.5. Modification of the gain λ through error feedback.

5.2. *Specific growth rate error feedback*

If an estimation of the specific growth rate is available, one can consider the feeding law:

$$F = \frac{y\mu}{s_i - s_r} xv + k_1 \frac{y(\mu - \mu_r)}{s_i - s_r} xv,$$

for some constant k_1 to be chosen. This can be easily rewritten as

$$\begin{aligned} F &= \frac{y\mu_r}{s_i - s_r} xv + (1 + k_1) \frac{y(\mu - \mu_r)}{s_i - s_r} xv \\ &= \lambda_0 \left(1 + k \frac{\mu - \mu_r}{\mu_r}\right) xv, \end{aligned}$$

i.e. the invariant control plus the correction using the error in μ multiplied by a new constant. Since the specific growth rate is bounded, there is implicitly a limitation on λ. Hence, for a properly chosen k we can make sure the system will always be in a given region. Actually, in [30] it is proved by other means that the system can be globally stabilized at any desired setpoint $s = s_r$ all along the (non-monotonic) kinetics.

In [10,11,32] similar laws are used, but they need the full state plus an estimation of the specific growth rate. In order to solve this problem, they are typically "approximated" taking the substrate concentration s equal to a reference concentration s_r. Besides, because of the inherent unstable properties of the system in the right flank of the Haldane-like kinetic functions, stabilization becomes a crucial problem in this operating region. In [10] the feedback gain switches according to the sign of the estimated growth rate derivative in order to globally stabilize the process. As a consequence of the high sensitivity to the noise corrupting the estimation, convergence to the desired set-point may be critically delayed in some circumstances. In [30], some points of the stability analysis are clarified. In particular, it is claimed that feedback discontinuity is actually a sufficient but not a necessary condition to accomplish global stability.

Defining $e_s \triangleq s - s_r^\star$, the substrate error dynamics are:

$$\begin{aligned}\dot{e}_s &= -\lambda_n x \left[1 + k\frac{p(s)}{s}\frac{\mu(s)}{\mu_r}\right] e_s + \frac{\delta\lambda_0}{s_i - s_r^\star} x \left[(s_i - s) - K\frac{p(s)}{s}\frac{\mu(s)}{\mu_r} e_s\right] \\ &= -\left[1 + k\frac{p(s)}{s}\frac{\mu(s)}{\mu_r}\right]\lambda_0 x e_s + \delta\lambda_0 x, \end{aligned} \tag{8.25}$$

where $\frac{p(s)}{s} \triangleq \frac{(s-\bar{s_r}^\star)(s_i-s)}{s}$ for Haldane kinetics and $\frac{p(s)}{s} \triangleq \frac{s_i-s}{s}$ for Monod ones. At steady state, the substrate error settles at

$$e_s = \frac{\delta\lambda_0}{\left[1 + k\frac{p(s)}{s}\frac{\mu(s)}{\mu_r}\right]\lambda_0}. \tag{8.26}$$

An analysis of the closed-loop dynamics reveals that for the case of Haldane kinetics a trade-off between steady state and dynamic performance has to be made, being more critical if the desired substrate concentration lies on the right flank of the kinetic function.

5.3. *Robust adaptation of the partial state feedback gain*

A robust adaptive controller is presented in this section. It is applied to processes with Monod-like and Haldane-like kinetic functions depending only on the substrate concentration. In the latter case, the results in Section 4 may be used to determine a saturation in the control action to ensure stability. In the following the theoretical derivation of the controller is shown.

Consider model (8.1). The kinetic function may be monotonic or non-monotonic. The basis for the new controller is the invariant control of Section 4. The goal manifold $\mathcal{Z}^\star$ associated to it plays an important role. Take a normalized off-the-manifold error as:

$$\sigma(z, v, \lambda) = \frac{z - z_{or}}{z} - \frac{\mu_r}{\lambda}\frac{v - v_{or}}{z}. \tag{8.27}$$

Normalization weights the errors at the beginning of the fermentation, as more important than at the end. Now the gain λ is not taken as a constant but as a variable, and $z = xv$ being x_{or}, v_{or}, z_{or} the initial conditions for a reference trajectory which can be generated by the exosystem (8.10).

In the sequel a law modifying λ is sought in order to compensate the effect of uncertainty. Consider the function

$$W = \frac{1}{2}\sigma^2. \tag{8.28}$$

It is intended to achieve $\sigma \to 0$. This could be achieved forcing

$$\dot{\sigma} = -\frac{1}{T_a}\sigma, \tag{8.29}$$

so that

$$\dot{W} = -\frac{1}{T_a}\sigma^2 \leq 0. \tag{8.30}$$

The derivative of σ with respect to time is

$$\dot{\sigma} = \mu(s)(1 - \sigma) - \mu_r + \dot{\lambda}\frac{\mu_r(v - v_{or})}{\lambda^2 z}. \tag{8.31}$$

From (8.29) and solving for the derivative of λ:

$$\dot{\lambda} = \frac{\lambda^2 z}{\mu_r(v - v_{or})}[\mu_r - \mu(s)(1 - \sigma) - \frac{1}{T_a}\sigma]. \tag{8.32}$$

In case an adequate on-line measure or estimation of $\mu(s)$ is not available, it should be substituted by a $\hat{\mu}$ with an *a priori* chosen value. Then:

$$\sigma\dot{\sigma} = (\sigma - \sigma^2)(\mu(s) - \hat{\mu}) - \frac{1}{T_a}\sigma^2. \tag{8.33}$$

Two elements, namely $\hat{\mu}$ and $\frac{1}{T_a}$, can be set so as to get

$$\sigma\dot{\sigma} \leq 0.$$

Choosing $\hat{\mu} = \mu_r$ and, for instance,

$$\frac{1}{T_a} = \begin{cases} 2\mu_m & for \quad |\sigma| \geq 1 \\ \dfrac{2\mu_m}{|\sigma|} & for \quad |\sigma| < 1 \end{cases}$$

the resulting controller would be:

$$\dot{\lambda} = \frac{\lambda^2 z}{\mu_r(v - v_{or})}[\mu_r - \frac{1}{T_a}]\sigma. \tag{8.34}$$

Clearly, practical implementation of this high gain scheme requires adding a dead zone δ. The following gain is suggested inside the dead zone:

$$\frac{1}{T_a} = \frac{2\mu_m}{\delta} \quad \text{for} \quad |\sigma| < \delta < 1 \tag{8.35}$$

which forces the system to get into it. In fact it can be checked that

$$\dot{W} < 0 \quad \text{for} \quad \frac{\delta}{2-\delta} < |\sigma| < \delta.$$

The resulting control structure is depicted in Figure 8.6.

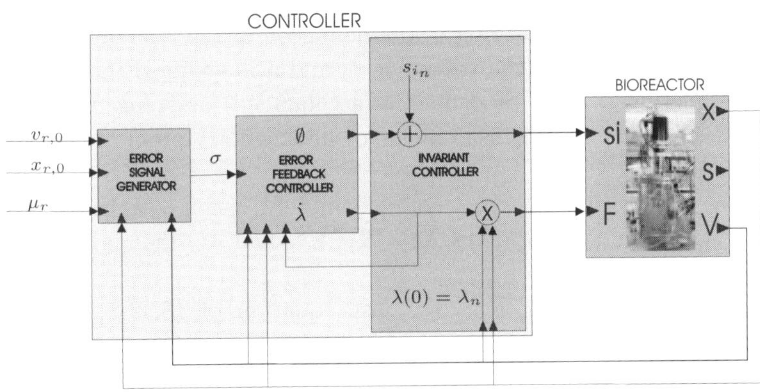

Figure 8.6. Modification of the gain λ through adaptation.

5.3.1. *Stability proof*

It has already been shown that state trajectories converge to (the close vicinity of) $\sigma = 0$. This section is devoted to prove that system trajectories on this sliding surface asymptotically converge to the goal manifold Z^*. In other words, we will prove that if the off-the-manifold error σ is maintained at zero, then $s \to s_r$ and the feedback gain tends to its nominal value λ_r given by (8.9). In the case of non-monotonous kinetics some precautions must be taken.

On the sliding manifold $\sigma = 0$, the closed-loop system dynamics can be rewritten as follows:

$$\Sigma_\sigma = \begin{cases} \dot{s} = \left[-y_s\mu(s) + \lambda(s_i - s)\right] x \\ \dot{\lambda} = \left[-\lambda^2 \dfrac{\mu(s) - \mu_r}{\mu_r} \dfrac{v}{v - v_{r,0}}\right] x \\ \dot{v} = [\lambda v]\, x, \end{cases} \tag{8.36}$$

where the equation for the evolution of λ has been obtained from (8.31) and the sliding mode existence condition ($\sigma = 0$, $\dot{\sigma} = 0$) [24]. Besides, the equation for the biomass concentration has been omitted to avoid redundancy. In fact, on the sliding manifold $\sigma = 0$, x is algebraically dependent on $\{\lambda, v\}$: $x = \frac{z_{r,0}}{v} + \frac{\mu_r}{\lambda v}(v - v_{r,0}) > 0$. Note also that the biomass concentration is bounded since v only can increase.

Let $\psi(v) = \frac{v}{v - v_{r,0}} : (v_{r,0}, \infty) \mapsto (\infty, 1)$. Note that $v = v_{r,0}\frac{\psi}{\psi - 1}$ and that the Frechet derivative $\psi\prime = -(\psi - 1)^2/v_{r,0}$. It must be taken into account that $v_{r,0}$ is a constant entering in the definition of the reference manifold. The initial condition for the volume is v_0 and it is assumed that $v_0 > v_{r,0}$. So, the function ψ could be defined as a (bounded) mapping from (v_0, ∞) into $(\psi_0, 1)$. A fact that is used in the complementary proof of asymptotic stability below.

Let ζ the partial state $\zeta = \mathrm{col}(s, \lambda)$ and $\zeta_r = \mathrm{col}(s_r, \lambda_r)$. Recall that $s \in \mathcal{S} = (0, s_i)$ and $\lambda \in \mathbb{R}^+$. Let $\mathcal{M} = \mathcal{S} \times \mathbb{R}^+$, and M_σ the region of $\sigma = 0$ such that $\zeta \in \mathcal{M}$.

Note that, replacing x in the last equation of (8.36), yields $\dot{v} = \mu_r(v - v_{r,0}) + \lambda z_{r,0}$, which confirms that, on $\sigma = 0$, the volume diverges exponentially. As a result, $\psi \overset{t\to\infty}{\longrightarrow} 1$.

On the other hand, for the partial system Σ_σ, define the candidate partial Lyapunov function:

$$V(\zeta, \psi(v)) = \psi \int_{s_r}^{s} \frac{\mu(\varsigma) - \mu_r}{\mu_r} \mathrm{d}\varsigma + +(s_i - s_r)\left[\ln\frac{\lambda}{\lambda_r} + \frac{(\lambda_r - \lambda)}{\lambda}\right]. \tag{8.37}$$

Its time derivative is

$$\dot{V}(\zeta, \psi(v)) = -\psi x \left[(\psi - 1)\lambda \int_{s_r}^{s} \frac{\mu(\varsigma) - \mu_r}{\mu_r} \mathrm{d}\varsigma + \right.$$
$$\left. \frac{y_s}{\mu_r}(\mu(s) - \mu_r)^2 + \frac{\lambda}{\mu_r}(\mu(s) - \mu_r)(s - s_r)\right]. \tag{8.38}$$

The equations for the evolution of s and λ on the sliding manifold $\sigma = 0$ are asymptotically independent of v. That is, as v diverges, $\psi \mapsto 1$ and

$x \mapsto \mu_r/\lambda$, thus approaching the system

$$\Sigma_{v\to\infty} = \begin{cases} \dot{s} = (-y_s\mu(s) + \lambda(s_i - s))\dfrac{\mu_r}{\lambda} \\ \dot{\lambda} = -\lambda(\mu(s) - \mu_r). \end{cases} \tag{8.39}$$

On the other hand it is possible to define positive definite functions $\overline{V}(\zeta) \triangleq V(\zeta, \psi_0)$ and $\underline{V}(\zeta) \triangleq V(\zeta, 1)$ such that

$$\underline{V}(\zeta) \leq V(\zeta, \psi) \leq \overline{V}(\zeta) \tag{8.40}$$

and an additional nonnegative definite function $W(\zeta) \triangleq -\dot{V}(\zeta, 1)$ such that

$$\dot{V}(\zeta, \psi) \leq -W(\zeta). \tag{8.41}$$

Consequently, ζ_r is a globally asymptotically stable partial equilibrium point for the partial system Σ_σ [27].

For non-monotonous kinetic functions, *e.g.* Haldane, the previous results about stability are only local. Actually, the system may present two equilibrium points. Let denote s_m the substrate concentration at which the growth rate is maximum, $s_r < s_m$ and $s^r > s_m$ the substrate concentrations satisfying $\mu(s_r) = \mu(s^r) = \mu_r$. Locally around s_r, the kinetic function behaves as a monotonous function. Then, $V(\zeta, \psi(v))$ is locally positive definite around ζ_r, whereas $\dot{V}(\zeta, \psi(v))$ is locally negative semi-definite and ζ_r is the largest invariant set for which $\dot{V} = 0$. Then, ζ_r is a locally asymptotically stable equilibrium point for the partial system Σ_σ, and the original system on $\sigma = 0$ locally asymptotically converges to the goal manifold $Z_{r,0}$.

Let $\mathcal{S}^r = \{s \in \mathcal{S} \mid s < s^r\}$, $\mathcal{L}^r = \left\{\lambda \in \mathbb{R}^+ \mid \lambda < \lambda^r = \frac{y_s\mu_r}{s_i - s^r}\right\}$, $\mathcal{M}^r = \mathcal{S}^r \times \mathcal{L}^r$ and M_σ^r the region of $\sigma = 0$ where $\zeta \in \mathcal{M}^r$.

It is clear from the previous expressions that M_σ^r is a domain of attraction of ζ_r on the sliding manifold $\sigma = 0$, that is a region of convergence towards $Z_{r,0}$ on $\sigma = 0$. Nevertheless, if the substrate concentration is initially very high, the system state might reach (the close vicinity of) the sliding manifold outside the domain of attraction, leading to undesired unstable dynamics.

A possible solution suggested here is to modify the adaptation law so that the trajectories are steered to reach the attractive region M_σ^r of the sliding manifold $\sigma = 0$. A natural way of avoiding the aforementioned undesired dynamics is limiting the feeding governed by λ and s_i. According to this, the adaptation law is modified by incorporating the saturation function

$$g(w_1, w_2) = \begin{cases} 0 & \text{if } w_1 \geq 0 \land w_2 \geq 0 \\ 1 & \text{otherwise} \end{cases} \tag{8.42}$$

where $w_1 = \lambda - \bar{\lambda}$ and $w_2 = -\sigma$. To complete the analysis we need to show that, despite saturation, all state trajectories finally reach the vicinity of $\sigma = 0$. In fact, during saturation, the derivative of the function (8.28) becomes

$$\dot{W} = -\mu\sigma^2 + (\mu - \mu_r)\sigma. \tag{8.43}$$

On one hand, if $\sigma > 0$, the saturation becomes inactive ($g(\cdot,\cdot) = 1$) and $\dot{\lambda}$ is negative. Thus, the inequality $\dot{W} < 0$ holds, *i.e.* the trajectory points towards $\sigma = 0$. On the other hand, if $\sigma < 0$, λ remains at its limit value. Therefore, to approach $\sigma = 0$, (8.43) should be negative whenever $\sigma < 0$. It can be shown that this is true, possibly except for an initial period of time. In fact, as λ is maintained fixed at $\bar{\lambda}$, the partial state ζ will finally reach $\mathcal{M}^r$, and moreover, will converge to $\bar{\zeta} = (\bar{s}, \bar{\lambda}) \in \mathcal{M}^r$, where $\bar{s}$ is the substrate concentration at which the solid line in (8.20) crosses the kinetic function (see Figure 8.4 (bottom)). Since $\mu(\bar{s}) > \mu_r$, $\dot{W}$ will, eventually, become negative. Consequently, trajectories will finally point towards $\sigma = 0$ from both sides, as desired.

Note that although it is not necessary to assure convergence toward $Z_{r,0}$, limiting λ may also be used in the case of Monod-like kinetic functions to improve the transient from certain initial conditions. A kind of windup effect may appear due to the saturation of Monod functions and the integrator implicit in the control law. Effectively, in order to reach and maintain the process state on the sliding manifold $\sigma = 0$, a large overshoot in λ may appear, leading to an excess of feeding and a large settling time. Limiting appropriately the value of λ, the substrate concentration is bounded hence avoiding strong saturation of the growth rate and the associated windup effect. Based on the previous analysis, convergence to the equilibrium point on the sliding manifold $\sigma = 0$ is still guaranteed despite the λ limitation, provided $\bar{\lambda} > \lambda_r$.

6. Results

6.1. *Simulation Results*

Simulation results for fed-batch processes with Haldane kinetics are presented to corroborate the attractive features of the proposed adaptive control strategy. The process and controller parameters used in simulations are listed in Table 8.1.

Figure 8.7 (top) shows (solid line) the response of the closed-loop system for initial conditions beyond its peak value ($s > s_m$): $(x_0, s_0, v_0) =$

Table 8.1. Parameters and test conditions.

Parameter	Value	Parameter	Value
μ_m [1/h]	0.22	v_f [L]	30
k_s [g/L]	0.14	μ_r [1/h]	0.1
y_s	1.43	s_i [g/L]	20
m [1/h]	0.05	δ	0.05
k_i [g/L]	4		
$x_{r,0}$ [g/L]	5	$v_{r,0}$ [L]	$\lesssim 1$
Meas. noise x	$+\{-0.1, 0, 0.1\}\, x + \mathcal{N}(0.5, 0.01)$		
Meas. noise V	$+\mathcal{N}(0, 0.001)$		

$(x_{r,0}, 2.5, v_{r,0}{}^{+})$. It is seen that the substrate concentration is rapidly reduced and, consequently, the growth rate converges to its desired value μ_r. Simulation results are also presented when a $\pm 10\%$ error in the biomass measurement is considered (dashed and dot-dashed lines). Note that the adjustable gain λ tends to different steady state values (Figure 8.7 (b, top)), to compensate for these large errors in the measured biomass (Figure 8.7 (c, top)). As a consequence, the growth rate is stabilized at μ_r, corroborating the robustness property against biomass measurement errors.

Finally, simulation analysis were conducted to validate the ability of the controller to reach the prescribed manifold from different initial conditions (Figure 8.7 (bottom)). The response from $(x_0, s_0, v_0) = (x_{r,0}, 0.1, v_{r,0}{}^{+})$ is drawn in solid line. The prescribed surface is immediately reached and the growth rate rapidly converges to μ_r. The response from a larger initial biomass concentration ($x_0 = 1.4 x_{r,0} = 7$g/L) is displayed in dashed line. The excess of biomass and the low incoming flow (low λ) necessary to reduce the normalized error σ lead to the undershoot in the growth rate observed in Figure 8.7 (a, bottom). After the prescribed surface is reached, the growth rate rapidly converges to its desired value. On the other hand, the response from an initial condition with negative error σ is depicted in dotted line. The incoming flow (*i.e.* λ) is increased to reduce the magnitude of this error. Unfortunately, when the prescribed manifold is reached, $\lambda \gg \lambda^r$, and the state trajectory is oriented in the opposite direction to the domain of attraction M_σ^r. Consequently, both the controller parameter λ and the substrate concentration s diverge. Conversely, the response of the system from the same initial condition, but bounding the gain $\lambda \leq \bar{\lambda}$, is shown in

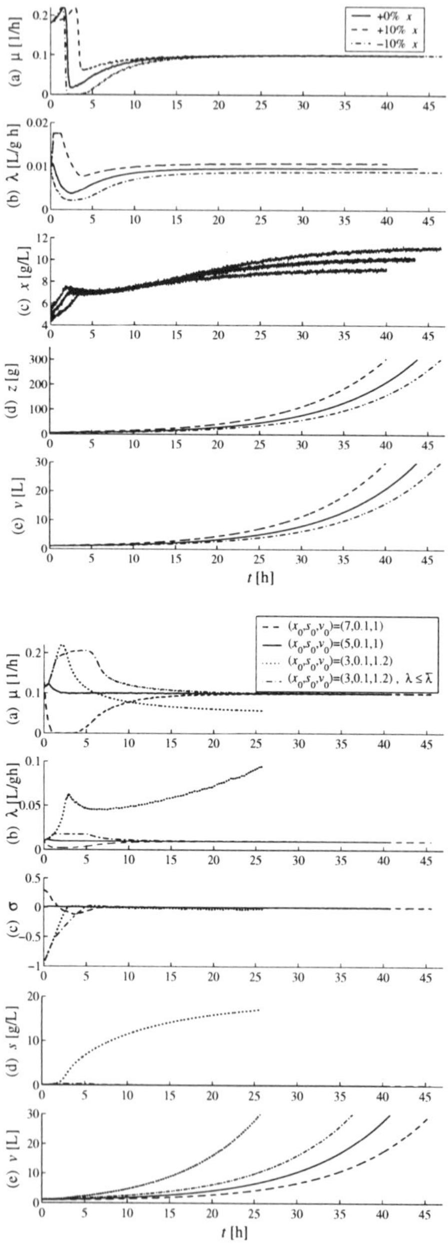

Figure 8.7. Simulation results for Haldane kinetics with biomass error measurement (top) and from different initial conditions (bottom).

dot-dashed line. At the cost of increasing the reaching time, this limitation of the feedback gain guarantees that the prescribed manifold is reached inside the domain of attraction of the equilibrium point for the partial state $\zeta = \zeta_r$.

The simulation results using large initial errors σ shown in this last example are intended to put in evidence the reaching and stabilizing properties of the proposed controller. In practice, however, the initial conditions for x and v of the process Σ and of the reference system Σ_r can be adjusted to avoid large transient responses.

6.2. *Experimental results*

The results shown in Figure 8.8 correspond to a fed-batch fermentation of *Saccharomyces cerevisiae* T_{73} on glucose. The product of biomass concentration by volume, *i.e.* the absolute biomass, is in logarithmic scale and a straight line with the slope corresponding to the reference specific growth rate is added to facilitate comparison. After an initial batch with a glucose concentration of 5g/L (Figure 8.8 (d)), the controlled fed-batch was switched on at $t_0 = 7.65$h, when the glucose in the medium was almost exhausted. The concentration of glucose in the feeding flow was set to 20g/L.

The constants of the goal manifold were set to $z_{r,0} = z(t_0)$ and $v_{r,0} = 0.9v(t_0)$, whereas the initial value of λ was set at $\lambda(t_0) = 1.3e-3L(gh)^{-1}$. Under these conditions, the initial value of the normalized off-the-manifold error is $\sigma(t_0) = -2.3$. Then, the control algorithm increases λ in order to approach the sliding surface $\sigma = 0$ (Figure 8.8 (c)). The long term variation in λ (Figure 8.8 (c)), which is commonly observed in all long experiments, can be explained as an adaptation to the varying yield coefficient y_s. For this reason, the control strategies that use *a priori* estimation of y_s usually fail to regulate the specific growth rate during the whole experiment.

The integral action inherent to the controller causes an initial overshoot in the specific growth rate (Figure 8.8 (a)) for approximately 4 hours. This transient overshoot could have been reduced by choosing $z_{r,0}$, $v_{r,0}$ and $\lambda(t_0)$ so that $\sigma(t_0) \cong 0$. Anyway, the large initial value of $\sigma(t_0)$ allows to corroborate the reaching properties towards $\sigma = 0$ of the algorithm (Figure 8.8 (c)). During the rest of the experiment, the specific growth rate μ keeps around the desired value, but for some periods of time (around $t = 20$h and $t = 25$h). At these periods, μ drops due to shortages in the oxygen supply, as seen in Figure 8.8 (b) looking at the decrease of pO_2 at $t = 20$h and the increase of the stirrer speed at $t = 25$h. This behavior occurs because

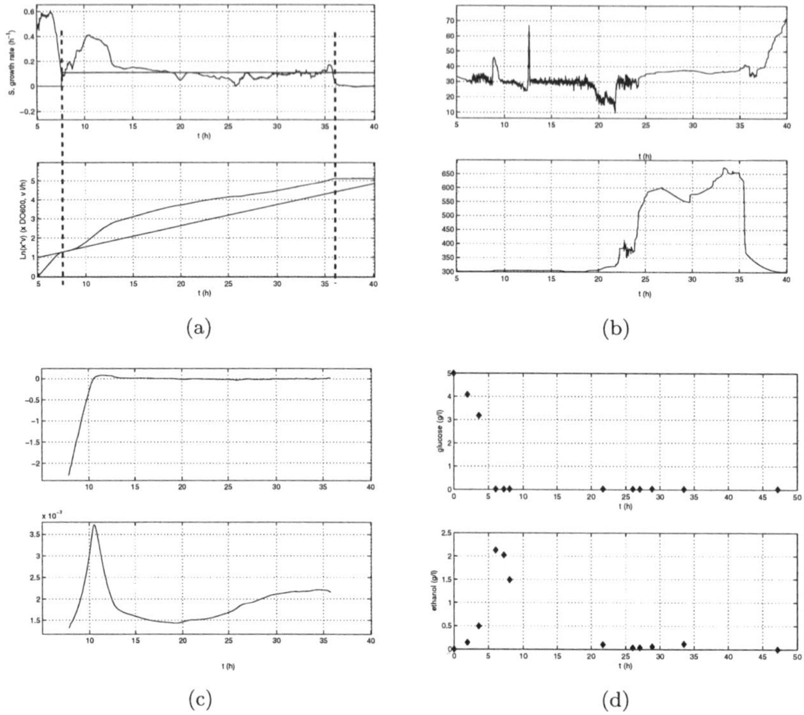

Figure 8.8. Fed-batch on *S. cerevisiae* T_{73}. (a) Specific growth rate, $\log(z)$ and line with slope μ_r. (b) pO_2 (%) and stirrer (r.p.m.). (c) σ and λ. (d) Off-line measurements of glucose and ethanol.

there was a deficient control loop for pO_2 in the experiment, and O_2 was not considered as a limiting substrate in the model. Actually, whenever this limitation appears, one should improve the oxygen transfer rate by means of the air supply and stirrer speed and/or demand for a lower specific growth rate.

Finally, it is important to stress the low values of glucose in the medium after the initial batch, which remained at around 0.023g/L throughout the experiment. Their order of magnitude is close to that of measurement noise. Therefore, a control strategy based somehow on measurements or estimation of the substrate is not feasible in practice. As for ethanol, the low specific growth rate permitted to avoid its formation.

7. Conclusions and Outlook

In this chapter the problem of defining a control law for the regulation of the specific growth rate in fed-batch bioreactors has been treated as that of defining an invariant, attractive goal manifold for the system. Within this framework, a closed-loop version of the exponential feeding law for fed-batch bioreactors has been derived, and the stability of the resulting closed loop analyzed as a problem of partial stability. The resulting invariant controller, a partial state feedback, has been the basis for further designs aiming at coping with process and measurement uncertainties. In particular, an adaptation of the controller gain using sliding mode theory has been introduced. It presents very interesting features, as seen through simulated and real experiments. No estimator for the specific growth rate is used. Additionally, the controller completely rejects actuator errors and is robust to process parameter uncertainties and bounded disturbances in environmental variables. Particularly, zero steady state error is achieved despite all these types of perturbations as well as despite a biomass measurement offset. The approach is valid for a set of models covering a large range of biotechnological applications.

Acknowledgments

This work has also been partially supported the Spanish Government (CICYT DPI2002-00525). The first three authors are with an Associated Unit to the Department of Biotechnology at the IATA (National Research Council, CSIC). The experiments were carried out at the facilities of Biopolis S.L.

Bibliography

1. G. Bastin and D. Dochain. *On-line Estimation and Adaptive Control of Bioreactors.* Elsevier, 1990.
2. P. Zlateva. Output stabilization of fermentation processes: a binary control system approach. *Bioprocess Eng.*, 23:81–87, 2000.
3. I. Dunn, E. Heinzle, J. Ingham, and J. Přenosil. *Biological Reaction Engineering.* Wiley-VCH Verlag, 2003.
4. A. Jobé, C. Herwig, M. Surzyn, B. Walker, I. Marison, and U. von Stockar. Generally applicable fed-batch culture concept based on the detection of metabolic state by on-line balancing. *Biotechnol. Bioeng.*, 82, 2003.
5. M. Henson and D. Seborg. Nonlinear control strategies for continuous fermenters. *Chemical Engineering Science*, 47:821–835, 1992.
6. S. Parulekar and H. Lim. Modeling, optimization and control of semi-batch bioreactors. *Advances in Biochemical Eng./Biotechnol.*, 32:207–258, 1985.

7. A. Johnson. The control of fed-batch fermentation processes – a survey. *Automatica*, 23:691–705, 1987.
8. J. Lee, S. Lee, S. Park, and A. Middelberg. Control of fed-batch fermentations. *Biotechnol. Advances*, 17:29–48, 1999.
9. K. Rani and V. Rao. Control of fermenters – a review. *Bioprocess Eng.*, 21:77–88, 1999.
10. I. Smets, G. Bastin, and J. van Impe. Feedback stabilization of fed-batch bioreactors: non-monotonic growth kinetics. *Biotechnol. Prog.*, 18:1116–1125, 2002.
11. I.Y. Smets, J.E. Claes, E.J. November, G.P. Bastin, and J.F. van Impe. Optimal adaptive control of (bio)chemical reactors: past, present and future. *J. Process Control*, 14(7):75–805, 2004.
12. J.R. Banga, C.G. Moles, and A.A. Alonso. Global optimization of bioprocesses using stochastic and hybrid methods. In *Frontiers in Global Optimization, C.A. Floudos and P.M. Pardalos (Eds.)*, Kluwer Academic Publishers, 2003.
13. J.F. van Impe and G. Bastin. Optimal adaptive control of fed-batch fermentation processes. *Control Engineering Practice*, 3(7):939–954, 1995.
14. T. Egli and M. Zinn. The concept of multiple-nutrient-limited growth of microorganisms and its application in biotechnological processes. *Biotechnol. Advances*, 22:35–43, 2003.
15. H.R. Bungay. Continuous cultivation of microorganisms. Available on-line *http://www.cape.canterbury.ac.nz/Archive/contin/working.htm.*
16. K. Schügerl and K.H. Bellgardt (Eds.). *Bioreaction Engineering: Modeling and Control.* Springer-Verlag, 2000.
17. J.M. Modak, H.C. Lim, and Y.J. Tayeb. General characteristics of optimal feed rate profiles for various fed-batch fermentation processes. *Biotechnol. Bioeng.*, 28:1396–1407, 1986.
18. J. Lee, S.Y. Lee, S. Park, and A.P.J. Middelberg. Control of fed-batch fermentations. *Biotechnol. Advances*, 17:29–48, 1999.
19. S.Y. Lee. High cell-density culture of Escherichia coli. *Trends Biotechnol.*, 14:98–105, 1996.
20. M.E. Gregory and C. Turner. Open-loop control of specific growth rate in fed-batch cultures of recombinant E. coli. *Biotechnol. Tech.*, 7(12):889–894, 1993.
21. D.M. Chang. The snowball effect in fed-batch bioreactions. *Biotechnol. Prog.*, 19:1064–1070, 2003.
22. T. Yamane, W. Hibino, K. Ishihara, Y. Kadotani, and M. Kominami. Fed-batch culture automated by uses of continuously measured cell concentration and culture volume. *Biotechnol. Bioeng.*, 39:550–555, 1992.
23. E. Pico-Marco. *Nonlinear robust control of biotechnological processes. Application to fed-batch bioreactors.* Ph.D. Thesis, Technical University of Valencia, 2004.
24. V. Utkin. *Sliding Modes in Control and Optimization.* Springer-Verlag, 1992.
25. J. Navarro, J. Pico, J. Bruno, E. Pico-Marco, and S. Valles. *On-line method and equipment for detecting, determining the evolution and quantifying a*

microbial biomass and other substances that absorb light along the spectrum during the development of biotechnological processes. Patent ES20010001757, EP20020751179, 2001.

26. E. Pico-Marco and J. Pico. Partial stability for specific growth rate control in biotechnological fed-batch processes. *Proc. IEEE Conf. Control Applications*, 2003.
27. V. Chellaboina and W. Haddad. A unification between partial stability and stability theory for time-varying systems. *IEEE Control Systems Magazine*, 6:66–75, 2002.
28. V.I. Vorotnikov. *Partial Stability and Control.* Birkhäuser, 1998.
29. V.I. Vorotnikov. Partial stability, stabilization and control: some recent results. *Proc. 15th IFAC World Congress, Barcelona, Spain*, 2002.
30. H. DeBattista, E. Pico-Marco, and J. Pico. On "Feedback stabilization of fed-batch bioreactors: non-monotonic growth kinetics". *Biotechnol. Prog.*, submitted.
31. R. Oliviera, R. Simutis, and S. Feyo de Azevedo. Design of a stable adaptive controller for driving aerobic fermentation processes near maximum oxygen transfer capacity. *J. Process Control*, 14:617–626, 2004.
32. M. Arndt and B. Hitzmann. Kalman filter based glucose control at small set points during fed-batch cultivation of Sacccharomyces cerevisiae. *Biotechnol. Prog.*, 20:377–383, 2004.
33. S. Valentinotti, B. Srinivasan, U. Holmberg, D. Bonvin, C. Cannizzaro, M. Rhiel, and U. von Stockar. Optimal operation of fed-batch fermentations via adaptive control of overflow metabolite. *Control Engineering Practice*, 11:665–674, 2003.
34. N.I. Marcos, M. Guay, D. Dochain, and T. Zhang. Adaptive extremum-seeking control of a continuous stirred tank bioreactor with Haldane's kinetics. *J. Process Control*, 14:317–328, 2004.
35. S. Velut and P. Hagander. Analysis of a probing control strategy. *Proc. American Control Conference, Denver, Colorado*, pages 609–614, 2003.

CHAPTER 9

Control of Fed-Batch Bioreactors. Part II

S. Velut and P. Hagander

Department of Automatic Control, Lund Institute of Technology
Box 118, SE 221 00 Lund, Sweden, E-mail: svelut@control.lth.se

This chapter addresses the control of glucose feeding in fed-batch reactors. The feeding strategy is based on a pulse technique, which does not require *a priori* knowledge of the process. A feedback algorithm optimizes the feed rate with respect to the constraints of aerobic conditions and overfeeding. Stability and performance analysis are performed on a simplified model of the reactor, using standard LMI optimization routines. Tuning guidelines that help the user for the design are also derived.

1. Introduction

Today many proteins are produced by genetically modified organisms. A common host is the bacterium *Escherichia coli*. It is well characterized and it can be quickly grown to high cell densities. A limiting factor is the formation of the by-product acetate that has been reported to reduce cell growth and protein production. Formation of acetate occurs under anaerobic conditions but also under fully aerobic conditions when the carbon source – often glucose – is in excess.

Aerobic conditions can be guaranteed by maintaining a sufficiently high dissolved oxygen concentration in the bioreactor. The main difficulty consists in finding the optimal feed rate. High feed rates result in short cultivation times, thereby high productivity, but may lead to overfeeding. The critical glucose concentration, above which the respiratory capacity of the cells saturates and acetate is produced, is unknown and may vary during a cultivation. Furthermore, accurate online measurement of such low glucose concentrations are not available. The time-varying and uncertain nature of the process makes the control task even more difficult. Feedback from an online cell density sensor is an interesting alternative. In Chapter 8 how to use that to achieve a predetermined reference growth rate μ, possibly time-varying, was investigated.

The probing feeding strategy described in [1] can be used to avoid acetate accumulation. The idea of the probing approach is to detect the saturation of the respiration by superimposing pulses in the feed. The control strategy does not require *a priori* knowledge of the process and is solely based on the dissolved oxygen measurement. The control algorithm has been implemented on real plants where good performance could be achieved, see for instance [2, 4]. The controller is able to adapt to process changes such as variations in the maximum oxygen uptake rate, which typically occurs when the foreign protein starts to be produced. Although the control strategy is simple, it results in a complex closed-loop system that is nonlinear, time-varying with continuous as well as discrete states. Rigorous analysis of the closed-loop system is valuable for a better understanding and tuning of the probing controller. The objective of this chapter is to study stability and performance of the closed-loop system. The analysis will be restricted to plants consisting of a piecewise linear static function followed by a linear time invariant system.

The outline of the chapter is as follows. In Section 2 a model of the fed-batch reactor is presented, similar to the ones in Section 2.2 of Chapter 8. The control problem, as well as the probing control strategy, are described in Section 3. In Section 4 a discrete and time-invariant representation of the closed-loop system is derived. Computational methods for stability and performance analysis, based on standard optimization routines are given in Section 5. Performance of the probing controller is measured by the ability to track a time-varying optimum. The analysis methods are illustrated on a numerical example in Section 6, where local analysis that helps the user for the design is also performed. Finally, the case of an input nonlinearity is studied in Section 7 and experimental results are shown in Section 8.

2. Process Description

2.1. *Stirred bioreactor*

We consider a bioreactor running in fed-batch mode. After an initial batch phase, glucose is continuously fed into the reactor at a limiting rate. The cell density grows exponentially and the feed rate is adjusted to meet the growing glucose demand. Air is sparged into the reactor and the dissolved oxygen concentration is controlled by manipulation of the agitation speed.

2.2. *Mass balances and metabolic relations*

The mass balance equations for the media volume V, the glucose concentration G, the cell concentration X and the dissolved oxygen concentration C_o are:

$$\begin{aligned} \frac{dV}{dt} &= F \\ \frac{d(VG)}{dt} &= FG_{in} - q_g(G)VX \\ \frac{d(VX)}{dt} &= \mu(G)VX \\ \frac{d(VC_o)}{dt} &= K_{La}(N)V(C_o^* - C_o) - q_o(G)VX, \end{aligned} \tag{9.1}$$

where F, G_{in}, K_{La} and μ are, respectively, the feed flow rate, the glucose concentration in the feed, the volumetric oxygen transfer coefficient and the specific growth rate. Further, C_o^*, q_o and q_g denote the oxygen concentration in equilibrium with the oxygen in gas bubbles, the specific oxygen uptake rate and the specific glucose uptake rate. The specific oxygen uptake rate is modeled by

$$q_o = \begin{cases} Y_{og}q_g & q_g < q_g^{crit} \\ q_o^{max} & q_g > q_g^{crit}. \end{cases} \tag{9.2}$$

The critical specific glucose uptake rate $q_g^{crit} = \frac{q_o^{max}}{Y_{og}}$ defines the limit for overfeeding. Above q_g^{crit} the respiratory capacity of the cells saturates and the byproduct acetate is produced.

Most sensors measure the dissolved oxygen tension O instead of the dissolved oxygen concentration C_o. They are related by Henry's law

$$O = HC_o. \tag{9.3}$$

The dynamics in the oxygen probe can also be taken into account and it is modeled by a first order system with time constant T_p

$$T_p\frac{dO_p}{dt} + O_p = O. \tag{9.4}$$

2.3. *Linearization*

During short periods of time, the volume V and the biomass X are approximately constant. The variations in the dissolved oxygen tension and in the glucose concentration are described by

$$T_g \frac{d\Delta G}{dt} + \Delta G = K_g \Delta F \tag{9.5}$$

$$T_o \frac{d\Delta O}{dt} + \Delta O = K_N \Delta N + K_o q_o(\Delta G), \tag{9.6}$$

where

$$\begin{aligned} T_g &= (X \frac{\partial q_g}{\partial G})^{-1} \qquad & K_g &= (\frac{VX}{G_i} \frac{\partial q_g}{\partial G})^{-1} \\ T_o &= (K_{La})^{-1} \qquad & K_o &= \frac{\partial K_{La}}{\partial N}(O^* - O)(K_{La})^{-1}. \end{aligned} \tag{9.7}$$

At the beginning of the fermentation, the cell density is low and the time constant T_g is large. The input dynamics are consequently dominant. At the end of the cultivation, the oxygen probe at the output provides the main dynamics.

3. Probing Control

3.1. *Control problem*

The control task is twofold. The first objective consists in maximizing the feed rate F with respect to the constraint of overfeeding ($q_g < q_g^{crit}$). This is related to extremum control problems. In extremum control, the optimal setpoint is not known and is often given by the extremum of a static input-output map. The classical approach to this problem consists in adding a known time-varying signal to the process input and correlating the output with the perturbation signal to get information about the nonlinearity gradient. The controller adjusts continuously the control signal towards the optimum. A good overview of extremum control is given in [8].

The second objective is to regulate the dissolved oxygen tension O at a constant level O_{sp} by manipulating the agitation speed. The closed-loop system is consequently a multivariable setup with two inputs, the feed rate F, the agitation speed N and one output, the dissolved oxygen concentration O.

3.2. *Probing control*

As in extremum control the probing strategy makes use of a perturbation signal to get information about the nonlinearity. The key idea of the probing approach is to detect the formation of acetate by superimposing short pulses in the feed rate, see Figure 9.1. The size of the pulse response, which depends

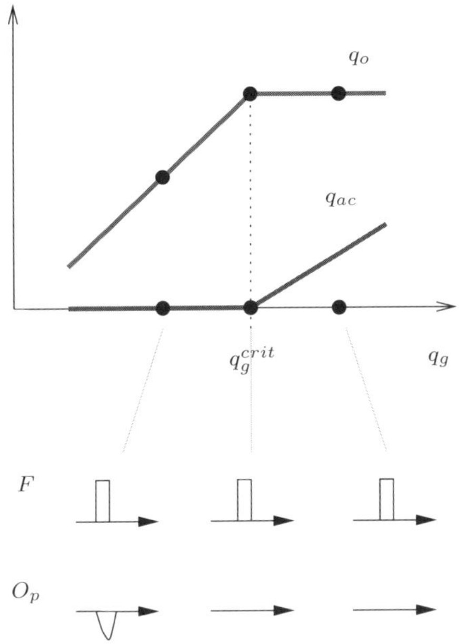

Figure 9.1. The superimposed pulses in the glucose feed rate F affect the glucose uptake rate q_g. When glucose is limiting, variations in the oxygen uptake q_o can be clearly seen in the dissolved oxygen measurement O_p. In this way acetate formation can be detected.

on the local gain of the nonlinearity, is used to adjust the control signal, namely the glucose feed rate.

The main difference with the classical scheme is the separation in time of the correlation phase and the control phase. Pulses are periodically introduced at the process input and a control action is taken at the end of every pulse. This also allows the regulation of the process output by manipulation of the agitation speed between two successive pulses, thereby maintaining aerobic conditions. Figure 9.2 shows the complete scheme.

4. Closed-Loop System Representation

The overall system is rather complex but analysis is possible when some dynamics are neglected. The periodic nature of the total controller, with one regulation phase followed by a probing phase, suggests a discrete-time description. We will restrict the analysis to static nonlinearities that are piecewise linear. Global stability can thus be investigated by using tools for discrete-time piecewise linear systems.

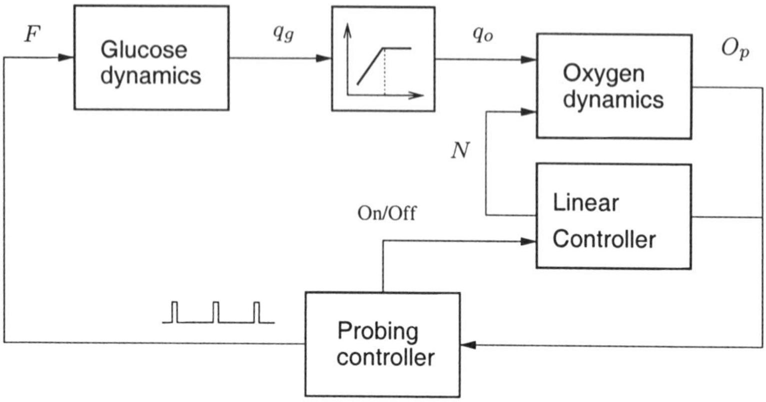

Figure 9.2. Block diagram of the closed-loop system. The size of the pulse responses in the dissolved oxygen O_p is used for feedback to adjust the feed rate F. The agitation speed N regulates O_p between the pulses in F and is kept constant during a pulse.

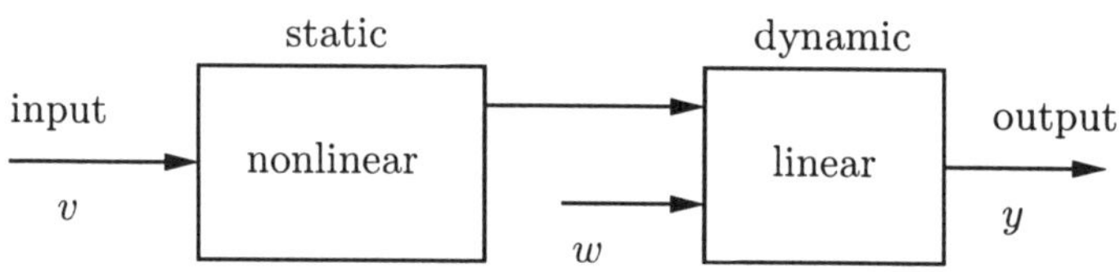

Figure 9.3. Hammerstein model.

We assume that the process is a Hammerstein model: a static nonlinearity followed by a dynamical linear process, see Figure 9.3. The case of input dynamics will be studied later. A state space representation of the process can be written as

$$\begin{aligned} \dot{x} &= Ax + B_1 f(v) + B_2 w \quad x \in R^n \\ y &= Cx. \end{aligned} \tag{9.8}$$

The control objective is to find and track the optimal point for which the gradient of f is small. The probing controller gets information about the nonlinearity from pulses that are periodically superimposed to the control signal. The input signal v to the process is the sum of the piecewise constant signal u_k and the perturbation signal $u_p(t)$:

$$v(t) = u_k + u_p(t) \qquad t \in [kT,\ kT + T]. \tag{9.9}$$

$u_p(t)$ is a pulse train with period T and amplitude u_p^0:

$$u_p(t) = \begin{cases} 0 & t \in [kT,\ kT + T_c) \\ u_p^0 & t \in [kT + T_c, (k+1)T). \end{cases} \tag{9.10}$$

T_c is the duration of the regulation phase, while T_p is the length of the probing pulse. We have the equality

$$T = T_p + T_c. \tag{9.11}$$

The piecewise constant control signal u_k is adjusted at the end of every pulse, depending on the size of the pulse response. Figure 9.4 illustrates the behavior of the probing controller.

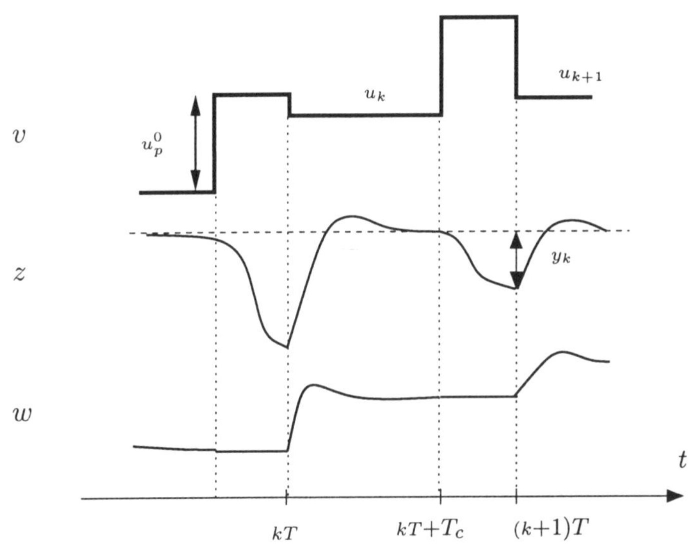

Figure 9.4. Illustration of the probing controller. A pulse in the input signal v leads to a response in the output z. The size y_k of the pulse response is used to compute the change $u_{k+1} - u_k$.

A cycle starts with the regulation phase: the process input v is kept constant while the second input w regulates the output y:

$$t \in [kT,\ kT + T_c) \qquad \begin{cases} \dot{x} = Ax + B_1 f(u_k) + B_2 w \\ \dot{x_c} = A_c x_c + B_c y \\ y = Cx \\ w = C_c x_c + D_c y, \end{cases} \tag{9.12}$$

where $x_c \in R^m$ is the controller state of the second loop. After integration between kT and $kT+T_c$, we get

$$\begin{bmatrix} x(kT+T_c) \\ x_c(kT+T_c) \end{bmatrix} = A_{d1} \begin{bmatrix} x(kT) \\ x_c(kT) \end{bmatrix} + B_{d1} f(u_k), \tag{9.13}$$

where A_{d1} and B_{d1} are given in Appendix A.

During the probing phase the control signal w is kept constant:

$$t \in [kT+T_c,\ kT+T) \quad \begin{cases} \dot{x} = Ax + B_1 f(u_k + u_p^0) + B_2 w(kT+T_c) \\ \dot{x}_c = 0 \\ y = Cx. \end{cases} \tag{9.14}$$

After integration between $kT+T_c$ and $kT+T$, we get

$$\begin{bmatrix} x(kT+T) \\ x_c(kT+T) \end{bmatrix} = A_{d2} \begin{bmatrix} x(kT+T_c) \\ x_c(kT+T_c) \end{bmatrix} + B_{d2} f(u_k + u_p^0) \tag{9.15}$$

where A_{d2} and B_{d2} are given in Appendix A.

From Eqs. (9.13) and (9.15) it is possible to express the response to a pulse $y_k = z((k+1)T) - z(kT+T_c)$ as the output of a discrete-time system with sample interval T:

$$\begin{aligned} X_{k+1} &= A_{d2} A_{d1} X_k + \begin{bmatrix} A_{d2} B_{d1} & B_{d2} \end{bmatrix} \begin{bmatrix} f(u_k) \\ f(u_k + u_p^0) \end{bmatrix} + a_d \\ y_k &= C(A_{d2} - I) A_{d1} X_k + C \begin{bmatrix} (A_{d2} - I) B_{d1} & B_{d2} \end{bmatrix} \begin{bmatrix} f(u_k) \\ f(u_k + u_p^0) \end{bmatrix}, \end{aligned} \tag{9.16}$$

where X_k is given by

$$X_k = \begin{bmatrix} x(kT) \\ x_c(kT) \end{bmatrix}.$$

For a better understanding of the probing controller consider the case of a very long regulation phase between the pulses. Assuming stability of (9.12), the influence of the state x_k on y_k vanishes when T_c goes to infinity and Eq. (9.16) reduces to a static input-output map:

$$y_k = C(e^{AT_p} - I) A^{-1} B \cdot (f(u_k + u_p^0) - f(u_k)). \tag{9.17}$$

Using the integrating feedback law

$$u_{k+1} = u_k + K(y_k - y_r) \tag{9.18}$$

with a desired pulse response $y_r = 0$, the equilibrium point u_∞ is such that $f(u_\infty + u_p^0) - f(u_\infty)$ vanishes, *i.e.* for a u_∞ that makes the gradient small.

This rough analysis indicates that a probing strategy using (9.18) can converge to the optimal point. What happens when the pulses are done more frequently? Do we have convergence to the optimal point when the dynamics in (9.12) are taken into account? Is the probing strategy able to track a time-varying nonlinearity? The chapter aims at answering those questions.

We will only consider functions f that are piecewise affine. The closed-loop equations described by (9.16) and (9.18) have consequently a piecewise affine structure. Stability analysis can be performed by searching for piecewise quadratic Lyapunov functions. Modifications of the existing methods are however necessary to cope with the integrator that the control law (9.18) introduces. In the next section, tools for stability as well as performance analysis of piecewise affine systems will be derived. Evaluation of the probing strategy will be performed in Section 6 using those tools on a bioreactor example.

5. Tools for Global Analysis

Consider the piecewise affine system

$$\bar{X}^+ = \bar{A}_i \bar{X} \qquad \bar{X} \in \bar{\mathbf{X}}_i \tag{9.19}$$

where

$$\bar{X} = \begin{bmatrix} X \\ 1 \end{bmatrix} \qquad \bar{A}_i = \begin{bmatrix} A_i & a_i \\ 0 & 1 \end{bmatrix},$$

$\bar{\mathbf{X}}_i \subseteq R^n$ is a partition of the state space into convex polyhedral regions.We assume that there is only one equilibrium point, and that it is located in the region with index $i = i_0$. The origin is shifted such that $a_{i_0} = 0$.

5.1. *Stability analysis*

The search for piecewise quadratic Lyapunov functions, as in [5, 6], is a powerful tool for stability analysis. Denote by V the Lyapunov function candidate:

$$V(\bar{X}) = \begin{cases} \bar{X}^T \bar{P}_i \bar{X} & \bar{X} \in \bar{\mathbf{X}}_i, \quad i = i_0 \\ \bar{X}^T \bar{P}_i \bar{X} & \bar{X} \in \bar{\mathbf{X}}_i, \quad i \neq i_0. \end{cases} \tag{9.20}$$

For V to be a Lyapunov function, one should have, for $i \neq i_0$,

$$\begin{aligned} &V(\bar{X}) = \bar{X}^T \bar{P}_i \bar{X} > 0 && \bar{X} \in \bar{\mathbf{X}}_i \\ &\Delta V(\bar{X}) = \bar{X}^T (\bar{A}_i{}^T \bar{P}_j \bar{A}_i - \bar{P}_i) \bar{X} < 0 && \bar{X} \in \bar{\mathbf{X}}_i, \ \bar{X}^+ \in \bar{\mathbf{X}}_j, \end{aligned} \tag{9.21}$$

and similarly for $i = i_0$. The search for the matrices $\bar{P}_i$ can be formulated as an optimization problem in terms of LMIs. The stability conditions (9.21) take the form:

$$\begin{aligned} &\bar{P}_i - \bar{R}_i > 0 \\ &\bar{A}_i{}^T \bar{P}_j \bar{A}_i - \bar{P}_i + \bar{S}_{ij} < 0 \end{aligned} \tag{9.22}$$

where $\bar{R}_i$ and $\bar{S}_{ij}$ are matrices used in the S-procedure. They express the fact that the inequalities are only required to hold for particular $\bar{X}$, *e.g.* $\bar{X}$ in $\bar{\mathbf{X}}_i$. More details on how to find such matrices can be found in [6].

A solution to (9.22) implies the existence of $\gamma > 0$ such that $\Delta V(\bar{X}) < -\gamma\|\bar{X}\|^2$ for all $\bar{X}$. When the state partition contains an unbounded region with an integrator, it may not be possible to bound ΔV quadratically in all directions although it is strictly negative. Modifications of Eqs. (9.22) for the regions with an integrator are therefore necessary. Our approach is similar to that in [3] for linear systems and consists in deriving a reduced LMI set after removal of the nullspace of $A_i - I$.

Consider a region $\bar{\mathbf{X}}$ of the state partition, where the dynamics contain an integrator. By a change of coordinates $\bar{Z} = \bar{T}^{-1}\bar{X}$, the dynamic equation in $\mathbf{X}$ can be put in the form:

$$\bar{Z}^+ = \underbrace{\left[\begin{array}{cc|c} A_s & 0_{n-1\times 1} & \multirow{2}{*}{$T^{-1}a$} \\ 0_{1\times n-1} & 1 & \\ \hline 0_{1\times n-1} & 0 & 1 \end{array}\right]}_{\bar{D}} \bar{Z}, \qquad \bar{Z} = \left[\begin{array}{c} z_s \\ z_u \\ \hline 1 \end{array}\right],$$

where A_s has all its eigenvalues in the open unit ball. Defining $\bar{Q} = \bar{T}^{-T}\bar{P}\bar{T}$, one can express $\Delta V(\bar{X})$ using the new coordinates $\bar{Z}$:

$$\begin{aligned} \Delta V(\bar{T}\bar{Z}) &= \bar{Z}^T(\bar{D}^T\bar{Q}\bar{D} - \bar{Q})\bar{Z} \\ &= \left[\begin{array}{c} z_s \\ z_u \\ \hline 1 \end{array}\right]^T \left[\begin{array}{cc|c} M_{ss} & M_{su} & M_s \\ M_{su}^T & 0 & m_u \\ \hline M_s^T & m_u & m \end{array}\right] \left[\begin{array}{c} z_s \\ z_u \\ \hline 1 \end{array}\right]. \end{aligned} \tag{9.23}$$

The absence of quadratic term in z_u is a consequence of the eigenvalue 1. Application of the S-procedure can introduce a negative quadratic term in z_u only if the region $\mathbf{X}$ is bounded in the z_u direction. When the state can pass to infinity along the eigendirection defined by z_u, the cell description can be rewritten as:

$$\bar{\mathbf{X}} = \{\bar{Z} \mid z_u < \begin{bmatrix} G_j & g_j \end{bmatrix} \begin{bmatrix} z_s \\ 1 \end{bmatrix}, \quad \text{for } j = 1 \cdots p\}. \tag{9.24}$$

Define matrices $\bar{N}_j, j = 1, \cdots, p$ such that for all $i \neq j$

$$\begin{bmatrix} z_s \\ 1 \end{bmatrix}^T \bar{N}_j \begin{bmatrix} z_s \\ 1 \end{bmatrix} > 0 \text{ when } \begin{bmatrix} G_i - G_j \; g_i - g_j \end{bmatrix} \begin{bmatrix} z_s \\ 1 \end{bmatrix} \prec 0.$$

The following result should be combined with (9.22) to investigate stability of piecewise affine systems.

Theorem 1: If there exist $\bar{Q}$ with $\bar{Q}_{su} = 0_{n-1\times 1}$ and $\bar{R}$ such that, for $j = 1, \cdots, p$,

$$\bar{T}^T \bar{Q} \bar{T}^{-1} - \bar{R} > 0$$

$$\begin{bmatrix} M_{ss} & M_s + m_u G_j \\ (M_s + m_u G_j)^T & m + m_u g_j \end{bmatrix} + \bar{N}_j < 0$$

$$m_u > 0$$

then $V(\bar{X}) > 0$ and $\Delta V(\bar{X}) < 0$ in $\bar{\mathbf{X}}$.

Proof: The first inequality in the statement guarantees positivity of V in $\bar{\mathbf{X}}$. From Eq. (9.23) we have

$$\Delta V(\bar{Z}) = \begin{bmatrix} z_s \\ 1 \end{bmatrix}^T \begin{bmatrix} M_{ss} & M_s^T \\ M_s & m \end{bmatrix} \begin{bmatrix} z_s \\ 1 \end{bmatrix} + 2 \begin{bmatrix} z_s \\ 1 \end{bmatrix}^T \begin{bmatrix} M_{su} \\ m_u \end{bmatrix} z_u.$$

M_{su} can be easily computed to be of the form

$$M_{su} = (A_s - I)\bar{Q}_{su}.$$

Since $\bar{Q}$ is such that $\bar{Q}_{su} = 0$, the cross-term $z_u z_s$ in ΔV vanishes, *i.e.*

$$\Delta V(\bar{Z}) = \begin{bmatrix} z_s \\ 1 \end{bmatrix}^T \begin{bmatrix} M_{ss} & M_s \\ M_s^T & m \end{bmatrix} \begin{bmatrix} z_s \\ 1 \end{bmatrix} + 2 m_u z_u.$$

By assumption, we have

$$m_u > 0,$$

hence it follows from (9.24) that

$$m_u z_u < m_u \begin{bmatrix} G_j \; g_j \end{bmatrix} \begin{bmatrix} z_s \\ 1 \end{bmatrix}, \quad j = 1, \cdots, p.$$

For less conservative bounds we add the relaxation term involving $\bar{N}_j$

$$m_u z_u < m_u \begin{bmatrix} G_j \; g_j \end{bmatrix} \begin{bmatrix} z_s \\ 1 \end{bmatrix} + \begin{bmatrix} z_s \\ 1 \end{bmatrix}^T \bar{N}_j \begin{bmatrix} z_s \\ 1 \end{bmatrix}, \quad j = 1, \cdots, p,$$

which leads to p upper-bounds for ΔV, namely

$$\Delta V(\bar{Z}) < \begin{bmatrix} z_s \\ 1 \end{bmatrix}^T \begin{bmatrix} M_{ss} & M_s + m_u G_j \\ (M_s + m_u G_j)^T & m + m_u g_j \end{bmatrix} \begin{bmatrix} z_s \\ 1 \end{bmatrix} + \begin{bmatrix} z_s \\ 1 \end{bmatrix}^T \bar{N}_j \begin{bmatrix} z_s \\ 1 \end{bmatrix}, \quad j = 1, \cdots, p,$$

and each of them is negative by hypothesis. □

Remark 1: The condition $\bar{Q}_{su} = 0$ is not restrictive but actually necessary for ΔV to be negative in the unbounded region.

Remark 2: Along the z_u direction, V is decreasing linearly and the inequality $m_u > 0$ imposes the correct sign of the slope to ΔV.

Remark 3: Since the search for Lyapunov functions is done in terms of $\bar{Q}$, *i.e.* in the new coordinate systems, it is easy to impose $\bar{Q}_{su} = 0$. $\bar{Q}$ is used in Eqs. (9.22) as $\bar{P}_i = \bar{T}^T \bar{Q} \bar{T}^{-1}$ for some i.

5.2. *Performance analysis*

The previous result provides a way to check global stability of piecewise affine systems with integrator, using standard LMI solvers. However, it does not guarantee a good behavior in presence of disturbances.

In [7], the authors propose a way to analyze the servo-problem for piecewise affine systems in continuous-time. Performance was evaluated by computing the L_2 gain between the derivative of the exogeneous input, $\dot{r}$, and the error, $x - x_r$, between the system trajectory x and a predetermined trajectory x_r.

An extension of the method to piecewise affine systems in discrete time will be performed. Consider the following piecewise affine system with input r:

$$x(k+1) = A_i x(k) + B_i r(k) + a_i. \tag{9.25}$$

The reference trajectory is determined by the sequence of equilibrium points $x_r(k)$:

$$x_r(k) = (I - A_{i_0})^{-1} B_{i_0} r(k) \tag{9.26}$$

and the performance is measured with the following cost function:

$$J(x, r) = \sum_{k=0}^{\infty} (x(k) - x_r(k))^T \bar{Q} (x(k) - x_r(k)).$$

Suppose that for any constant $r \in \mathbb{R}$ the piecewise linear system has a unique equilibrium point located in $\mathbf{X}_{i_0}$.

Define

$$\bar{A}_i = \begin{bmatrix} A_i & a_i & B_i + (A_i - I)(I - A_0)^{-1} B_0 \\ 0 & 1 & 0 \\ 0 & 0 & 1 \end{bmatrix}$$

$$\bar{B}_i = \begin{bmatrix} -(I - A_0)^{-1} B_0 \\ 0 \\ 1 \end{bmatrix}, \quad \bar{I} = \mathrm{diag}(I, 0, 0)$$

and the matrices $\bar{S}_i$ such that

$$\begin{bmatrix} x - x_r \\ 1 \\ r \end{bmatrix}^T \bar{S}_i \begin{bmatrix} x - x_r \\ 1 \\ r \end{bmatrix} > 0 \text{ for } x \in \mathbf{X}_i, r \in \mathbb{R}.$$

We then have the following statement.

Theorem 2: If there exist $\gamma > 0$ and $P_i > 0$ such that $\bar{P}_i = \mathrm{diag}\{P_i, 0\}$ satisfies

$$\begin{bmatrix} \bar{A}_i{}^T \bar{P}_j \bar{A}_i - \bar{P}_i + \bar{Q} + \bar{S}_i & (\bar{B}_i{}^T \bar{P}_j \bar{A}_i)^T \\ \bar{B}_i{}^T \bar{P}_j \bar{A}_i & \bar{B}_i{}^T \bar{P}_j \bar{B}_i - \gamma^2 \end{bmatrix} < 0, \qquad i, j \neq i_0$$

and similarly for $i = j = 0$, then every trajectory defined by (9.25) with $x(0) = 0$ satisfies

$$J(x, r) < \gamma^2 \sum_{k=0}^{\infty} (r(k+1) - r(k))^2.$$

The proof is similar to that in the continuous case [7] and a sketch of the proof is given in Appendix C.

6. Case Study

In this section we will illustrate the performance of the probing controller on an example based on Section 2. It is desirable to control the process to a saturation instead of an extremum. The process is modeled by a first order system:

$$\begin{aligned} \dot{x} &= -ax + f(v) + w \qquad a > 0 \\ v &= u_k + u_p(t). \end{aligned}$$

The static nonlinearity f that models the saturation in the cell respiration system is taken to be a min function, *i.e.*

$$f(v) = \min(v, r_k) = \begin{cases} v, & v \le r_k \\ r_k, & v > r_k. \end{cases} \tag{9.27}$$

To start with we will assume that the saturating point r_k is constant $r_k = 0$. Time varying r_k will be considered in Section 6.3 for performance evaluation of the controller.

For the sake of simplicity, we assume that the output regulation between the pulses is performed using state feedback. In that way, no additional state is introduced by the controller and symbolic computation for local analysis is tractable

$$w = \begin{cases} Lx & t \in [kT,\ kT + T_c) \\ Lx(kT + T_c) & t \in [kT + T_c, (k+1)T). \end{cases} \tag{9.28}$$

The closed-loop system described by (9.16) and (9.18) is a discrete piecewise linear system with a state space partitioned into 3 regions:

$$\begin{aligned} \mathbf{X}_1 &= \{X_k \in R^2, u_k < -u_p^0\} \\ \mathbf{X}_2 &= \{X_k \in R^2, -u_p^0 < u_k < 0\}\ , \quad X_k = \begin{bmatrix} x(kT) \\ u_k \end{bmatrix}. \\ \mathbf{X}_3 &= \{X_k \in R^2, u_k > 0\} \end{aligned}$$

The system equations can be written as

$$X_{k+1} = A_i X_k + a_i, \quad \text{for} \quad X_k \in \mathbf{X}_i, \quad i \in I = \{1, 2, 3\},$$

where the matrices A_i and a_i are given in Appendix B for the case $L = 0$.

6.1. *Simulations*

A simulation of the closed-loop system has been carried out and the results are shown in Figure 9.5. The input v to the process starting at 0 is gradually increased by the controller. At time $t \approx 15$, the saturation is reached and no pulse response is visible. As a consequence, the controller output is decreased. At steady state, the pulse response is equal to the desired value of $y_r = 2$ and the controller output u is slightly below the saturation.

The probing control strategy has a few parameters to be chosen. Some tuning guidelines are necessary for the strategy to work well.

Figure 9.6 shows how the choice of the probing controller gain K influences the closed-loop performance. Large K values give fast convergence but may lead to instability.

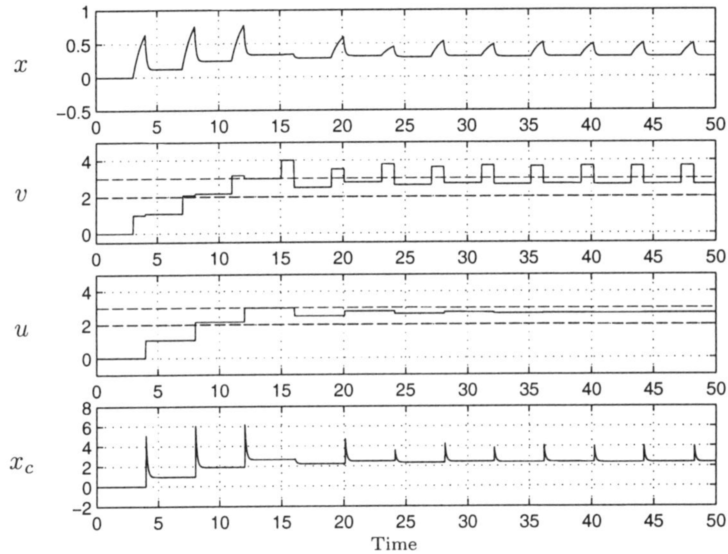

Figure 9.5. Simulation of the closed-loop system using the probing controller. The controller output u is gradually increased until the response to the pulses equals the setpoint $y_r = 0.2$. At steady state, u is just below the saturation and the process output x is regulated around 0. The dashed lines represent the cell borders.

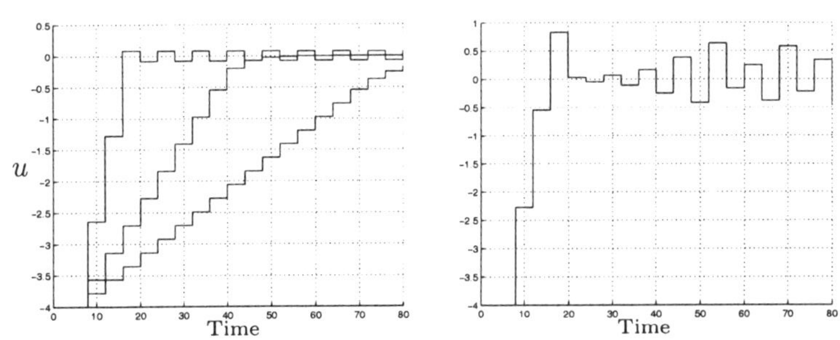

Figure 9.6. Control signal u from the probing controller for several gains: K = 0.5, 1, 3.1 (left) and 4 (right). Small gains result in a sluggish convergence while large gains may give overshoots or instability.

6.2. *Local analysis*

Local analysis in the different state space regions can be performed to derive tuning guidelines. The probing controller uses the size of the pulse response given by the dynamical system (9.16) for feedback. The static response of this system gives an idea of the pulse responses one can expect. The steady-state amplitude of the pulse response is plotted in Figure 9.7 where the constant β is a function of the plant dynamics

$$\beta = (1 - e^{-aT_c})(1 - e^{-aT})^{-1}(1 - e^{-aT_p})a^{-1}.$$

The static pulse response indicates that the integrating feedback law (9.18) with $0 < y_r < \beta u_p^0$ can drive the input u_k close to the saturation. The reference value y_r for the pulse response affects the steady-state distance to the saturating point.

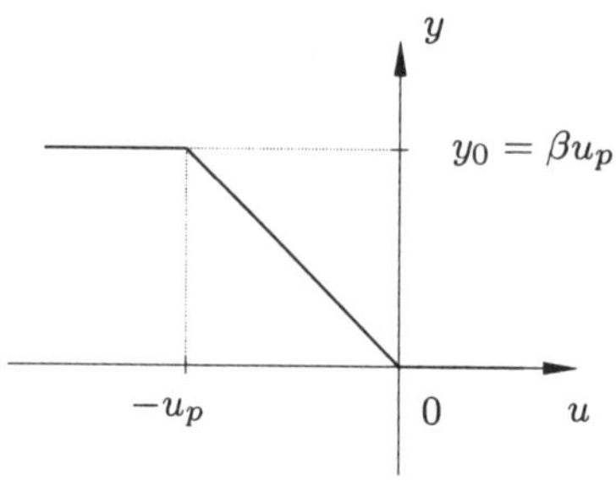

Figure 9.7. Stationary amplitude of the pulse response y as a function of u. Small responses indicate that u is close to the saturating point $r = 0$.

Since integrators are always present in the extreme regions, the equilibrium point of the closed-loop system, if it exists, is located in the middle region $\mathbf{X}_2$. It is easy to derive a necessary condition for A_2 to be Hurwitz

$$0 < K < \frac{2}{\beta}\frac{1 + e^{-aT}}{1 - e^{-aT}} \xrightarrow[T_c \to \infty]{} \frac{2a}{1 - e^{-aT_p}}. \tag{9.29}$$

The presence of the integrator in the extreme regions may give rise to situations where the state vector tends to infinity along the critically stable eigendirections. The vector field should therefore be oriented towards the middle region on these directions. Inspection of the vector field leads to the following inequality:

$$0 < \frac{y_r}{u_p^0} < \beta \tag{9.30}$$

and

$$0 < K < \frac{e^{aT_c} - 1}{\beta}. \tag{9.31}$$

Equation (9.30) relates the size of the desired pulse response to the dynamics of the open-loop: y_r should not exceed the largest pulse response βu_p^0 that one gets at steady state, see Figure 9.7. Equations (9.29) and (9.31) give bounds on the controller gain K and therefore limit the convergence speed to the middle region. Figure 9.8 shows the stability region in the parameter space (K, T_c) in two cases: with output feedback between the pulses ($L = 10$) and without output regulation ($L = 0$). It is interesting to notice that with a fast output regulation there remains only one constraint on K, which is $K < \frac{2}{\beta}$, and smaller T_c are allowed.

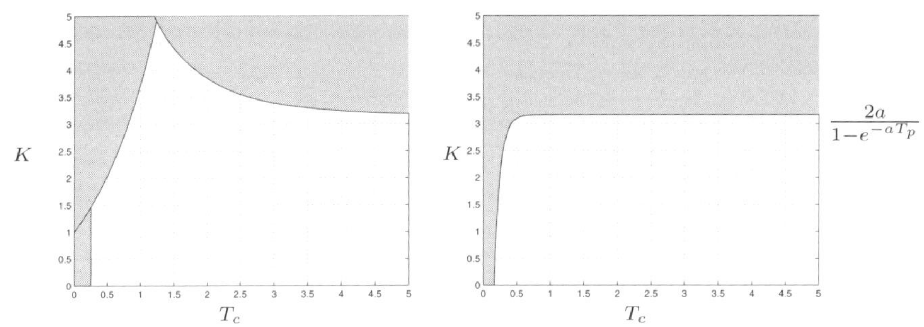

Figure 9.8. Stability region in the parameter space for $L = 0$ (left) and $L = 10$ (right). Pairs (K, T_c) in the shaded zone gives instability.

6.3. *Global analysis*

The stability conditions derived in the previous section are useful for the design of the probing controller, but they are not sufficient for global stability. Since the equilibrium point is located close to a cell border, the validity of the local analysis in $\mathbf{X}_2$ is rather limited.

For numerical computations we take

$$a = 1, \quad u_p^0 = 1, \quad T_p = 1.$$

Conditions (9.29) and (9.31) impose constraints on K and T_c. We choose $T = 4$ and $K = 2$ to get a fast convergence and some robustness margin.

Equation (9.30) provides an interval in which y_r should be: $0 < y_r < 0.612$. We choose $y_r = 0.3$ to get a symmetric behavior above and below the saturation.

The parameters of the probing controllers have now been chosen such that all necessary conditions are fulfilled. Global stability can be investigated using Theorem 1.

6.3.1. *Global stability*

The LMIs are implemented and solved using *Matlab*. Stability of the closed-loop system can be proved for gains $K < 3.1$, which is very close to the upper bound from the local analysis, Eq. (9.29), when $T = 4$. Level curves of the piecewise quadratic Lyapunov function as well as the phase plane are shown in Figure 9.9. Convergence of u to a neighborhood of the saturation can be guaranteed for all initial values of u and x.

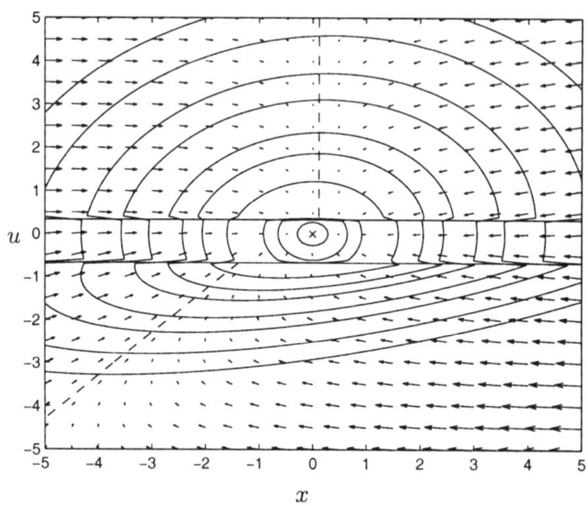

Figure 9.9. Phase plane with some level curves of the Lyapunov function. The solid lines represent the cell borders and the dashed line are the eigendirections corresponding to the eigenvalue 1.

6.3.2. *Performance*

For a better understanding of the probing controller a simulation with a particular trajectory r_k has been performed. The result is shown in Figure 9.10. The probing controller succeeds to track the time-varying saturation by using the pulse responses for feedback. Theorem 2 from last section can be used to quantify the performance of the closed-loop system for all variations of $r_k \in [-5, 5]$. It can be checked that for these r_k the equilibrium point is always in the middle region defined by $r_k < u_k < r_k - u_p^0$. The integrator in the control law is replaced by a pole close to 1 to simplify calculations.

The matrix $\bar{Q}$ that penalizes the state deviation $x - x_r$ from its equilibrium point is taken to be $\bar{Q} = \text{diag}\{1, 1, 0, 0\}$. The LMIs from Theorem 2 turn out to be feasible. Minimizing γ subject to the constraints, one obtains $\gamma \approx 30$.

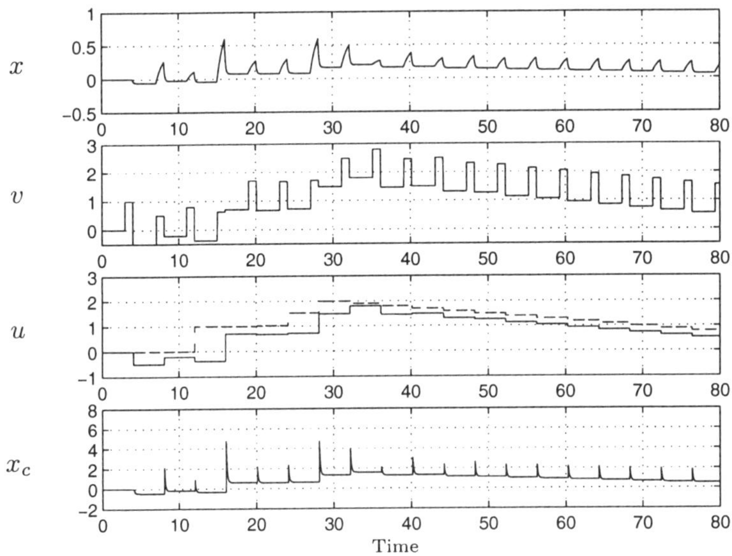

Figure 9.10. Simulation of the probing controller with a time-varying optimal point. Top: the input signal v with superimposed pulses is tracking the reference trajectory defined by r_k. Bottom: output signal z and deviation $y_k - y_r$ of the pulse responses from the reference value.

Minimization of γ has been performed for different values of the gain K. The result is plotted in Figure 9.11 together with the gain obtained by simulation with a particular r. A better agreement between the gain obtained by simulations and γ can be achieved by looking for a worst case disturbance r. The graph is however helpful for design purposes. The slow convergence speed for small K values is indicated by the large γ values. The plot suggests a K value of about 1.5. Larger values of K do not improve much the performance and may give poor robustness properties, as it is seen in simulations.

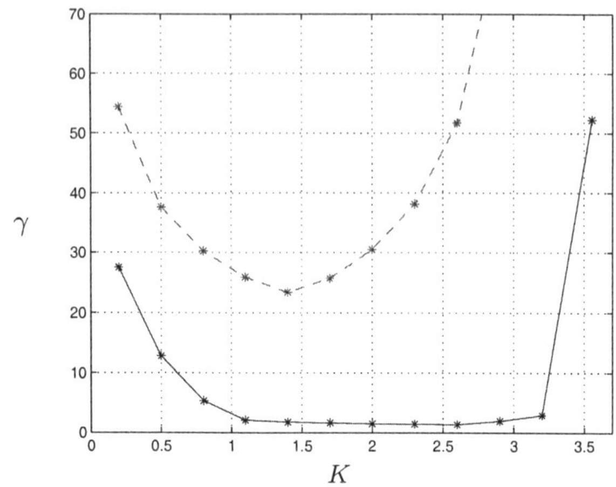

Figure 9.11. Performance measurement for different values of the probing controller gain K. The dashed line represents the γ values obtained by numerical computations whereas the solid line is the result of simulations.

7. Input Versus Output Dynamics

As it was mentioned in Section 2, the Hammerstein system is a good description of the fed-batch process in the later part of a cultivation where the oxygen dynamics are dominating over the glucose one. At the beginning of the fermentation, the situation is the opposite and the Wiener system is a better description of the process. In spite of their similar static responses the two configurations are significantly different. Figure 9.12 illustrates the difference between the two configurations.

Different behavior can therefore be expected when applying the probing strategy to the two different systems. It will be shown in this section that the case of an output nonlinearity is more complex and that similar performance cannot be achieved.

As in Section 4, it is possible to describe the closed-loop system by a discrete-time system. The response y_k to a pulse is given by the output of a nonlinear system of the form:

$$\begin{aligned} x_{k+1} &= A_{d2}A_{d1}x_k + (A_{d2}B_{d1} + B_{d2})u_k + a_d \\ y_k &= f(Cx_{k+1}) - f(C(A_{d1}x_k + B_{d1}u_k)). \end{aligned} \tag{9.32}$$

When f is a saturation function and the control law is given by Eq. (9.18), the closed-loop system is a piecewise affine system with a state space divided into 4 regions. There are still two regions associated with dynamics containing an integrator and a region where the equilibrium point is lo-

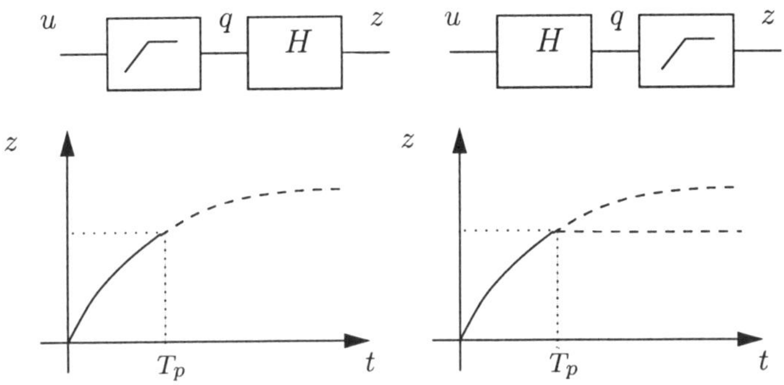

Figure 9.12. The effect of an input nonlinearity is to restrict the range of the input values to the linear dynamics. After a step change in u the output z will smoothly increase and it is possible to predict the future output from a finite time experiment. In the output nonlinearity case (right), the output z is not smooth and it is not possible to predict the output z after a short time observation.

cated. The additional region describes the situation where the output q of the linear process is above the saturation at the beginning of a pulse and below the saturation at the end of the same pulse. The dynamics associated with this state region are unstable.

Global stability analysis can be performed as in Section 5, but it will not be performed here. A good insight into the fundamental differences between the configurations can be obtained by considering an approximated system. When the time T_c between two successive pulses is large, Eq. (9.32) can be approximated as in (9.17) by

$$y_k = f(-CA^{-1}Bu_k + C(e^{AT_p} - I)A^{-1}Bu_p^0) - f(-CA^{-1}Bu_k). \tag{9.33}$$

The condition (9.29) for local stability is changed to

$$0 < K < 2a\frac{e^{-aT} + 1}{1 - 2e^{-aT_c} + e^{-aT}} \xrightarrow[T_c \to \infty]{} 2a. \tag{9.34}$$

From Eq. (9.34) it can be seen that the plant dynamics have a strong influence on the maximal gain K. Contrary of the input saturation case, the maximal gain K can take values below 2.

The dynamics of the two systems below the saturation are identical and lead to a convergence speed V to the saturation given by

$$V = K\frac{1 - e^{-aT_p}}{a(T_c + T_p)}u_p^0.$$

The constraints (9.29) and (9.34) on the gain K for local stability lead however to different maximal speeds:

$$\begin{aligned} V^{max}_{Hammerstein} &= \frac{2}{T_c + T_p} u^0_p \\ V^{max}_{Wiener} &= 2\frac{1 - e^{-aT_p}}{T_c + T_p} u^0_p. \end{aligned} \tag{9.35}$$

From Eq. (9.35), the effect of the plant dynamics and the pulse duration on the performance are clear. In the Hammerstein system case arbitrary short pulses could be performed in order to increase the convergence speed to the saturation. The process dynamics have no impact on the achievable convergence speed. In practice, the minimal pulse duration is related to the output noise level. In the Wiener system case the situation is much different. Both the plant dynamics and the pulse duration play an essential role. Slow plant dynamics have a direct impact on the convergence speed to the saturation. The influence of the pulse duration is illustrated in Figure 9.13. The convergence speed presents an optimum for $T_p \approx 1.7$. Small T_p result in small pulse responses, which in turn are fed back for small feed adjustments. Long pulses imply larger pulse periods and reduce the convergence speed, too. There exists an optimal length for the pulses that is a trade-off between frequent updates and long probing.

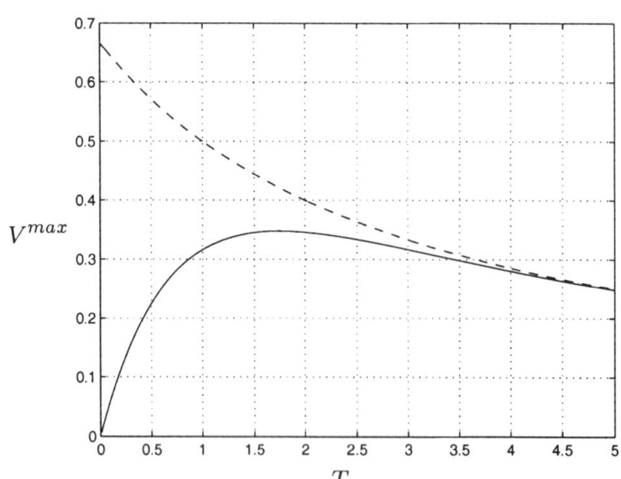

Figure 9.13. Maximal convergence speed as a function of the pulse duration T_p when the saturation is at the input (dashed line) and output (solid line). The numerical values $T_c = 3,\ a = 1,\ u^0_p = 1$ have been used in both cases.

8. Experimental Results

The probing strategy has been evaluated on many different platforms, from laboratory to large scale bioreactors, see for instance [2, 4]. The tuning of the probing controller is done for every reactor setup. Figure 9.14 shows a part of a fed-batch experiment with *E. coli* in a 3 liters bioreactor. The following parameters were used.

- A pulse length T_p of 90 s was chosen, which is a bit larger than the glucose time constant estimated to about one minute at this stage of the cultivation.
- The length T_c of the output regulation phase was chosen to be 6 minutes. The dissolved oxygen control, achieved by PID control of the agitation speed, sets a limit on the minimal length of the regulation phase.
- The normalized gain K of the probing controller was taken to be 1.2 in agreement with the analysis from Section 6.
- A rather low y_r value of 3 % was taken in order to achieve a fast increase in the feed during the exponential growth phase.

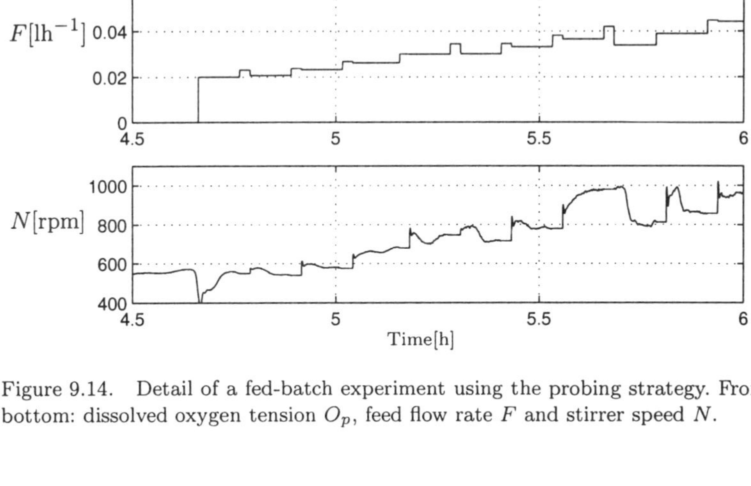

Figure 9.14. Detail of a fed-batch experiment using the probing strategy. From top to bottom: dissolved oxygen tension O_p, feed flow rate F and stirrer speed N.

The feed started after depletion of the initial glucose, detected by a peak in the dissolved oxygen signal. The feed is thereafter increased by the feedback algorithm to meet the glucose demand of the growing biomass. At two occasions, the feed is not increased because of the absence of clear pulse response. This is a good illustration of the sensitivity of the cells to glucose feeding and it demonstrates the ability of the strategy to adapt to the time-varying demand in glucose.

9. Conclusion

A probing strategy used for feed rate control in fed-batch fermentations has been analyzed. After neglecting some dynamics in the plant, the problem can be formulated in a piecewise affine framework. Numerical algorithms have been proposed to study stability and performance of the closed-loop system. The efficiency of the methods have been illustrated on an example where it is desirable to track a saturating point. Finally, the difference between the input and output saturation cases has been discussed. The efficiency of the probing technique has also been demonstrated by an experiment.

Bibliography

1. M. Åkesson and P. Hagander. A simplified probing controller for glucose feeding in Escherichia coli cultivations. *Proc. 39th IEEE Conf. Decision and Control*, pages 4520–4525, 2000.
2. M. Åkesson, P. Hagander, and J.P. Axelsson. Avoiding acetate accumulation in Escherichia coli cultures using feedback control of glucose feeding. *Biotechnol. Bioeng.*, 73(3):223–230, 2001.
3. S. Boyd, L. El Ghaoui, E. Feron, and V. Balakrishnan. *Linear Matrix Inequalities in System and Control Theory.* SIAM, Philadelphia, 1994.
4. L. de Maré, L. Andersson, and P. Hagander. Probing control of glucose feeding in Vibrio cholerae cultivations. *Bioprocess and Biosystems Eng.*, 25:221–228, 2003.
5. G. Ferrari-Trecate, F.A. Cuzzola, D. Mignone, and M. Morari. Analysis and control with performance of piecewise affine and hybrid systems. *Proc. American Control Conference, Arlington, Virginia*, 2001.
6. M. Johansson. *Piecewise Linear Control Systems.* Ph.D. Thesis, Department of Automatic Control, Lund Institute of Technology, Sweden, 1999.
7. S. Solyom and A. Rantzer. The servo problem for piecewise linear systems. *Proc. 15th Int. Symp. Math. Theory of Networks and Systems, Notre Dame, Indiana*, 2002.
8. J. Sternby. Extremum control systems – an area for adaptive control? *Preprints Joint American Control Conference, San Francisco, California*, 1980.

9. S. Velut, L. de Maré, J.P. Axelsson, and P. Hagander. Evaluation of a probing feeding strategy in large scale cultivations. Technical Report ISRN LUTFD2/TFRT--7601--SE, Department of Automatic Control, Lund Institute of Technology, Sweden, 2002.
10. S. Velut and P. Hagander. Analysis of a probing control strategy. *Proc. American Control Conference, Denver, Colorado*, pages 609–614, 2003.
11. S. Velut and P. Hagander. A probing control strategy: stability and performance. *Proc. 43rd IEEE Conf. Decision and Control, Paradise Island, Bahamas*, 2004.

Appendix A

From Eq. (9.12) we get

$$\begin{bmatrix} \dot{x} \\ \dot{x}_c \end{bmatrix} = A_{c1} \begin{bmatrix} x \\ x_c \end{bmatrix} + B_{c1} f(u_k),$$

where A_{c1} and B_{c1} are given by

$$A_{c1} = \begin{bmatrix} A + B_2 D_c C & B_2 C_c \\ B_c C & A_c \end{bmatrix} \qquad B_{c1} = \begin{bmatrix} B_1 \\ 0 \end{bmatrix}.$$

The matrices A_{d1} and B_{d1} are then

$$A_{d1} = e^{A_{c1} T_c} \qquad B_{d1} = \int_0^{T_c} e^{A_{c1}(T_c - \sigma)} d\sigma \; B_{c1}.$$

One can compute A_{d2} and B_{d2} similarly and get

$$A_{d2} = e^{A_{c2} T_p} + \int_0^{T_p} e^{A_{c2}(T_p - \sigma)} d\sigma \begin{bmatrix} B_2 \begin{bmatrix} D_c C & C_c \end{bmatrix} \\ 0 \end{bmatrix}$$

$$B_{d2} = \int_0^{T_p} e^{A_{c2}(T_p - \sigma)} d\sigma \; B_{c2},$$

where A_{c2} and B_{c2} are

$$A_{c2} = \begin{bmatrix} A & 0 \\ 0 & 0 \end{bmatrix} \qquad B_{c2} = \begin{bmatrix} B_1 \\ 0 \end{bmatrix}.$$

Appendix B

The closed-loop dynamics in the three regions of the state partition are given by the matrices

$$
\begin{aligned}
A_1 &= \begin{bmatrix} e^{AT} & (e^{AT}-I)A^{-1}B \\ KC(e^{AT}-e^{AT_c}) & 1+KC(e^{AT}-e^{AT_c})A^{-1}B \end{bmatrix} \\
a_1 &= \begin{bmatrix} 0 \\ -Ky_r \end{bmatrix} + \begin{bmatrix} I \\ KC \end{bmatrix} (e^{AT_p}-I)A^{-1}Bu_p^0 \\
A_2 &= \begin{bmatrix} e^{AT} & (e^{AT}-e^{AT_p})A^{-1}B \\ KC(e^{AT}-e^{AT_c}) & 1+KC(e^{AT_p}-I)(e^{AT_c}-I)A^{-1}B \end{bmatrix} \\
a_2 &= \begin{bmatrix} 0 \\ -Ky_r \end{bmatrix} \\
A_3 &= \begin{bmatrix} e^{AT} & 0 \\ KC(e^{AT}-e^{AT_c}) & 1 \end{bmatrix} \qquad a_3 = \begin{bmatrix} 0 \\ -Ky_r \end{bmatrix}.
\end{aligned}
$$

Appendix C

The state error can be written as

$$
\begin{aligned}
x_{k+1} - x_{rk+1} &= A_i(x_k - x_{rk}) + A_i(I-A_{i_0})^{-1}B_{i_0}r_k \\
&\quad + B_i r_k + a_i - (I-A_{i_0})^{-1}B_{i_0}r_{k+1} \\
&= A_i(x_k - x_{rk}) - (I-A_{i_0})^{-1}B_{i_0}(r_{k+1}-r_k) \\
&\quad + \underbrace{B_i + (A_i - I)(I-A_{i_0})^{-1}B_{i_0}r_k + a_i}_{=0 \text{ for } i=i_0}.
\end{aligned}
$$

Consider now the system with input $r_{k+1} - r_k$, which describes the error $x_k - x_{rk}$

$$
\begin{bmatrix} x_{k+1} - x_{rk+1} \\ 1 \\ r_{k+1} \end{bmatrix} = \underbrace{\begin{bmatrix} A_i & a_i & B_i + (A_i - I)(I-A_{i_0})^{-1}B_{i_0} \\ 0 & 1 & 0 \\ 0 & 0 & 1 \end{bmatrix}}_{\bar{A}_i} \begin{bmatrix} x_k - x_{rk} \\ 1 \\ r_k \end{bmatrix}
+ \underbrace{\begin{bmatrix} -(I-A_{i_0})^{-1}B_{i_0} \\ 0 \\ 1 \end{bmatrix}}_{\bar{B}_i} (r_{k+1} - r_k).
$$

If the following LMI is feasible

$$\begin{bmatrix} \bar{A}_i{}^T \bar{P}_j \bar{A}_i - \bar{P}_i + \bar{I} + \bar{S}_{ij} & * \\ \bar{B}_i{}^T \bar{P}_j \bar{A}_i & \bar{B}_i{}^T \bar{P}_j \bar{B}_i - \gamma^2 \end{bmatrix} < 0$$

then the L_2 gain from $r_{k+1} - r_k$ to $x_k - x_{rk}$ is less than γ, see [5].

CHAPTER 10

A Compensator Design Framework for Attenuation of Wave Reflections in Long Cable Actuator-Plant Interconnections

A. de Rinaldis[1], R. Ortega[1] and M. W. Spong[2]

[1] *Laboratoire des Signaux et Systèmes, Supélec, 91192 Gif-sur-Yvette, France, E-mail: {derinaldis,Romeo.Ortega}@lss.supelec.fr*

[1] *Coordinated Science Laboratory, University of Illinois, Urbana, IL 61801, USA, E-mail: m-spong@uiuc.edu*

The wave reflection phenomenon that appears when actuator and plant are connected through long cables is studied in this chapter. In several applications, the perturbation induced by the presence of these reflected waves is non-negligible and seriously degrades the performance of the control and the operativity of the system. Standard compensation schemes are based on matching impedances at specific frequencies and are realized with the addition of linear RLC filters. Impedance matching is clearly ineffective if there is no single dominant frequency in the system and/or the plant is highly uncertain. In this work a novel compensator design framework applicable for the latter scenario is proposed. In contrast with the standard schemes the compensators are *active* and require for their implementation regulated sources placed either on actuator or plant side. A port-interconnection viewpoint is adopted and the cable is modeled via the transmission line equations in their scattering representation. Under the assumptions of plant linearity and piecewise approximation of the signals – which is reasonable if the line propagation delay is small – a family of current decoupling compensators, that requires only knowledge of the line parameters, and ensures stability and asymptotic tracking for all (unknown) plants with *passive impedance*, is proposed. An adaptive version that estimates the line characteristic impedance is also presented. Some simulation results in a benchmark example of voltage overshoot suppression in AC drives are shown.

1. Introduction

In this chapter, we are interested in the problem of compensation of the wave effects that appear when the controlled plant, with non-negligible dynamic impedance, is coupled to the actuator through long feeding cables. The connecting cables behave as a transmission line inducing a wave reflection that deforms the transmitted signals and degrades the quality of the control. In some applications, including the classical power distribution and digital communications problems, attention can be centered in one dominant frequency at which the overall system operates. In these cases, and assuming the plant is linear and known, it is possible to design linear time-invariant (LTI) RLC filters that will match – *at that particular frequency* – the load impedance to the impedance of the (compensated) line, hence avoiding the wave reflection problem. If the plant parameters are uncertain adaptive implementations are needed, see the second example in Section 3 for an illustration. Furthermore, if the plant is nonlinear the overall concept of impedance matching is far from clear.

There are some applications where there is no single dominant frequency and/or the plant is highly uncertain. A prototypical example is the overvoltage problem in high-performance AC drives [12], where the actuator is a PWM inverter that sends through the long cables a fast rising pulse that should be reproduced without distortion on the motor side, whose linear approximation changes dramatically with the operating point. The reflecting waves generate high voltage spikes at the motor terminals that can produce potentially destructive stress on the motor insulation, constituting a serious practical problem. Since impedance matching *for all frequencies* is possible only in the case when the load is purely resistive, it is clear that an alternative perspective should be adopted when the load high-frequency dynamics cannot be neglected – like in the case of induction motors.

A novel framework for the design of active compensators to reduce the wave reflections, when the plant is unknown and there is no single dominant frequency in the system, was suggested in [7]. The qualifier "active" is important since we depart from the standard RLC LTI filter implementations and assume that regulated sources can be placed either on the actuator or the plant side. The central objectives of this chapter are to elaborate in detail the material briefly sketched in [7] and to present some new results stemming from the use of the control design framework. The main novelties of the proposed approach are

(1) the adoption of the port interconnection viewpoint for the controller design, and
(2) the use of the scattering variables representation of the transmission line.

Using port representations of the four components – actuator, compensator, transmission line and plant – allows to formulate this (non-standard) controller design problem in terms of achievable behaviors between the terminals of the first and the third ports, without the *knowledge* of the plant. Furthermore, assuming that the plant is *passive*, stability can be enforced restricting to behaviors such that the operator seen from the plant is also *passive*.

On the other hand, the scattering representation relates voltages and currents at the line extremes via a simple delay transfer matrix, with the delay being the line propagation. Henceforth, the interconnection of the line with a linear discrete-time compensator will also be an LTI system and the characterization of the achievable (actuator-to-plant) behaviors becomes an algebraic problem.

The remaining of this chapter is organized as follows. In Section 2 we present the model of the system under consideration, including the compensator configuration, and discuss the limitations for *performance improvement* of RLC LTI compensators. Two motivating practical examples are given in Section 3. Section 4 contains the scattering representation and the ideal full-decoupling scheme of [7]. In Section 5 we present the compensator design framework that aims at characterizing all behaviors that correspond to causal and well-posed interconnections and ensure stability. The well-posedness analysis requires the additional assumptions of plant linearity and piecewise approximation of the signals – which is reasonable if the line propagation delay is small – and is carried out in Section 6. This analysis reveals that any full-decoupling scheme, as well as any voltage-decoupling one, will yield ill-posed interconnections. This motivates the consideration in Section 7 of a current-decoupling controller, for which a complete stability analysis, based on the passivity considerations described above, is possible. An adaptive version of the scheme is given in Section 8. Finally, in order to test the compensation schemes performances, Section 9 contains some simulation results of a benchmark example of voltage overshoot suppression in an AC drive consisting of a PWM inverter and an induction motor. We wrap up the chapter with some concluding remarks and open problems in Section 10, where we discuss in particular the technological implications of our approach.

Notation: We define the differentiation and advance-delay operators, acting on signals $x : \mathbb{R} \to \mathbb{R}$, as $(p^k x)(t) \triangleq \frac{d^k}{dt^k} x(t)$ and $(q^{\pm k} x)(t) = x(t \pm kd)$, respectively, where $d \in \mathbb{R}_+$ and $k \in \mathbb{Z}_+$. Their Laplace transform counterparts, which are used to define transfer functions, are s and $z = e^{ds}$, respectively.

2. Systems Configuration and Limitations of Current Practice

2.1. *System model*

To model the plant connected to the actuator through long cables we consider the configuration shown in Figure 10.1, where we model the connecting cables as a two-port system whose dynamics are described via the Telegrapher's equations

$$C\frac{\partial v(t,x)}{\partial t} = -\frac{\partial i(t,x)}{\partial x}, \quad L\frac{\partial i(t,x)}{\partial t} = -\frac{\partial v(t,x)}{\partial x}, \tag{10.1}$$

where $v(t,x)$, $i(t,x)$ represent the line voltage and current, respectively, $x \in [0,\ell]$ is the spatial coordinate, with $\ell > 0$ the cable length and $C, L > 0$, which are assumed constant, are the capacitance and inductance of the line, respectively. The actuator is modeled as a one-port whose port variables, $(v(t,0), i(t,0))$, are directly connected to the line. It consists of a voltage source, $v_s(t)$, connected in series with a resistor R_a.

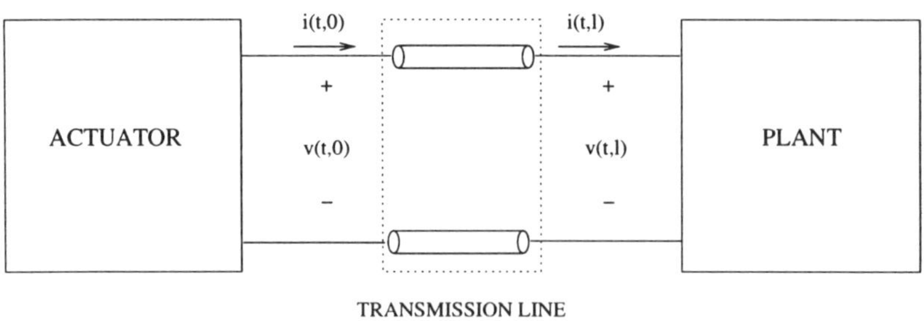

Figure 10.1. Uncompensated systems configuration.

The transmission line is terminated by the plant, which is a one-port, with port variables $(v(t,\ell), i(t,\ell))$. If we assume the plant is LTI[a] the dy-

[a]Neither the compensator design framework nor the stability analysis, presented in Section 5, require this assumption, however, the well-posedness analysis is presented only for LTI plants.

namics of the overall system is described by (10.1) together with

$$\begin{aligned} v(t,0) &= -R_a i(t,0) + v_s(t) \\ v(t,\ell) &= Z_p(p) i(t,\ell) \end{aligned} \tag{10.2}$$

where $Z_p(s) \in \mathbb{R}(s)$ is the plant impedance – that we assume is strictly stable but otherwise *unknown*. In Appendix A it is shown that the mapping from the source voltage to the plant voltage is given by the linear *delay-differential* operator

$$v(t,\ell) = \frac{1}{2}[1 + K_p(p)](1 - K_a)\frac{q^{-1}}{1 - K_a K_p(p) q^{-2}} v_s(t) + \epsilon_t, \tag{10.3}$$

where $d \triangleq \ell\sqrt{LC}$ is the propagation delay, ϵ_t is an exponentially decaying term, that will be omitted in the sequel, and

$$K_a \triangleq \frac{R_a - Z_0}{R_a + Z_0}, \quad K_p(s) \triangleq \frac{Z_p(s) - Z_0}{Z_p(s) + Z_0}, \tag{10.4}$$

are the so-called reflection coefficients, with $Z_0 \triangleq \sqrt{\frac{L}{C}}$ the line characteristic impedance. For further developments it will be assumed that $K_p(s)$ is also strictly stable. Notice that if these coefficients are nonzero the delayed signal $K_a K_p(p) v(t - 2d, \ell)$ appears in the dynamics. This term captures the physical phenomenon of wave reflection that deforms the transmitted signals and degrades the quality of the control.

Compensators are introduced to attenuate the wave reflections. The compensator may be placed at the actuator or plant sides leading to the configurations shown in Figure 10.2 and Figure 10.3, respectively.

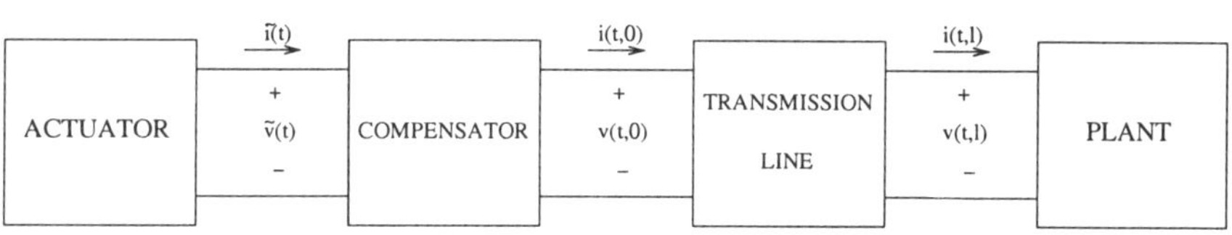

Figure 10.2. Port representation of the system with compensator on the actuator side.

2.2. *Limitations of impedance matching*

As explained in the introduction, the standard way to attenuate the wave reflections is to introduce RLC LTI filters to match the load impedance to the impedance of the (compensated) line. At this point it is convenient to

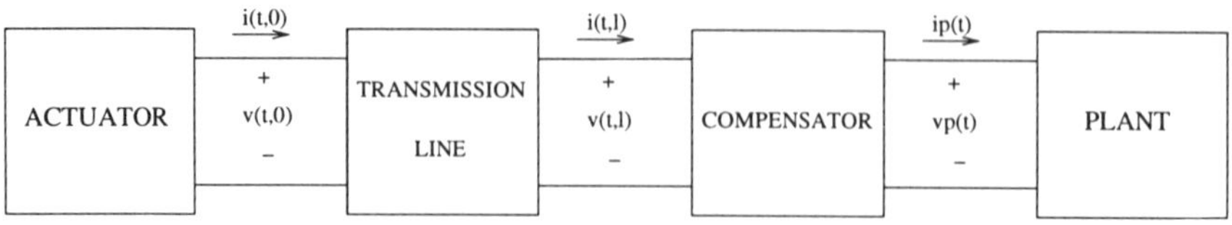

Figure 10.3. Port representation of the system with compensator on the actuator side.

review this approach to underscore its intrinsic limitation for the problem at hand.

Under the assumption of LTI plant, an LTI filter can be placed between the line and the plant, Figure 10.3, to create a new impedance $\tilde{Z}_p(s)$, which for any given frequency ω_0 satisfies $\tilde{Z}_p(j\omega_0) = Z_0$. This makes the reflection coefficient $K_p(j\omega_0) = 0$ and reduces the transfer function (10.3) to a simple (scaled) delay, hence eliminating the wave reflections for a system *operating at this frequency.* This property is known as impedance matching and is well documented in the transmission lines literature. See the first example below for a practical illustration of this idea. (Clearly, a similar effect can be achieved, but now modifying the reflection coefficient K_a, if the compensator is placed as shown in Figure 10.2.)

There are at least three drawbacks of the impedance matching approach. First, the parameters of the plant dynamics must be *exactly known* – notice that both, phase and amplitude of $Z_p(j\omega_0)$ must be changed. Second, as the concept of impedance matching is poorly understood for nonlinear systems, it also heavily relies on the assumption of linearity of the plant. Third, and more importantly, in applications where the signals contain a wide spectrum this approach is effective only if the plant high-frequency dynamics can be neglected. Indeed, impedance matching *for all frequencies* is possible only in the case when the load is purely resistive, when we can make $\tilde{Z}_p = Z_0$.

The interest in this chapter is in cases, like the overvoltage problem described in the second example below, where there is no dominant operating frequency and the plant is unknown with high-frequency dynamics that cannot be approximated by a constant – in these cases, an alternative perspective to the problem should be adopted.

2.3. *Limitations of RLC LTI filtering*

Due to power considerations, in the standard compensator configuration used in the impedance matching, only *shunt* RLC LTI filters are used, which as shown now severely restricts our ability to change the dynamic behavior of the system. In Figure 10.4 is shown the electrical configuration of a shunt RLC filter placed on the actuator side. Applying Thevenin's rules, it turns out that

$$v(t,0) = \frac{Z_c(p)}{Z_c(p)+R_a} v_s(t), \tag{10.5}$$

where $Z_c(p)$ is the equivalent impedance of the RLC filter. Equation (10.5) reveals the critical role played by the coefficient R_a, called actuator surge impedance, on the achievable behaviors. Indeed, the effect of a shunt filter for small values of R_a is negligible, and totally disappears in the limit as $R_a \to 0$, *i.e.* when $\tilde{v}(t) \to v_s(t)$.

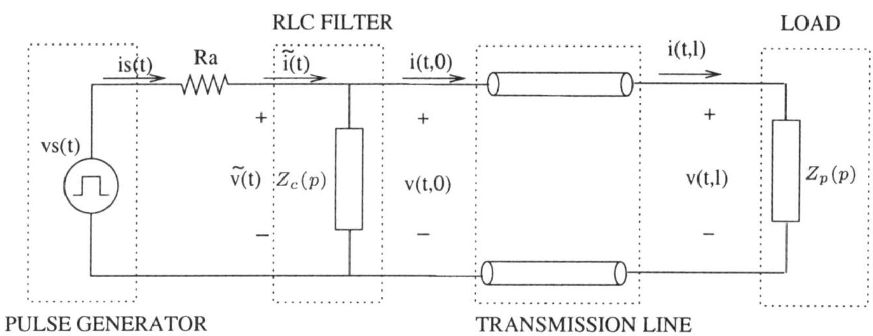

Figure 10.4. Electrical scheme of an Actuator Output RLC filter placed in shunt.

To overcome this limitation we will, following [7], assume that *active regulated sources* can be placed either on the actuator or the plant side. This is an important departure from the standard approach whose technological implications are discussed in Section 10.

3. Two Motivating Examples

Before presenting our proposal let us illustrate with two examples the issues described above.

3.1. *Microwave heating*

Electromagnetic energy in the microwave and radio-frequency (RF) portions of the spectrum can be used to heat or defrost (thaw) foodstuffs [4]. In this example, we will consider a fixed frequency RF oven, where a matching circuit is used to ensure that the combined impedance of the oven and the food (the load) matches the impedance of the generator and the transmission line, in order to avoid reflection power from the load, see Figure 10.5. The system is highly nonlinear and there are significant sources of uncertainty in the model of the process. While the characteristic impedance of the transmission line Z_0 is not matched, the resonant frequency and quality factor of the circuit is altered by the presence of wave reflections, which in turn, changes the power transfer to the load. The change in load impedance may also move the operating frequency of the circuit outside the agreed limits, which can result in interference with radio communications.

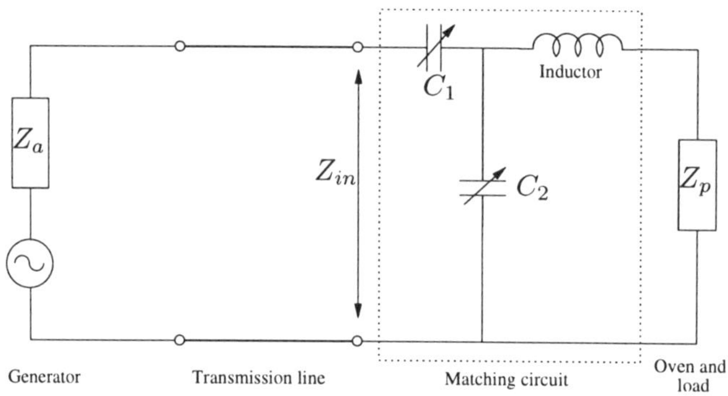

Figure 10.5. Arrangement of generator, transmission line, matching circuit and load.

In this case the actuator is simply a sinusoidal voltage source series-connected to a resistor Z_a. As far as the frequency is fixed, the load Z_p can be expressed as a complex number. Moreover, the coaxial cable, through which the electrical power is supplied, is modeled as lossless transmission line where the characteristic impedance is Z_0. Then, the problem is how to tune the capacitors parameters C_1, C_2 in order to get $Z_{in} = Z_0$. If $Z_{in} = R_{in} + jX_{in}$ denotes the impedance matching circuit, oven and load then the expression for Z_{in} can be simplified by defining $Y_L = a + jb$ as the admittance of the load and inductor in series, $j\alpha$ as the impedance of capacitor C_1, and $j\beta$ as the admittance of capacitor C_2.

The relation between the load impedance and the tuning parameters α and β is then

$$Z_{in} = j\alpha + \frac{1}{j\beta + Y_p}.$$

Differentiating this to obtain the change of Z_{in} with respect to time gives

$$\dot{Z}_{in} = j\dot{\alpha} - (Z_{in} - j\alpha)^2(j\dot{\beta} + \dot{Y}_p).$$

Because the dynamics in a heating process have time constants of the order of hundreds of seconds whereas α and β can be moved over their full range in a few seconds, the load admittance, Y_p, changes slowly compared to the changes in α and β. s a result, the control problem can be considered as adjusting α and β to bring Z_{in} to the desired value, *i.e.* 1+j0, and then to maintain it at this value in the presence of slowly varying changes in Y_p.

In this chapter we are interested in applications where there is no single dominant frequency. In the previous section we argue that impedance matching *for all frequencies* is possible only in the case when the load is purely resistive, hence if the load high-frequency dynamics cannot be neglected the impedance matching approach is inadequate. This scenario arises when the actuator is a fast switching device that generates pulses that excite the high frequency modes of the plant, for instance in the overvoltage problem in high-performance AC drives [12] that we explain below.

3.2. *Overvoltage in AC electrical drives*

In modern AC drive applications the use of fast switching actuators (typically PWM inverters based on IGBT technology) induces high voltage spikes at the motor terminals which can produce potentially destructive stress on the motor insulation. The motor cables represent an impedance to the PWM voltage pulses from the drive. These cables contain significant values of inductance and capacitance that are directly proportional to their length. The peak value and rise time of the reflected voltage waveform can have significant impact on the insulation inside the motor, which invariably exhibits mechanical stress cracks in the enamel wire insulation and microscopic voids in the insulation coating. These holes and cracks can become insulation failure points when voltage peaks are impressed on the stator winding by the reflected wave phenomenon.

The model proposed in Section 2 is often adopted in this application, where the inverter is modeled as an ideal PWM voltage source, $v_S(t)$, plus a series resistor, R_a. To represent the high frequency terms in the rising and falling edges of the PWM pulses R_a usually takes large values. Indeed, for such high frequency components, the inverter contains large stray inductance, skin effects, and RF emission losses. Such losses are typically included in R_a that is called surge impedance. The transmission line is modeled by (10.1) while the motor is assumed to be LTI and modeled by a high frequency R-C circuit in parallel with a low frequency R-L branch.

Figure 10.6 shows the pulse train response of the system (10.3) with a dynamic load that approximates the high-frequency behavior of an induction motor and leads to the reflection coefficients $K_a = -0.97$ and

$$K_p(s) = \frac{0.34s^2 + 2.98 \times 10^6 s - 2.39 \times 10^9}{s^2 + 2.99 \times 10^6 s + 2.48 \times 10^9}.$$

Inductance, capacitance and resistance of the cable are chosen as $L = 0.97\mu\text{H/m}$, $C = 45p\text{F/m}$, $R = 50m\Omega/\text{m}$, and the length of the cable is set to be $\ell = 100\text{m}$. These values are calculated from the measured S-parameters by a network analyzer in the experimental facility of POSTECH, see [10]. Further, they are confirmed by a high speed digital sampling oscilloscope which is used to measure transport delay ($d = 0.66\mu\text{s}$) and characteristic impedance ($Z_0 = 146.8\Omega$). As for the cable, the motor parameters are calculated from the measured values, using an impedance analyzer. The simulation clearly exhibits the undesirable ringing due to the wave propagation.

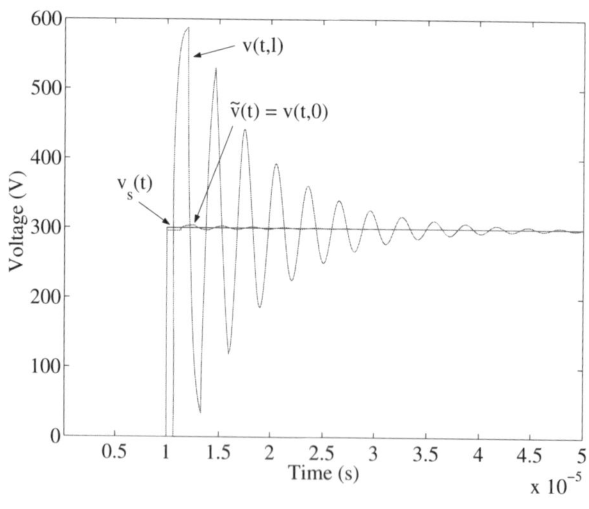

Figure 10.6. Pulse train response of the system (10.3) for an approximate model of an induction motor load, with zero initial conditions.

4. Scattering Representation

As will become clear in the next section the use of scattering representation of the transmission line is instrumental for the compensator design. This is contained in the following well-known lemma [13] which is essential for our further developments and whose proof is given for completeness.

Lemma 1: *Consider the transmission line Eqs. (10.1). Then,*

$$\begin{bmatrix} v(t,\ell) \\ i(t,\ell) \end{bmatrix} = W(q) \begin{bmatrix} v(t,0) \\ i(t,0) \end{bmatrix}, \tag{10.6}$$

where the transfer matrix

$$W(z) \triangleq T^{-1} \begin{bmatrix} z^{-1} & 0 \\ 0 & z \end{bmatrix} T, \qquad T \triangleq \begin{bmatrix} 1 & Z_0 \\ 1 & -Z_0 \end{bmatrix},$$

with $Z_0 = \sqrt{\frac{L}{C}}$ *the line characteristic impedance, and* $d = \ell\sqrt{LC}$ *the propagation delay.*

Proof: Define the so-called *scattering variables*[b]

$$\begin{bmatrix} s_+(t,x) \\ s_-(t,x) \end{bmatrix} = T \begin{bmatrix} v(t,x) \\ i(t,x) \end{bmatrix}. \tag{10.7}$$

It can easily be shown that Eqs. (10.1) can be written as

$$\frac{\partial s_+(t,x)}{\partial t} = -\frac{1}{\sqrt{LC}} \frac{\partial s_+(t,x)}{\partial x}, \qquad \frac{\partial s_-(t,x)}{\partial t} = \frac{1}{\sqrt{LC}} \frac{\partial s_-(t,x)}{\partial x}. \tag{10.8}$$

It follows that $s_+(t,x) = m(t - \sqrt{LC}x)$, $s_-(t,x) = m(t + \sqrt{LC}x)$ are solutions of (10.8) for any C^1 function $m(\cdot)$.[c] Noting that

$$\begin{aligned} s_+(t,0) &= m(t), \qquad s_+(t,\ell) = m(t-d) \\ s_-(t,0) &= m(t), \qquad s_-(t,\ell) = m(t+d) \end{aligned}$$

where we have used the definition of d above, we establish the well-known relation for the scattering variables

$$\begin{bmatrix} s_+(t,\ell) \\ s_-(t,\ell) \end{bmatrix} = \begin{bmatrix} q^{-1} & 0 \\ 0 & q \end{bmatrix} \begin{bmatrix} s_+(t,0) \\ s_-(t,0) \end{bmatrix}. \tag{10.9}$$

The proof is completed using (10.7) in (10.9) again and noting that T is a full-rank matrix. □

---

[b]The functions $s_+(t,x)$ and $s_-(t,x)$ have the interpretation as left and right traveling waves, respectively.

[c]This is, of course, the well-known D'Alembert solution of the wave equations, that may be found on any elementary text of partial differential equations, *e.g.* [13].

5. Compensator Design Problem

Two distinguishing features of our control problem are the *lack of knowledge* of the plant dynamics and the fact that specifications are given in terms of *transient performance* improvement – namely, reducing the overshoot of the step response – instead of stabilization. (As a matter of fact, stabilization is not an issue here because the uncompensated system is stable and, under the reasonable assumption that the plant is passive, any passive compensator – for instance, RLC LTI filters – will preserve stability.)

To handle these two aspects of the problem the compensator design framework proposed here proceeds in the following steps:

S1 Characterize the behaviors that can be assigned to the compensator-transmission line subsystem, that is, to the mappings

$$(\tilde{v}(t), \tilde{i}(t)) \to (v(t,\ell), i(t,\ell))$$

for the configuration of Figure 10.2, and

$$(v(t,0), i(t,0)) \to (v_p(t), i_p(t))$$

for the one of Figure 10.3.[d] In this step, the main issues to be addressed concern properness and well-posedness.

S2 As we will be using active elements in the compensator stability is no longer ensured, to enforce this *sine-qua-non* condition we then restrict to behaviors such that the operator seen from the plant is *passive*.

Motivated by the scattering representation of the transmission line (10.6) we consider discrete-time compensators of the form[e]

$$\begin{bmatrix} v(t,0) \\ i(t,0) \end{bmatrix} = C(q) \begin{bmatrix} \tilde{v}(t) \\ \tilde{i}(t) \end{bmatrix}, \tag{10.10}$$

with $C(z) \in \mathbb{R}^{2\times 2}(z)$ – not necessarily proper. Connected with the transmission line in the configuration of Figure 10.2 yields

$$\begin{bmatrix} v(t,\ell) \\ i(t,\ell) \end{bmatrix} = M(q) \begin{bmatrix} \tilde{v}(t) \\ \tilde{i}(t) \end{bmatrix}, \tag{10.11}$$

where we have defined the transfer matrix

$$M(z) \triangleq W(z)C(z) \in \mathbb{R}^{2\times 2}(z), \tag{10.12}$$

---

[d]Throughout the rest of the chapter we will consider the control configuration of Figure 10.2. Totally analogous arguments will apply to the one of Figure 10.3.

[e]Clearly, the realization of this controller assumes knowledge of the line propagation delay d.

which characterizes the mappings alluded to in point S1 above.

Assuming known the line characteristic impedance Z_0, Eq. (10.12) parameterizes the compensator in term of the free matrix $M(z)$. This parameterization is convenient to reformulate steps S1 and S2 above as follows.

5.1. *Problem formulation*

Identify matrices $M(z)$ such that, for all plants with strictly positive real impedance $Z_p(s)$ and all source voltage references satisfying $\lim_{t\to\infty} v_s(t) = V_s$, $V_s \in \mathbb{R}$, we achieve:

W *(Well-posedness)* The compensator $C(z) = W^{-1}(z)M(z)$ admits a causal realization and the overall interconnected system is well-posed.

S *(Stability)* The compensator-transmission line subsystem asymptotically converges to a lossless steady state, that is,

$$\lim_{t\to\infty} [\tilde{v}(t)\tilde{i}(t) - v(t,\ell)i(t,\ell)] = 0.$$

Furthermore, internal stability of the overall system and asymptotic regulation of the terminal voltage is ensured, *e.g.* all internal signals are bounded and

$$\lim_{t\to\infty} [\tilde{v}(t) - v(t,\ell)] = 0.$$

The stability objective can be easily expressed in terms of constraints on $M(z)$ – even without the assumption of plant linearity. Indeed, from (10.11) it is clear that, if the steady state exists, the asymptotic stability conditions will be ensured imposing $M(1) = I$. Internal stability, established with the passivity argument explained in Step S2, imposes some degree and parametric restrictions on $M(z)$. On the other hand, establishing well-posedness for a delay-differential system, even for a linear plant, seems to be a formidable task and a key "discretization" assumption will be imposed. Under this assumption, we will show below that the well-posedness restriction will translate into some structural constraints for $M(z)$, specifically some non-decoupling and relative degree conditions.

5.2. *An ideal full-decoupling compensator*

Before concluding this section let us discuss, in the light of the remarks above, the ideal full-decoupling compensator proposed in [7]. The compensator was inspired by the codification scheme for teleoperator manipulators of [1] that transforms the pure delays introduced by the (two-directional) communication channel into a transmission line. The motivation in [1] was to avoid the destabilizing effects of the delays in force-reflecting tasks by exploiting the passivity of the transmission line, which is terminated in both extremes by passive ports – the human operator controlling the master and the contact environment of the slave. The codification scheme is depicted in Figure 10.7 where

$$C^{-1}(z) \triangleq T \begin{bmatrix} 1 & 0 \\ 0 & z^{-2} \end{bmatrix} T^{-1}.$$

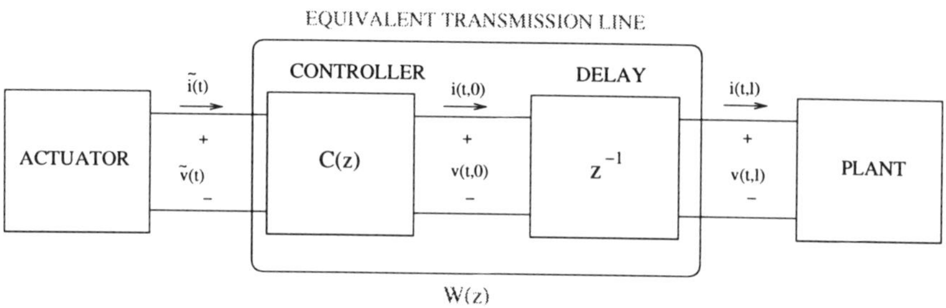

Figure 10.7. Block diagram representation of the teleoperator codification scheme of [1]

Inspired by this idea, one is tempted to try in the problem at hand the "inverse" operation, that is to undo the scattering and transform the transmission line into a pure delay decoupled system. Indeed, from (10.6) we see that the compensator (10.10) with $C(z)$ as given above yields

$$M(z) = W(z)C(z) = \begin{bmatrix} z^{-1} & 0 \\ 0 & z^{-1} \end{bmatrix}, \tag{10.13}$$

that is, it ensures

$$v(t,\ell) = \tilde{v}(t-d), \quad i(t,\ell) = \tilde{i}(t-d).$$

Compare Figure 10.7 with Figure 10.8, where a block diagram input-output representation of the compensator action is depicted.

The relations above indicate that we can arbitrarily assign the voltage and current to the plant one-port, independently of the nature of the plant. Since this is, clearly, physically impossible some further analysis is needed to identify the flaw in our derivations. Interestingly, we can prove that the compensator (10.10) admits a *causal* realization using active regulated voltage and current sources. Although there are several theoretically admissible

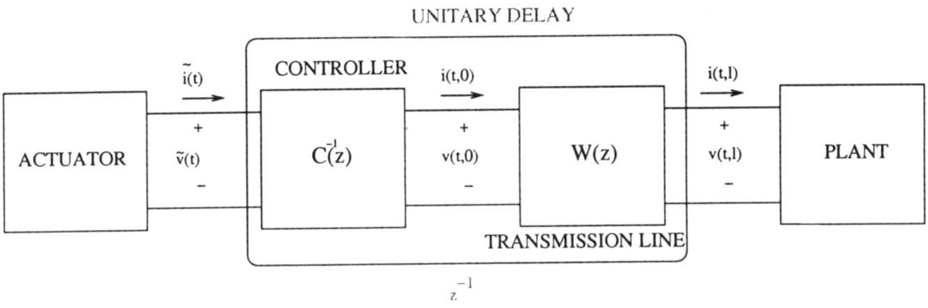

Figure 10.8. Block diagram representation of the full-decoupling ideal compensator.

configurations, for technological reasons to be explained in Section 10, we propose the one shown in Figure 10.9. Indeed, from

$$C(z) = \frac{1}{2}\begin{bmatrix} 1+z^{-2} & Z_0(1-z^{-2}) \\ \frac{1}{Z_0}(1-z^{-2}) & 1+z^{-2} \end{bmatrix}, \tag{10.14}$$

and writing the controller equations – which are given in the (t-parameter) representation – using the equivalent inverse hybrid representation (see Table 19.1 of [5]) we get

$$\begin{bmatrix} \tilde{i}(t) \\ v(t,0) \end{bmatrix} = H(q)\begin{bmatrix} \tilde{v}(t) \\ i(t,0) \end{bmatrix}, \tag{10.15}$$

where

$$H(z) = \frac{1}{1+z^{-2}}\begin{bmatrix} \frac{-1}{Z_0}(1-z^{-2}) & 2 \\ 2z^{-2} & Z_0(1-z^{-2}) \end{bmatrix}.$$

Since $H(z)$ is proper the regulated current source $\tilde{i}(t)$ and the regulated voltage source $v(t,0)$ can be causally generated as linear combinations of (delayed and undelayed) measurable signals $\tilde{v}(t)$, $i(t,0)$.

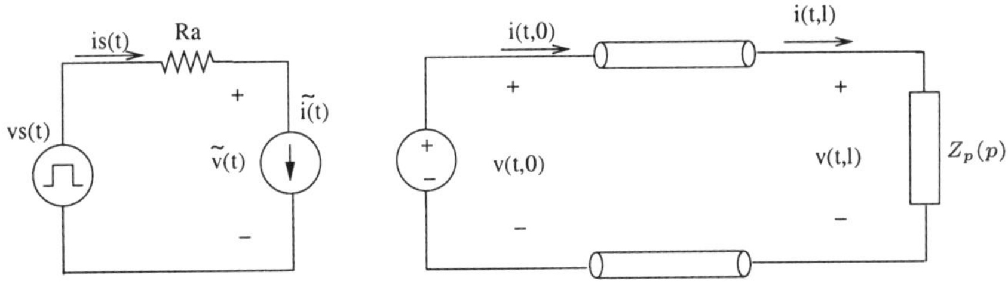

Figure 10.9. One possible circuit realization of the compensation schemes.

Once we have shown that the ideal scheme is causal it remains to study the well-posedness of the interconnection – point where the flaw will be revealed – that is carried out in the next section.

6. Discrete-Time Representation and Well-Posedness Analysis

To make the well-posedness analysis tractable we introduce the key assumption that the transmission delay is sufficiently small so that the signals can be suitably described by their piecewise approximation. More precisely, we need the following assumption:

Assumption 1: The plant current $i(t,\ell)$ verifies

$$i(t,\ell) = i(kd,\ell), \quad \forall\, t \in [kd,(k+1)d), \quad k \in \mathbb{Z}_+$$

where $d = \ell\sqrt{LC}$ is the propagation delay.

Under this assumption, the plant voltage $v(t,\ell)$ can be approximated as

$$v(t,\ell) \approx v(kd,\ell) = Z_p^*(q)i(kd,\ell), \quad \forall\, t \in [kd,(k+1)d), \quad k \in \mathbb{Z}_+ \quad (10.16)$$

with $Z_p^*(z) \in \mathbb{R}(z)$ the pulse transfer function representation (with sampling time d) of the plant impedance.

A natural question that arises at this point is the validity of this approximation and the ability of the approximated system to capture the dynamics of interest. Obviously, the pertinence of Assumption 1 is determined by the order relation between d and the frequency content of $i(t,\ell)$ – that is, if d is sufficiently small in comparison to the rate of change of the signal.

Sampling the signals every d units of time is done only for simplicity, and the sampling period can be taken as $\frac{d}{N}$ for any $N \in \mathbb{Z}_+$, making the approximation even better. Unfortunately, taking a smaller sampling period generates repeated poles of $W(z)$ in the unit disk that makes the subsequent stability analysis (which is based on passivity) inapplicable.

As a consequence of our assumption the behavior of the overall delay-differential dynamics is described, at the sampling instants kd, by a purely discrete-time system, for which the well-posedness analysis follows standard lines.

We recall at this point Definition 3.9 of [3]. As indicated in [3] this definition is needed for a practical design – see also [2]. Indeed, the standard definition that looks only at the overall transfer matrix is *not sufficient* to avoid the presence of "internal" improper loops.

Definition 1: Let every subsystem of a composite system be describable by a rational transfer function. Then the composite system is said to be *well-posed* if the transfer function of every subsystem is proper and the closed-loop transfer function from any point chosen as an input terminal to every other point along the directed path is well defined and proper.

We present the well-posedness analysis for general LTI compensators that admit a difference equation representation of the form (10.15), where $H(z) \in \mathbb{R}^{2\times 2}(z)$ is *proper*. This class contains the ideal scheme (10.15), the classical passive RLC filters – which, invoking the approximation described above, can be represented by their pulse transfer functions – as well as some delay-differential systems. From the transmission line Eqs. (10.6) and the plant Eq. (10.2) we can establish the relation

$$i(t,0) = -P(q)v(t,0), \tag{10.17}$$

where

$$P(z) \triangleq -\frac{1}{Z_0}\frac{z^2 - K_p(z)}{z^2 + K_p(z)}, \tag{10.18}$$

and we recall that the reflection coefficient is defined as

$$K_p(z) = \frac{Z_p^*(z) - Z_0}{Z_p^*(z) + Z_0}.$$

We will need in the sequel the following assumption:

Assumption 2:

$$\beta_p + Z_0 \neq 0, \tag{10.19}$$

where $\beta_p \in \mathbb{R}$ is the high-frequency gain of the discretized plant impedance.

Under this (generic) assumption it is easy to prove that – even if $Z_p^*(z)$ is improper – $P(z)$ is well-defined, has zero relative degree and its high-frequency gain is $-\frac{1}{Z_0}$.

In the presence of a compensator, the actuator Eq. (10.2) becomes

$$\tilde{v}(t) = -R_a\tilde{i}(t) + v_s(t). \tag{10.20}$$

The overall system can be represented with the block diagram of Figure 10.10, where we have explicitly kept the three, potentially troublesome, feedback loops, which after some elementary operations can be reduced to the form depicted in Figure 10.11.

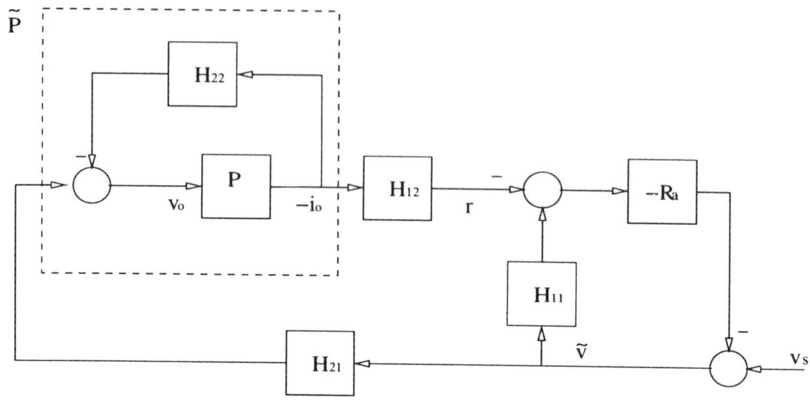

Figure 10.10. Block diagram representation of the closed-loop system.

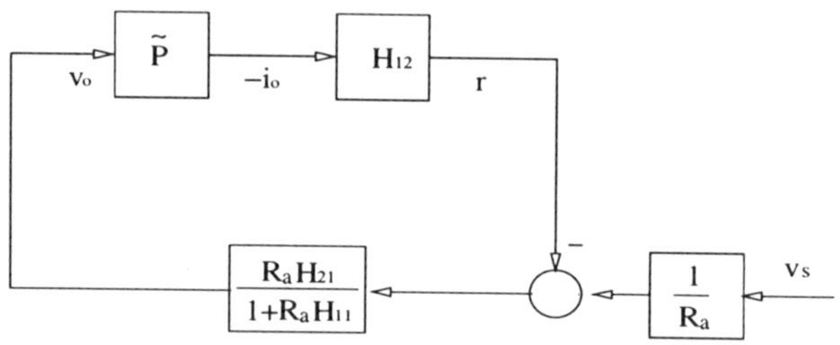

Figure 10.11. Reduced block diagram representation of the closed-loop system.

From properness of $H(z)$ and $P(z)$ the proposition below can be established via direct application of Definition 1 to the block diagram of Figure 10.11.

Proposition 1: *Consider the system depicted in Figure 10.2 where the actuator is described by (10.20), the compensator by (10.15), with* $H(z) \in \mathbb{R}^{2\times 2}(z)$ *proper, the transmission line by the Telegraphers Eq. (10.1) and the plant by* $v(t,\ell) = Z_p(p)i(t,\ell)$, *where* $Z_p(s) \in \mathbb{R}(s)$ *is strictly stable. Suppose Assumptions 1 and 2 hold. Then the overall system is well-posed, if and only if*

$$\begin{gathered} H_{11}(\infty) \neq -\tfrac{1}{R_a} \\ H_{22}(\infty) \neq Z_0 \\ \left[\tfrac{P}{1+PH_{22}} H_{12} \tfrac{R_a H_{21}}{1+R_a H_{11}}\right](\infty) \neq 1. \end{gathered} \tag{10.21}$$

Let us illustrate our derivations with the ideal compensator (10.10). Towards this end, we attract the readers attention to the first feedback loop around $P(z)$ in Figure 10.10 and recall that, under the standing Assumption 2, $P(\infty) = -\frac{1}{Z_0}$. Some simple calculations proceeding from (10.10) yield

$$H_{22}(z) = Z_0 \frac{z^2 - 1}{z^2 + 1}$$

for which we have $H_{22}(\infty) = Z_0$ that violates condition (10.21), confirming that the overall system is not well posed. Actually, we can establish the following stronger result.

Proposition 2: *All compensators $C(z)$ leading to a voltage-decoupled dynamics $(\tilde{v}(t), \tilde{i}(t)) \mapsto (v(t,\ell), i(t,\ell))$ – that is, such that the matrix*

$$W(z)C(z) = M(z) = \begin{bmatrix} m_{11}(z) & 0 \\ m_{21}(z) & m_{22}(z) \end{bmatrix}$$

where $m_{ij}(z) \in \mathbb{R}(z)$ are arbitrary and possibly improper – lead to ill-posed interconnections.[f]

Proof: The compensator transfer matrix is defined as

$$\begin{aligned} C(z) &= W^{-1}(z)M(z) \\ &= \frac{1}{2}\begin{bmatrix} (z + z^{-1}) & Z_0(z - z^{-1}) \\ \frac{1}{Z_0}(z - z^{-1}) & (z + z^{-1}) \end{bmatrix} \begin{bmatrix} m_{11}(z) & 0 \\ m_{21}(z) & m_{22}(z) \end{bmatrix} \\ &= \frac{1}{2z}\begin{bmatrix} m_{11}(z^2+1) + Z_0 m_{21}(z^2-1) & Z_0 m_{22}(z^2-1) \\ m_{11}(z^2-1) + Z_0 m_{21}(z^2+1) & Z_0 m_{22}(z^2+1) \end{bmatrix}. \end{aligned}$$

To use Proposition 2 we must express the compensator in the t-parameter representation $H(z)$. After some simple calculations we get, that independently of $M(z)$,

$$H_{22}(z) = Z_0 \frac{z^2 - 1}{z^2 + 1}$$

for which we have $H_{22}(\infty) = Z_0$ that violates condition (10.21). □

The proposition above shows that, to ensure well-posedness, we cannot aim at *voltage-decoupling*. Henceforth, the scheme that will be proposed in Section 7 will aim at current decoupling.

[f] Since $m_{21}(z)$ may be taken equal to zero, the proposition covers also the case of full-decoupled dynamics. We recall from (10.13) that the ideal compensator corresponds to $M(z) = \frac{1}{z}I$.

7. A Class of Provably Stable Compensators

In this section we present the main result of our work: a family of compensators – parameterized by one tuning coefficient – that satisfies the requirements of properness, well-posedness and stability imposed in the problem formulation of Section 5. After a brief discussion motivating our choice of current decoupling we express in terms of the achievable behaviors, *e.g.* of the matrices $M(z)$, these three requirements.

7.1. *Motivations*

In Proposition 2 we have shown that all compensators that achieve decoupling of voltage are not well-posed. We pursue then the objective of current decoupling, hence we select

$$M(z) = \begin{bmatrix} m_{11}(z) & m_{12}(z) \\ 0 & m_{22}(z) \end{bmatrix}.$$

Our motivation to aim at current decoupling is two-fold. On one hand, we will show in the next section that, due to the signal decoupling, it is possible to design *adaptive* versions of the resulting compensators, reducing the required prior knowledge on the transmission line. On the other hand, we prove at the end of this section that, thanks to the triangular structure, it is possible to carry out a complete *stability analysis* for these compensators. Indeed, terminating the current-decoupled system with the plant dynamics we obtain the transfer function

$$v(t,\ell) = \frac{m_{11}(q)}{1 - \frac{m_{12}(q)}{m_{22}(q)} \frac{1}{Z_p(q)}} \tilde{v}(t). \tag{10.22}$$

That admits the block diagram representation of Figure 10.12, from which we see that stability of $m_{11}(z)$ and *positive realness* of $-\frac{m_{22}(z)}{m_{12}(z)}$ ensure stability for all strictly positive real plants. These conditions will be imposed on our design below.

Throughout the rest of the chapter we fix $m_{11}(z) = \frac{1}{z}$. Given the voltage tracking objective this is the most reasonable choice which furthermore leads, with a slight loss of generality, to a considerable simplification in our derivations.

7.2. *Properness conditions*

We start here by imposing properness to $H(z)$.

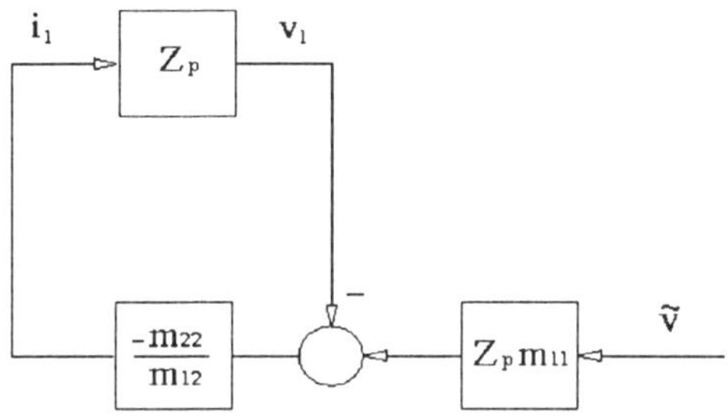

Figure 10.12. Closed-loop behavior of the current-decoupled system.

Lemma 2: (Compensator properness) *Consider the family of current-decoupling compensators with $m_{11}(z) = \frac{1}{z}$ and m_{12}, m_{22} defined as follow*

$$m_{12}(z) = \alpha z^{\rho_1} + \sum_{i=1-\rho_1}^{\infty} a_i z^{-i}$$
$$Z_0 m_{22}(z) = \beta z^{\rho_2} + \sum_{i=1-\rho_2}^{\infty} b_i z^{-i}, \tag{10.23}$$

where $\rho_i \in \mathbb{Z}$ and $\alpha, \beta, a_i, b_i \in \mathbb{R}$. Then $H(z)$ is proper if and only if $\rho_1 \geq -1$ or $\rho_2 \geq -1$.

In the particular case[g] *where*

(i) $\rho_1 = \rho_2 = \rho$,
(ii) $\alpha = -\beta$ *and* $a_{1-\rho} = -b_{1-\rho}$

$H(z)$ is proper if, and only if, $\rho \geq 1$

Proof: From $C(z) = W^{-1}(z)M(z)$ we have

$$H(z) = \begin{bmatrix} -\frac{m_{11}(z^2-1)}{z^2(m_{12}+Z_0m_{22})-(m_{12}-Z_0m_{22})} & \frac{2Z_0z}{z^2(m_{12}+Z_0m_{22})-(m_{12}-Z_0m_{22})} \\ \frac{2Z_0m_{11}m_{22}z}{z^2(m_{12}+Z_0m_{22})-(m_{12}-Z_0m_{22})} & \frac{Z_0\left[z^2(m_{12}+Z_0m_{22})+(m_{12}-Z_0m_{22})\right]}{z^2(m_{12}+Z_0m_{22})-(m_{12}-Z_0m_{22})} \end{bmatrix}, \tag{10.24}$$

where we omit the arguments of the transfer functions $m_{ij}(z)$ for brevity. Replacing $m_{11}(z) = \frac{1}{z}$ in (10.24) it is easy to see that the first line of $H(z)$

[g] As shown next, this is the case of interest when we additionally impose well-posedness.

is proper if and only if $\rho_1 \geq -1$ or $\rho_2 \geq -1$. On the other hand, the terms of the second row are proper for all values of ρ_1 and ρ_2.

For the particular case where $\alpha = -\beta$ and $a_{1-\rho} = -b_{1-\rho}$, the first line of $H(z)$ becomes

$$[-\frac{z^2 - 1}{a_{2-\rho}z^{\rho+1} + b_{2-\rho}z^{\rho+1} + 2\beta z^{\rho+1} + O^i}, \frac{2Z_0 z}{a_{2-\rho}z^{\rho} + b_{2-\rho}z^{\rho} + 2\beta z^{\rho} + O^{i-1}}]$$

where O^i represent all the terms having a degree $i \leq \rho$. We establish the claim noticing that, for both terms of this vector, properness is verified if and only if $\rho \geq 1$. □

7.3. *Well-posedness conditions*

We proceed now to characterize the matrices $M(z)$ ensuring *well-posedness* of the interconnection and properness of $H(z)$. Towards this end we recall that for fixed $M(z)$ the compensator transfer matrix $H(z)$ is uniquely determined by the transmission line. With this observation in mind and referring to conditions (10.21), we note that the only critical one is $H_{22}(\infty) \neq Z_0$ – as the other conditions do not involve the line parameters and are therefore satisfied for almost all R_a. We will concentrate our attention then on the critical (non-generic) condition.

Lemma 3: (Well-posedness) *Let m_{11}, m_{12} and m_{22} be defined as in Lemma 2. Then,*

(i) $\rho_1 \neq \rho_2 \Rightarrow H_{22}(\infty) = Z_0$.

(ii) $\rho_1 = \rho_2 = \rho$ *with* $\rho \geq 1$ *ensures the properness of* $H(z)$ *and* $H_{22}(\infty) \neq Z_0$ *provided*

$$\alpha = -\beta \neq 0, \quad a_{1-\rho} = -b_{1-\rho}.$$

Proof: For $\rho \geq 1$ we know that $H(z)$ is proper from Lemma 2. From (10.24), with $m_{11}(z) = \frac{1}{z}$, we compute

$$H_{22}(z) = Z_0 \frac{z^2(m_{12} + Z_0 m_{22}) + (m_{12} - Z_0 m_{22})}{z^2(m_{12} + Z_0 m_{22}) - (m_{12} - Z_0 m_{22})}.$$

The expression above reveals that some cancellations in the factor $m_{12}(z) + Z_0 m_{22}(z)$ are required to change the high-frequency gain of $H_{22}(z)$ from the critical value Z_0. To enforce these cancellations we must have $\rho_1 = \rho_2$ and satisfy the coefficients conditions, which establishes the claim. □

7.4. *Stability conditions*

To simplify the derivations for the second (stability enforcement) step of the compensator design procedure, and considering Lemma 3, we will fix $\rho = 1$ and select the lowest order form for $M(z)$ – that will in its turn yield the lowest order controller.

As indicated before, stability will be enforced invoking a passivity argument, which as is well-known is equivalent to positive realness of the transfer function in LTI systems. Even though positive realness is well-understood and widely documented in continuous-time, the discrete-time version has been studied less. Actually, some widely accepted statements (dealing with fundamental issues) reported in textbooks have recently been proven incorrect – see [6]. For completeness we recall below the necessary background material taken from [6].

Definition 2: $H(z) \in \mathbb{R}(z)$ is called *positive real* if:

(i) $H(z)$ is analytic in $|z| \geq 1$,

(ii) $H(z^*) + H(z) \geq 0$ for all $|z| \geq 1$.

Lemma 4: *A real rational function $H(z)$, analytic in $|z| > 1$, is positive real if and only if:*[h]

(i) all poles of $H(z)$ on $|z| = 1$ are simple,

(ii) $\mathrm{Re}\{H(e^{j\theta})\} \geq 0$ *for all real θ at which $H(e^{j\theta})$ exists,*

(iii) if $z_0 = e^{j\theta_0}$, $\theta_0 \in \mathbb{R}$, is a pole of $H(z)$, and if r_i is the residue of $H(z)$ at $z = z_0$, then $e^{-j\theta_0} r_i \geq 0$.

Equipped with Lemmas 2, 3 and 4 we can proceed with our characterization of admissible behaviors with guaranteed stability properties.

Lemma 5: (Compensator positive realness) *Consider the transfer functions (10.23) satisfying the well-posedness and properness conditions of Lemmas 2, 3 with $\rho_1 = \rho_2 = 1$ and $b_i = a_i = 0$ for all $i > 1$. Fix*

$$a_1 = -\alpha - a_0, \quad -b_1 = \alpha + Z_0 + a_0,$$

[h] We draw the readers attention to condition *(iii)* which, as pointed out in [6], is given erroneously in several textbooks.

and select α and a_0 such that

$$0 \leq a_0 < -2\alpha$$
$$\max_{\theta\in[0,\pi]} \mathrm{Re}\left\{\frac{1}{e^{j\theta}+1+\frac{a_0}{\alpha}}\right\} \leq 1+\frac{a_0}{Z_0}-\frac{2\alpha}{Z_0}.$$

Then the steady-state conditions

$$m_{12}(1)=0, \quad m_{22}(1)=1,$$

are satisfied, and the transfer function $-\frac{m_{22}(z)}{m_{12}(z)}$ is positive real. In particular, the conditions are satisfied if $a_0=0$ and $\alpha \leq -\frac{1}{2}Z_0$.

Proof: Under the conditions of the lemma the transfer functions (10.23) become

$$\begin{aligned} m_{12}(z) &= \alpha z + a_0 + \frac{a_1}{z} \\ -Z_0 m_{22}(z) &= \alpha z + a_0 + \frac{b_1}{z}. \end{aligned} \tag{10.25}$$

Imposing $m_{12}(1)=0$, $m_{22}(1)=1$, reduces the number of free parameters to two, that we select as α and a_0. This yields

$$H_c(z) \triangleq -Z_0\frac{m_{22}(z)}{m_{12}(z)} = \frac{z^2+\eta z+\gamma}{z^2+\eta z-(\eta+1)}$$

where, to simplify the derivations, we have introduced the parameterization $\eta=\frac{a_0}{\alpha}, \gamma=\frac{b_1}{\alpha}$. The partial fraction expansion of the right hand side is

$$H_c(z)=1+\left(\frac{1}{z-1}-\frac{1}{z+\eta+1}\right)\frac{1+\eta+\gamma}{2+\eta}.$$

We notice that the residues have opposite signs. Since the residue associated with the fixed pole at $z=1$ is $r_1=1$, $e^{-j\theta_1}r_1$ should be positive, that implies $1+\eta+\gamma>0$. The second pole can be strictly inside the unit disk or on it[i], hence $-2<\eta\leq 0$. Finally, in the first case, some simple calculations prove that the real part of this transfer function is positive if

$$\max_{\theta\in[0,\pi]} \mathrm{Re}\{\frac{1}{e^{j\theta}+1+\eta}\} \leq \frac{1}{2}\left(\frac{3+\eta-\gamma}{1+\eta+\gamma}\right).$$

To prove the last claim we note that for $a_0=0$ we have $\eta=0$, hence the other pole is also on the unitary disk. To complete the proof compute

$$H_c(z)=1+\left(\frac{1}{z-1}-\frac{1}{z+1}\right)\frac{1+\gamma}{2}. \tag{10.26}$$

[i]Recall that each pole on the unitary disk should be simple, then $\eta \neq -2$.

On one hand, we have to enforce the positivity of $\mathrm{Re}\{H_c(e^{j\theta})\}$, that implies $\alpha \leq -\frac{1}{2}Z_0$. On the other hand, we have also to verify condition (iii) of Lemma 4, that is $e^{-j\theta_2}r_2 \geq 0$, where $r_2 = -1$ is the second residue related to the pole $z = -1$. Remarking that $e^{-j\theta_2} = -1$, we establish the claim. □

As a closing remark of this subsection notice that, setting $\gamma = 0$ which corresponds to the case considered in [7], the transfer function (10.26) becomes

$$H_c(z) = 1 + \left(\frac{1}{z-1} - \frac{1}{z+1}\right)\frac{1}{2}.$$

As shown above this transfer function is positive real. However, using the "standard" definition of residues we would conclude that *it is not* positive real – see Appendix B for more details.

7.5. *Main result*

We are now in position to present the main result of the chapter.

Proposition 3: *Consider the system depicted in Figure 10.2 where the actuator is described by (10.20) the transmission line by the Telegraphers equation (10.1) and the plant by $v(t,\ell) = Z_p(p)i(t,\ell)$. Suppose Assumptions 1 and 2 hold, $Z_p(s) \in \mathbb{R}(s)$ is strictly positive real and $\lim_{t\to\infty} v_s(t) = V_s$, $V_s \in \mathbb{R}$. Let the compensator be realized as shown in Figure 10.9, where the voltage and current of the regulated sources are defined by*

$$\begin{bmatrix} \tilde{i}(t) \\ v(t,0) \end{bmatrix} = \begin{bmatrix} \frac{q^2-1}{2\alpha(q^2-1)-Z_0(q^2+1)} & \frac{-2Z_0 q}{2\alpha(q^2-1)-Z_0(q^2+1)} \\ \frac{2[\alpha(q^2-1)-Z_0]}{2\alpha(q^2-1)-Z_0(q^2+1)} & -\frac{Z_0[(2\alpha+Z_0)(q^2-1)]}{2\alpha(q^2-1)-Z_0(q^2+1)} \end{bmatrix} \begin{bmatrix} \tilde{v}(t) \\ i(t,0) \end{bmatrix}, \quad (10.27)$$

where $\alpha \leq -\frac{1}{2}Z_0$.

Then

P1 the overall system is well-posed and internally stable,

P2 the compensator-transmission line subsystem, with $\alpha = -Z_0$, is described by

$$\begin{bmatrix} v(t,\ell) \\ i(t,\ell) \end{bmatrix} = \begin{bmatrix} \frac{1}{q} & -Z_0(q-\frac{1}{q}) \\ 0 & q \end{bmatrix} \begin{bmatrix} \tilde{v}(t) \\ \tilde{i}(t) \end{bmatrix},$$

P3 the following asymptotic behavior is ensured

$$\lim_{t\to\infty}[\tilde{v}(t)\tilde{i}(t) - v(t,\ell)i(t,\ell)] = 0$$
$$\lim_{t\to\infty}[\tilde{v}(t) - v(t,\ell)] = 0.$$

Proof: The current-decoupling compensator (10.27) is obtained fixing $m_{11}(z) = \frac{1}{z}$, using $a_0 = 0$ and $\alpha \leq -\frac{1}{2}Z_0$ in (10.25), and replacing in (10.24). Considering Lemma 2, matrix $H(z)$ is proper, while Lemma 3 ensures well-posedness. The proof is completed with Lemma 5, that establishes the internal and asymptotic stability properties. □

We close this section with the observation that setting $\alpha = -Z_0$ the compensator (10.27) takes the simple form

$$\begin{bmatrix} \tilde{i}(t) \\ v(t,0) \end{bmatrix} = \frac{1}{1-3q^2} \begin{bmatrix} \frac{1}{Z_0}(q^2-1) & -2q^2 \\ 2q^2 & Z_0(q^2-1) \end{bmatrix} \begin{bmatrix} \tilde{v}(t) \\ i(t,0) \end{bmatrix}, \tag{10.28}$$

which is the one reported in [7].

8. Adaptive Compensators

In this section we prove that, using standard discrete-time adaptive control techniques [9], it is possible to design an adaptive version of the proposed compensator (10.27) that estimates the line impedance Z_0, but still assumes the transmission delay is known. For ease of presentation we treat here the case $\alpha = -Z_0$ that leads to the simpler controller expression (10.28) – the general case follows *mutatis mutandis* from these derivations.

To set up an error model upon which we can base the adaptation we assume measurable the current at the terminal point of the line, *i.e.* $i(t,\ell)$. We sample the signals every d units of time and, for brevity, denote

$$i_k^\ell = i(kd,\ell), \quad v_k^0 = v(kd,0), \quad i_k^0 = i(kd,0),$$

$$\tilde{i}_k = \tilde{i}(kd), \quad \tilde{v}_k = \tilde{v}(kd)$$

where $k \in \mathbb{Z}$. Using this notation we can write the sampled version of the second line equation of (10.6) as

$$2i_l(k) = \theta[v_0(k-1) - v_0(k+1)] + i_0(k-1) + i_0(k+1) \tag{10.29}$$

where we have defined the unknown parameter $\theta \triangleq \frac{1}{Z_0}$. (Notice that, if d is known, this is the only parameter needed in (10.28).) A certainty equivalent adaptive version of (10.28) is obtained replacing the unknown parameter θ by its current estimate, that we denote $\hat{\theta}_k$, to yield

$$\tilde{v}(k) = -\tilde{v}(k-2) + 2v_0(k) + \frac{1}{\hat{\theta}(k)}(\tilde{i}(k) - \tilde{i}(k-2)) \tag{10.30}$$

$$i_0(k) = -i_0(k-2) + 2\tilde{i}(k) + \hat{\theta}(k)(v_0(k) - v_0(k-2)). \tag{10.31}$$

Defining the parameter error $\tilde{\theta}(k) \triangleq \hat{\theta}(k) - \theta$, shifting (10.31) and replacing it in (10.29) yields, after some simple derivations, the error equation

$$e(k) = \tilde{\theta}(k)\phi(k), \tag{10.32}$$

where we have defined the measurable quantities

$$e_k \triangleq i_l(k-1) - \tilde{i}(k), \quad \phi(k) \triangleq \frac{1}{2}(v_0(k) - v_0(k-2)). \tag{10.33}$$

The error equation (10.32) suggests the parameter update law [9]

$$\hat{\theta}(k+1) = P\left\{\hat{\theta}(k) - \frac{\gamma\phi(k)}{1+\phi(k)^2 e(k)}\right\} \tag{10.34}$$

with $2 > \gamma > 0$ an adaptation gain and $P\{\cdot\}$ a projection operator that keeps the estimate bounded away from zero. Replacing (10.32) in (10.34) we have that, for almost all sampled instants k,

$$\tilde{\theta}(k+1) = \tilde{\theta}(k) - \frac{\gamma\phi(k)^2}{1+\phi(k)^2}\tilde{\theta}(k).$$

Hence,

$$\tilde{\theta}(k+1)^2 - \tilde{\theta}(k)^2 = \gamma\left[-2 + \frac{\gamma\phi(k)^2}{1+\phi(k)^2}\right]\frac{e(k)^2}{1+\phi(k)^2},$$

where we have used (10.32) again. From the latter, taking into account that $2 > \gamma > 0$, we immediately conclude that

$$\lim_{k\to\infty} \frac{e(k)}{\sqrt{1+\phi(k)^2}} = 0.$$

Hence, if $v_s(t)$ is such that the sequence $\phi(k)$ is bounded, we have that $\lim_{k\to\infty} e(k) = 0$, that is, $\lim_{k\to\infty}(i_l(k-1) - \tilde{i}(k)) = 0$. Now, if $\lim_{k\to\infty} \phi(k) \neq 0$, then we can also conclude from (10.32) that the estimated parameter converges to the true value.

The choice of a *decoupling controller* is essential to generate the necessary error equations. Indeed, since the interconnected dynamics depends on the unknown parameter Z_0, the derivation of direct adaptive versions of other controllers, at least along lines of the present derivation, does not seem obvious. Alternatively, we can take an *indirect* adaptive control perspective an estimate Z_0 from the line Eq. (10.29). Proceeding in this way we also get the error Eq. (10.32) but now with

$$e_k \triangleq i_0(k) + i_0(k-2) - 2i_l(k-1), \quad \phi(k) \triangleq v_0(k) - v_0(k-2),$$

and the same update law (10.34) can be used to generate the parameter estimates.

9. Simulation Results

In order to show the performances of the proposed compensators and their adaptive version, we consider the benchmark example introduced in Section 3, *i.e.* an induction motor drive connected to a fast-sampling actuator through long cables.

The transmission line was modeled by (10.6). Inductance, capacitance and resistance of the cable are chosen as $L = 0.97\mu$H/m, $C = 45p$F/m, $R = 50m\Omega$/m, and the length of the cable is set to be $\ell = 100$m. These values are calculated from the measured S-parameters by a network analyzer in the experimental facility of POSTECH, see [10]. Further, they are confirmed by a high speed digital sampling oscilloscope which is used to measure transport delay ($d = 0.66\mu$s) and characteristic impedance ($Z_0 = 146.8\Omega$). The motor is modeled by a high frequency R-C ($C_{hf} = 750p$F, $R_{hf} = 300\Omega$), in parallel with a low frequency R-L model ($R_{lf} = 2.5\Omega$, $L_{lf} = 180m$H). As for the cable, the motor parameters are calculated from the measured values, using an impedance analyzer.

Several approaches could be used to simulate the behavior of the overall system, as far as many software are available on the market. Our simulations have been performed with Matlab-Simulink and 20-sim, which relies on a port-based description of the system. In any case, no significant differences between the two results were observed. It is, however, of interest to indicate that for the 20-sim simulations the compensator was represented in the equivalent form indicated in Figure 10.13, obtained applying the Kirchoff's laws to the electrical scheme of the compensator reported in Figure 10.9, that is $\widetilde{i}(t) = i_c + i(t,0)$ and $v(t,0) = \widetilde{v}(t) - v_c$ with $\widetilde{i}(t) = i_s(t)$.

Some simulations results of the uncompensated and the compensated system are depicted in the Figures 10.14 and 10.15 below.

10. Conclusions and Outlook

In this chapter we have given rigorous theoretical foundations for the compensator design framework proposed in [7], and explicitly formulated in

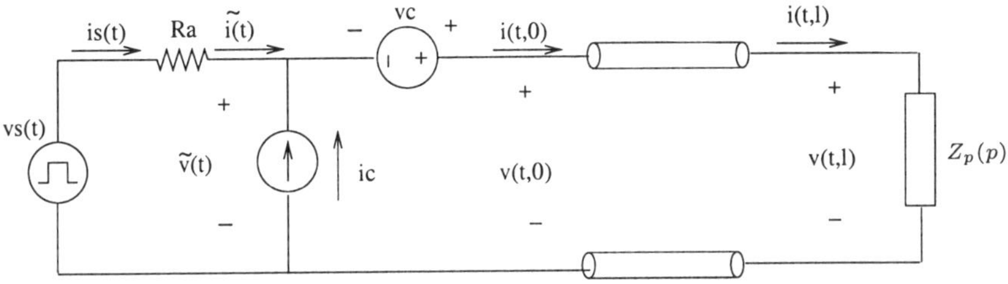

Figure 10.13. Equivalent electrical scheme of compensated system used for 20-sim simulations.

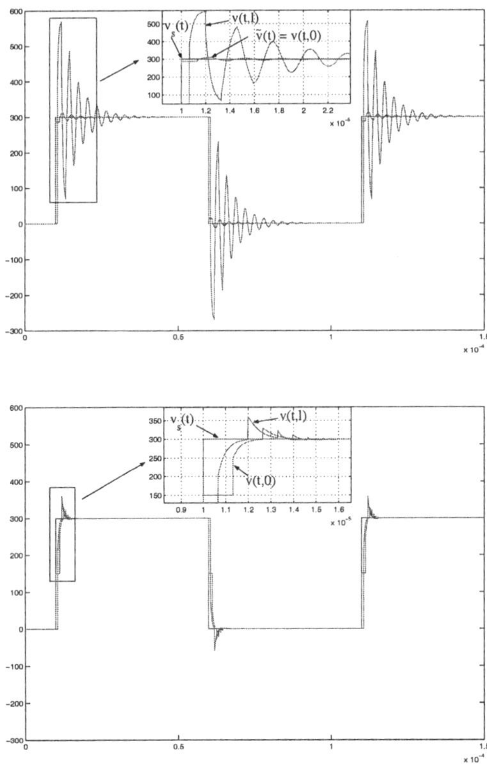

Figure 10.14. Top graph: Pulse response of the uncompensated system. Bottom graph: Pulse response with compensator (10.28).

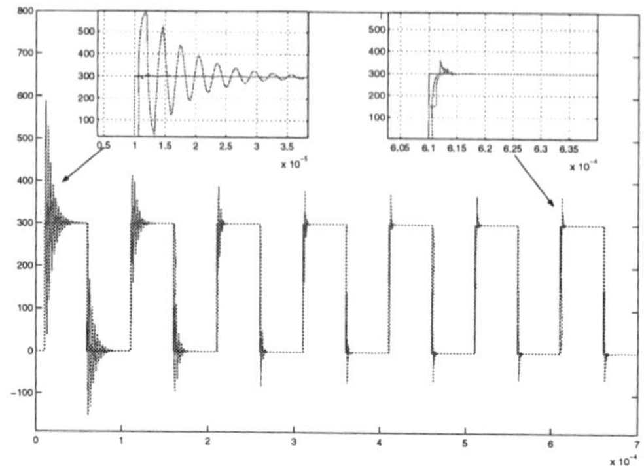

Figure 10.15. Adaptive version of the compensator (10.28).

Section 6. In particular,

- using the adequate definition of well-posedness we have completely characterized the achievable compensator-line behaviors that lead to proper compensators with well-defined interconnections;
- we have shown that all voltage-decoupling compensators lead to ill-posed systems;
- we have identified a family of current-decoupling (well-posed and proper) schemes that ensure asymptotic stability for all strictly positive real plants;
- a provably stable adaptive version of the compensator, that identifies the line characteristic impedance, is given.

These issues were not properly addressed in [7], where inadequate definitions of well-posedness and positive realness led to overly conservative conditions and no clear explanation – besides simulation evidence – to the interest of current-decoupling. Of particular relevance is the new stability analysis given here which was mentioned as an open problem in [7].

The proposed scheme relies on active elements – regulated sources – hence it differs from standard practice. A practical device that can be used for implementation is the hybrid active filter depicted in Figure 10.16[j], see [8] for some preliminary results. Hybrid filters are now systematically used for various applications and they seem to be a feasible solution for low-power

[j]The authors thank Gerardo Escobar for this pertinent suggestion.

applications. However, for overvoltage problems in AC drives their use is currently stymied due to power and switching times considerations. Recent advances in power electronics may change this situation in the future.

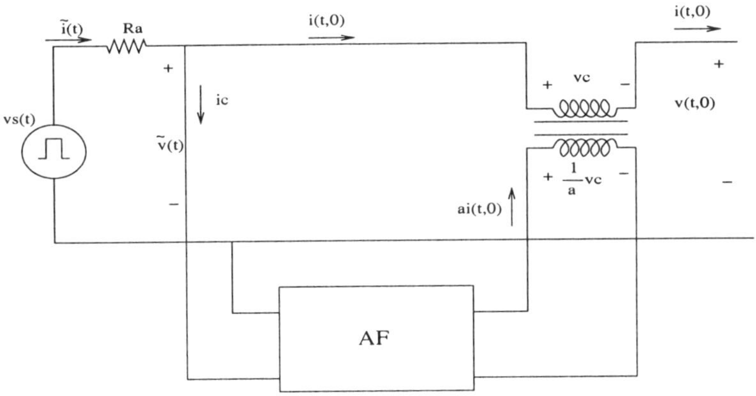

Figure 10.16. Hybrid active filter. Note the placement between actuator and transmission line, and the action as compensator in terms of voltage and current.

On the other hand, power considerations are conspicuously absent in VLSI applications. We believe that the theory developed here, will lead to filters which can be used in VLSI circuit design to minimize the distortion of the pulses traveling on the interconnects.

Our current research proceeds along three different lines:

- We have adopted in this work a robustness perspective – specifically, a passivity argument – to cope with plant uncertainty. It is well-known that passivity-based schemes, although robust, may lead to below par performances. The natural candidate to overcome this problem is, of course, adaptation. From (10.3) it is easy to see that applying to the transmission line the voltage

$$v(t,0) = \frac{1}{2}\left[1 + \frac{Z_0}{Z_p(p)} + \left(1 - \frac{Z_0}{Z_p(p)}\right) q^{-2}\right] \widetilde{v}(t)$$

also achieves perfect suppression of the reflected wave, that is $v(t,\ell) = \widetilde{v}(t-d)$. This solution, however, has the big drawback that it requires the exact knowledge of the plant impedance. Current research is under way to develop suitable estimation algorithms.

- For AC drive applications an approximation of the proposed active compensator with a shunt passive LTI filter would lead to a workable design. It is possible to show that a continuous-time approximation, *e.g.* with a Pade approximation of the delay, of the compensator (10.28) is not positive real – hence not realizable with RLC circuits. However, some preliminary computations for the more general scheme (10.27) suggests the existence of an interval for the free parameter α for which positive realness is ensured. The outcome of this research will be reported elsewhere.
- The construction at the University of Illinois of an experimental, low-power rig to test the proposed algorithms is also under investigation.

Acknowledgments

This research was also partially supported by the National Science Foundation under grant INT-0128656, the U.S. Office of Naval Research under Grant N00014-02-1-0011 and the European project GEOPLEX (code IST-2001-34166).

Bibliography

1. R. Anderson and M. Spong. Bilateral control of teleoperators with time delay. *IEEE Trans. Automatic Control*, 34(5):494–501, 1989.
2. F.M. Callier and C.A. Desoer. *Multivariable Feedback Systems.* Springer-Verlag, New York, 1982.
3. C.T. Chen. *Linear System Theory and Design.* Saunders-HBC, 1984.
4. C. Cottee and S. Duncan. Design of matching circuit controllers for radio-frequency heating. *IEEE Trans. Control Systems Technology*, 11(1):91–100, 2003.
5. R. DeCarlo and P.M. Lin. *Linear Circuit Analysis.* Oxford University Press, New York, 2001.
6. C. Xiao and D.J. Hill. Generalization and new proof of the discrete-time positive real lemma and bounded real lemma. *IEEE Trans. Circuits and Systems*, 4(6):740–743, 1999.
7. R. Ortega, A. de Rinaldis, M.W. Spong, S. Lee, and K. Nam. On compensation of wave reflections in transmission lines and applications to the overvoltage problem in AC motor drives. *IEEE Trans. Automatic Control*, 49(10):1757–1763, 2004.
8. R. Ortega, A. de Rinaldis, G. Escobar, and M. Spong. A hybrid active filter implementation of an overvoltage suppression scheme. *Proc. IEEE Int. Symp. Industrial Electronics*, pages 1141–1146, 2004.
9. G. Goodwin and K. Sin. *Adaptive filtering prediction and control.* Prentice-Hall, 1984.

10. S.C. Lee and K.H. Nam. An overvoltage suppression scheme for AC motor drives using a half DC-link voltage level at each PWM transition. *IEEE Trans. Industrial Electronics*, 49(3):549–557, 2002.
11. R.J. Kerkman. Twenty years of PWM AC drives: when secondary issues become primary concerns. *Proc. IEEE Int. Conf. Industrial Electronics, Control, and Instrumentation (IECON), Taipei, Taiwan*, pages 57–63, 1996.
12. E. Persson. Transient effects in application of PWM inverters to induction motors. *IEEE Trans. Industry Applications*, 28(5):1095–1101, 1992.
13. P. Berg and J. McGregor. *Elementary Partial Differential Equations.* McGraw-Hill, New York, 1966.

Appendix A. Derivation of (10.3) and (10.5)

Using the scattering representation it is straightforward to derive (10.3). Indeed, replacing (10.2) in (10.7) we have on the actuator side

$$s_+(t,\ell) = v(t,\ell) + Z_0 i(t,\ell) = \left(1 + \frac{Z_0}{Z_p(p)}\right) v(t,\ell).$$

Now, using from (10.9) the relationship $s_+(t,\ell) = s_+(t-d,0)$, and the fact that

$$\begin{aligned} s_+(t,0) = v(t,0) + Z_0 i(t,0) &= v(t,0) + \frac{Z_0}{R_a}[v_s(t) - v(t,0)] \\ &= \left(1 - \frac{Z_0}{R_a}\right) v(t,0) + \frac{Z_0}{R_a} v_s(t), \end{aligned}$$

we obtain

$$\left(1 + \frac{Z_0}{Z_p(p)}\right) v(t,\ell) = \left(1 - \frac{Z_0}{R_a}\right) v(t-d,0) + \frac{Z_0}{R_a} v_s(t-d). \tag{10.35}$$

Proceeding analogously for $s_-(t,\ell)$, using this time $s_-(t,\ell) = s_-(t+d,0)$, one gets

$$\left(1 - \frac{Z_0}{Z_p(p)}\right) v(t-2d,\ell) = \left(1 + \frac{Z_0}{R_a}\right) v(t-d,0) - \frac{Z_0}{R_a} v_s(t-d). \tag{10.36}$$

The delay-differential equation (10.3) is finally obtained eliminating $v(t-d,0)$ from (10.35), (10.36), operating on both sides of the equation with $\frac{Z_p(p)}{Z_p(p)+Z_0}$, which we assumed is stable, and using the operator q^k. Due to the filtering operation the additive exponentially decaying term ϵ_t appears in (10.3).

Appendix B. Equivalent formulation of (iii) condition for the DTPR lemma

Consider the following general rational function

$$H(z^{-1}) = \frac{b_0 + b_1 z^{-1} + ... + b_n z^{-n}}{a_0 + a_1 z^{-1} + ... + a_m z^{-m}},$$

where $a_i, b_i \in \mathbb{R}$. Operating the partial fractional expansion we should distinguish two cases:

(i) if $n < m$, then

$$H(z^{-1}) = \sum_{i=1}^{m} \frac{r_i^*}{1 - p_i z^{-1}};$$

(ii) else, if $n \geq m$, then

$$H(z^{-1}) = F(z^{-1}) + \sum_{i=1}^{m} \frac{r_i^*}{1 - p_i z^{-1}};$$

where the degree of the polynomial $F(z^{-1})$ is $\rho = n - m$. In both cases the residues r_i related to the poles p_i can be evaluated by the summatory

$$\sum_{i=1}^{m} \frac{r_i^*}{1 - p_i z^{-1}}. \tag{10.37}$$

An equivalent proper representation of the rational function $H(z^{-1})$ is the following

$$H(z) = z^{m-n} \frac{b_0 z^n + b_1 z^{n-1} + ... + b_n}{a_0 z^m + a_1 z^{m-1} + ... + a_m}.$$

Now, if $b_0 \neq 0$ and operating the same partial fractional expansion we get

$$H(z) = \frac{a_0}{b_0} + \sum_{i=1}^{m} \frac{r_i}{z - p_i} \tag{10.38}$$

obtaining, obviously, different values of the residues, i.e, $r_i^* \neq r_i$. They are, anyway, co-related. In particular, considering r_i^* of a simple pole $p_i = e^{j\theta_0}$ on the unitary disc, from (10.37) and (10.38) we have

$$e^{-j\theta_0} r_i = r_i^*,$$

then, we have two ways to check condition (iii) of DTPR lemma:

(i) write the rational function in z, i.e, $H(z)$ and verify that $e^{-j\theta_0} r_i \geq 0$,
(ii) write the rational function in z^{-1}, i.e, $H(z^{-1})$ and verify that $r_i^* \geq 0$.